P9-ARB-295

Recent Titles in This Series

131 **V. V. Sharko,** Functions on manifolds: Algebraic and topological aspects, 1993
130 **V. V. Vershinin,** Cobordisms and spectral sequences, 1993
129 **Mitsuo Morimoto,** An introduction to Sato's hyperfunctions, 1993
128 **V. P. Orevkov,** Complexity of proofs and their transformations in axiomatic theories, 1993
127 **F. L. Zak,** Tangents and secants of algebraic varieties, 1993
126 **M. L. Agranovskiĭ,** Invariant function spaces on homogeneous manifolds of Lie groups and applications, 1993
125 **Masayoshi Nagata,** Theory of commutative fields, 1993
124 **Masahisa Adachi,** Embeddings and immersions, 1993
123 **M. A. Akivis and B. A. Rosenfeld,** Élie Cartan (1869–1951), 1993
122 **Zhang Guan-Hou,** Theory of entire and meromorphic functions: Deficient and asymptotic values and singular directions, 1993
121 **I. B. Fesenko and S. V. Vostokov,** Local fields and their extensions: A constructive approach, 1993
120 **Takeyuki Hida and Masuyuki Hitsuda,** Gaussian processes, 1993
119 **M. V. Karasev and V. P. Maslov,** Nonlinear Poisson brackets. Geometry and quantization, 1993
118 **Kenkichi Iwasawa,** Algebraic functions, 1993
117 **Boris Zilber,** Uncountably categorical theories, 1993
116 **G. M. Fel′dman,** Arithmetic of probability distributions, and characterization problems on abelian groups, 1993
115 **Nikolai V. Ivanov,** Subgroups of Teichmüller modular groups, 1992
114 **Seizô Itô,** Diffusion equations, 1992
113 **Michail Zhitomirskiĭ,** Typical singularities of differential 1-forms and Pfaffian equations, 1992
112 **S. A. Lomov,** Introduction to the general theory of singular perturbations, 1992
111 **Simon Gindikin,** Tube domains and the Cauchy problem, 1992
110 **B. V. Shabat,** Introduction to complex analysis Part II. Functions of several variables, 1992
109 **Isao Miyadera,** Nonlinear semigroups, 1992
108 **Takeo Yokonuma,** Tensor spaces and exterior algebra, 1992
107 **B. M. Makarov, M. G. Goluzina, A. A. Lodkin, and A. N. Podkorytov,** Selected problems in real analysis, 1992
106 **G.-C. Wen,** Conformal mappings and boundary value problems, 1992
105 **D. R. Yafaev,** Mathematical scattering theory: General theory, 1992
104 **R. L. Dobrushin, R. Kotecký, and S. Shlosman,** Wulff construction: A global shape from local interaction, 1992
103 **A. K. Tsikh,** Multidimensional residues and their applications, 1992
102 **A. M. Il′in,** Matching of asymptotic expansions of solutions of boundary value problems, 1992
101 **Zhang Zhi-fen, Ding Tong-ren, Huang Wen-zao, and Dong Zhen-xi,** Qualitative theory of differential equations, 1992
100 **V. L. Popov,** Groups, generators, syzygies, and orbits in invariant theory, 1992
99 **Norio Shimakura,** Partial differential operators of elliptic type, 1992
98 **V. A. Vassiliev,** Complements of discriminants of smooth maps: Topology and applications, 1992
97 **Itiro Tamura,** Topology of foliations: An introduction, 1992

(Continued in the back of this publication)

Functions on Manifolds
Algebraic and Topological Aspects

Translations of
MATHEMATICAL MONOGRAPHS

Volume 131

Functions on Manifolds
Algebraic and Topological Aspects

V. V. Sharko

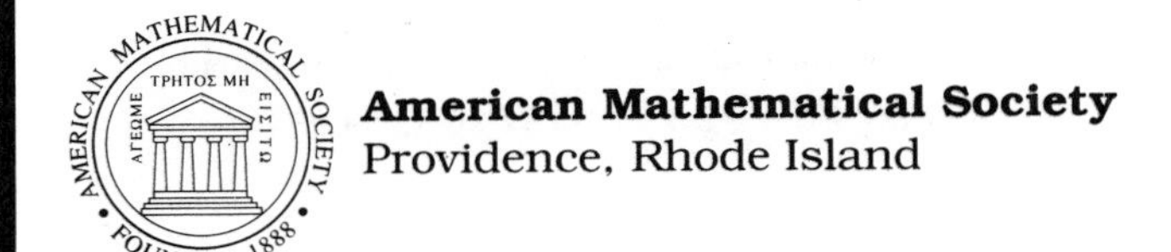

American Mathematical Society
Providence, Rhode Island

В. В. Шарко
ФУНКЦИИ НА МНОГООБРАЗИЯХ
(АЛГЕБРАИЧЕСКИЕ И ТОПОЛОГИЧЕСКИЕ АСПЕКТЫ)

«НАУКОВА ДУМКА» КИЕВ, 1990

Translated from the Russian by V. V. Minachin
Translation edited by Simeon Ivanov

1991 *Mathematics Subject Classification.* Primary 57R45, 57R70, 58E05;
Secondary 57N20, 58B05.

Library of Congress Cataloging-in-Publication Data

Sharko, V. V. (Vladimir Vasil′evich)
[Funkt͡sii na mnogoobrazii͡akh. English]
Functions on manifolds: algebraic and topological aspects
V. V. Sharko; [translated from the Russian by V. V. Minachin].
p. cm. — (Translations of mathematical monographs, ISSN 0065-9282; v. 131)
Includes bibliographical references.
ISBN 0-8218-4578-0 (acid-free)
1. Differentiable mappings 2. Morse theory. 3. Manifolds. 4. Singularities (Mathematics)
I. Title. II. Series.
QA614.58.S5313 1993
515′.352—dc20
93-25901
CIP

♾ The paper used in this book is acid-free and falls within the guidelines
established to ensure permanence and durability.
♻ Printed on recycled paper.
This volume was typeset using $\mathcal{A}\mathcal{M}\mathcal{S}$-TEX,
the American Mathematical Society's TEX macro system.

10 9 8 7 6 5 4 3 2 1 98 97 96 95 94 93

Contents

Preface

The idea of studying smooth manifolds by means of level curves of functions defined on them goes back to Poincaré, and even to Möbius. But its systematic development is due to Marston Morse who first observed that the number of critical points of different indices of a smoth function on a manifold can be made use of in order to study the geometric properties of the manifold. Morse proved certain inequalities connecting the number of critical points with the ranks and torsion orders of the homology groups of the manifold [102]. At various times, the development of this line of topology has been carried forward by L. A. Lyusternik, L. G. Snirel′man, G. S. Chogoshvili, L. E. Elsholz, E. Pitcher, and G. Reeb [78, 20, 21, 44, 113, 121].

In 1960, Smale showed that on any smooth simply connected manifold M^n $(n \geq 5)$ there exists a Morse function with the minimal number of critical points of each index. Among many corollaries thereto we mention the generalized Poincaré conjecture and the h-cobordism theorem [143].

Morse theory plays an important role in modern topology. Morse surgery, the theory of Smale's handles, provides flexible tools for the analysis of differentiable manifolds. The effectiveness of the theory was repeatedly demonstrated in the works of Milnor, Kervaire, S. P. Novikov, and Browder on the classification of manifolds [108]. On the other hand, despite the intensive studies of non-simply-connected manifolds made by Novikov, Wall, Farrell, Cappell, A. S. Mishchenko, Siebenmann and others, the question of the existence of a minimal Morse function on a non-simply-connected manifold remained open [105, 108, 155].

Another aspect of the Morse functions application should be mentioned. While studying families of smooth functions on simply connected manifolds, Cerf proved that the equivalence relations of isotopy and pseudo-isotopy for simply connected smooth manifolds of dimension greater than five coincide.

A new impulse was given to Morse theory in 1981 by the work of Novikov about multivalued Morse functions. The situation in this case differs qualitatively from the classical one. This area is now being investigated very actively.

The present monograph considers Morse functions on finite-dimensional smooth manifolds.

Chapter I includes the necessary material from the theory of Fréchet spaces and manifolds, which is then applied to the stratification of smooth functions on a manifold. The main result of Chapter II is a necessary and sufficient condition for two minimal Morse functions on a simply connected smooth manifold of dimension greater than 5 to be homotopic, isotopic, and conjugate. Chapter III is devoted to the algebraic technique used to construct minimal chain complexes over s-rings. It also contains a necessary and sufficient condition for the existence of a minimal chain complex in a given homotopic type.

Chapter IV considers the homotopic theory of chain complexes which is applied to the study of Morse functions on non-simply-connected cobordisms. The existence of minimal Morse functions for a wide class of non-simply-connected cobordisms is proved in Chapter V. New numerical invariants of manifolds are introduced resulting in the substantial improvement of the Morse inequalities. Chapter VI contains results on the homotopic properties of cell complexes needed for the analysis of Morse functions on closed manifolds. Our attention is centered on investigations concerning homotopic systems in the sense of Whitehead.

The question of the existence of minimal Morse functions on closed manifolds and on manifolds with a single component of the boundary is studied in Chapter VII. Chapter VIII considers so-called round Morse functions and includes recent developments in this area.

The author is sincerely grateful to S. P. Novikov for constructive advice and stimulating discussions, as well as to A. T. Fomenko and M. A. Shtanko whose counsel contributed to a deeper understanding of the problems discussed in the monograph. The author also appreciates the help and support he received in writing this book from many employees of the Institute of Mathematics of the Ukrainian Academy of Sciences.

CHAPTER I

Fréchet Manifolds

It is known that for any two finite-dimensional manifolds M^n and N^k the structure of a Banach manifold is introduced into the space $C^r(M^n, N^k)$ of all C^r-mappings from M^n into N^k. This manifold and its submanifolds are the most important examples of Banach manifolds. The space $C^\infty(M^n, N^k)$ involving derivatives of any order gives rise to Fréchet manifolds, i.e., those manifolds whose model spaces are Fréchet spaces (i.e., complete locally convex spaces with countably many norms).

As a rule, finite-dimensional results are transferred to Banach manifolds without much difficulty, but similar generalizations to Fréchet manifolds encounter serious obstacles. This is connected, in the first place, with the fact that the inverse function theorem does not hold for Fréchet spaces, so the use of a more complicated technique is required. In particular, one has to introduce additional structures (smoothing operators). Significant progress was made in this area and a number of issues were clarified; hopefully, the theory of Fréchet manifolds will gradually become as developed as that of Banach manifolds.

The first section of this chapter presents basic facts on Fréchet manifolds. The second section considers the inverse function theorem for Fréchet spaces. Here we follow the line of Nash-Moser-Sergereart-Hamilton. The chapter concludes with the analysis of the stratification of the Fréchet space of smooth functions on a compact manifold for which the inverse function theorem plays an important role.

§1. Brief summary

A seminorm on a vector space F is a mapping $\| \, \|: F \to \mathbb{R}$ satisfying the conditions:

$$\|f + g\| \leq \|f\| + \|g\|, \tag{1}$$

$$\|\alpha f\| = |\alpha| \, \|f\| \quad \text{for all } \alpha \in \mathbb{R}. \tag{2}$$

A collection of seminorms $\{\| \, \|_n : n \in \mathbb{N}\}$ on F is said to separate points if the condition $\|f\|_n = 0$ for all $n \in \mathbb{N}$ implies that $f = 0$. By definition, a locally convex space is a vector space F with a collection of seminorms

$\{\| \ \|_n : n \in \mathbb{N}\}$ * that separates points. The natural topology on a locally convex space is the weakest topology with respect to which all the seminorms $\| \ \|_n$ and the operation of addition in F are continuous. Note that occasionally one does not require that the collection of seminorms appearing in the definition of a locally convex space must separate points. The significance of this condition is in that the natural topology of a locally convex space (under our definition) is Hausdorff. In the natural topology a complete system of neighborhoods at zero is given by the sets $\{U_{n_1, \dots, n_k; \varepsilon} | n_1, \dots, n_k \in \mathbb{N}; \varepsilon > 0\}$, where $U_{n_1, \dots, n_k; \varepsilon} = \{f | \ \|f\|_{n_i} < \varepsilon, i = 1, \dots, k\}$. Thus $f_m \to f$ if and only if $\|f_m - f\|_n \to 0$ for all $n \in \mathbb{N}$. A sequence $\{f_m\}$ in a locally convex space F is called a Cauchy sequence if for any $\varepsilon > 0$ and any seminorm $\| \ \|_n$ there exists a positive integer k_0 such that $\|f_k - f_l\|_n < \varepsilon$ for all positive integers $k, l > k_0$. The space F is said to be complete if every Cauchy sequence converges. It is known that if a locally convex space F is metrizable, then the topology on F is defined by a countable collection of seminorms $\{\| \ \|_n : n \in \mathbb{N}\}$. Therefore, if the metric on F is defined by the inequality

$$\rho(f, g) = \sum_{n=1}^{\infty} 2^{-n} \left[\frac{\|f - g\|_n}{1 + \|f - g\|_n} \right],$$

then it induces the same topology on F as $\{\| \ \| : n \in \mathbb{N}\}$.

DEFINITION 1.1. A complete metrizable locally convex space F is called a Fréchet space.

We now present some known examples of Fréchet spaces. Clearly, any Banach space is a Fréchet space.

Let $\mathbb{R}^\infty$ be the vector space of all sequences $\{a_j\}$ of real numbers. Let

$$\|\{a_j\}\|_n = \sum_{j=0}^{n} |a_j|, \qquad n = 0, 1, 2, \dots.$$

Then $\mathbb{R}^\infty$ is a Fréchet space.

Denote by $C^\infty[0, 1]$ the space of all infinitely differentiable functions on the segment $[0, 1]$. A locally convex topology is usually introduced into $C^\infty[0, 1]$ by the collection of norms

$$\|f\|_n = \sum_{j=0}^{n} \sup_x |D^j f(x)|, \qquad n = 0, 1, 2, \dots.$$

The space $C^\infty[0, 1]$ is a Fréchet space.

Let $C^\infty(\Omega)$ be the vector space of infinitely differentiable functions on an open set Ω in $\mathbb{R}^n$. Choose an increasing sequence of compact sets K_n such

---

*Editor's note. This choice of the index set makes "all" locally convex spaces metrizable (see below). $\mathbb{N}$ should be replaced by "some" index set.

that $\Omega = \bigcup_{n=1}^{\infty} K_n$. The collection of seminorms

$$\|f\|_n = \sup_{x \in K_n} \sum_{j=0}^{n} |D^j f(x)|, \qquad n = 0, 1, \ldots,$$

turns $C^\infty(\Omega)$ into a Fréchet space.

Let M^n be a compact closed manifold of dimension n. Then the vector space of infinitely differentiable functions $C^\infty(M^n, \mathbb{R})$ is a Fréchet space. Indeed, let us cover M^n by finitely many open sets U_α such that the closure of each of them is contained in a coordinate neighborhood. The compactness of M^n implies that this is always possible. Let $g: M^n \to \mathbb{R}$ be a smooth function. Set

$$\|g\|_n^{U_\alpha} = \sup_{x \in U_\alpha} \sum_{j=0}^{n} |D^j g(x)|.$$

The sequence of seminorms is defined by the equality

$$\|g\|_n = \sum_\alpha \|g\|_n^{U_\alpha}, \qquad n = 0, 1, \ldots.$$

One can easily check that this collection of seminorms defines the structure of a Fréchet space on $C^\infty(M^n, \mathbb{R})$.

Consider a more general construction. Let $p: V \to M^n$ be a vector bundle over a smooth manifold M^n. Then the vector space $C^\infty(M^n, V)$ of the smooth sections of this bundle is a Fréchet space (we assume that M^n is compact). Cover the manifold M^n by finitely many open sets U_α such that the restriction $V|_{U_\alpha}$ of the bundle is trivial for each α. Choose a Riemannian metric on M^n. Then each fiber of the bundle acquires the structure of a Euclidean space. For a fixed section $f: M^n \to V$, making use of the preceding example, we define $\|f\|_n^{U_\alpha}$ as the maximum of $\|f_i\|_n^{U_\alpha}$, where the f_i are the coordinate functions defining the section f. Set

$$\|f\|_n = \sum_\alpha \|f\|_n^{U_\alpha}, \qquad n = 0, 1, \ldots.$$

Then $C^\infty(M^n, V)$ is a Fréchet space. It is easy to see that the norms corresponding to different coverings of this form are equivalent and do not depend on the choice of the Riemannian metric.

A closed subspace of a Fréchet space is also a Fréchet space, as is the quotient of a Fréchet space by a closed subspace. The direct sum of Fréchet spaces is a Fréchet space. However, the space of linear mappings of one Fréchet space into another is not, in general, a Fréchet space.

The Hahn-Banach theorem holds for Fréchet spaces. Thus, if F is a Fréchet space and f is a nonzero vector in F, one can find a continuous linear functional $l: F \to \mathbb{R}$ such that $l(f) = 1$. Hence, if $l(f) = l(g)$ for all continuous linear functionals $l: F \to \mathbb{R}$, then $f = g$, $f, g \in F$.

Also, the open mapping theorem holds for Fréchet spaces. In other words, if F and G are Fréchet spaces and $L: F \to G$ is a linear, continuous, and invertible mapping, then the inverse mapping $L^{-1}: G \to F$ is also continuous.

Let $f(t)$ be a continuous curve in a Fréchet space. Then its derivative is defined by the equality

$$f'(t) = \lim_{\Delta t \to 0} [f(t + \Delta t) - f(t)]/\Delta t.$$

We say that the path $f(t)$ is C^∞ if all its derivatives $f^{(n)}(t)$ exist and are continuous.

Let F and G be Fréchet spaces, U an open subset of F, and $P: U \subset F \to G$ a continuous (nonlinear) mapping. The derivative of P at a point $f \in U$ in the direction of a vector $h \in F$ is defined by the equalities

$$DP(f)h = \lim_{t \to 0} [P(f + th) - P(f)]/t.$$

The mapping P is said to be continuously differentiable in U if this limit exists for all $f \in U$ and all $h \in F$ and if the mapping $DP: (U \subset F) \times F \to G$ is continuous. It is known that if the mapping $P: U \subset F \to G$ is continuously differentiable and $h_1, h_2 \in F$, $f \in U$, then

$$DP(f)(h_1 + h_2) = DP(f)h_1 + DP(f)h_2.$$

Also the chain rule holds. For continuous mappings $U \times U \to G$ in two or more variables the partial derivatives are defined by

$$D_f P(f, g)h = \lim_{t \to 0} [P(f + th, g) - P(f, g)]/t,$$
$$D_g P(f, g)k = \lim_{t \to 0} [P(f, g + tk) - P(f, g)]/t.$$

The following statement holds: the partial derivatives $D_f P$ and $D_g P$ exist and are continuous if and only if the derivative $DP(f, g)(h, k)$ exists and is continuous. In this case

$$DP(f, g)(h, k) = D_f P(f, g)h + D_g P(f, g)k.$$

The second derivative of the mapping $P: U \subset F \to G$ is defined by the equality

$$D^2 P(f)\{h, k\} = \lim_{t \to 0} [DP(F + tk)h - DP(f)h]/t.$$

The mapping P is of class C^2 if the derivative DP is continuously differentiable and $D^2 P$ exists and is continuous. One can show that if P is of class C^2, then the mapping $D^2 P(f)\{h, k\}$ is bilinear. The higher derivatives $D^n P(f)\{h_1, h_2, \dots, h_n\}$ are defined by induction in an obvious manner. The mapping P is of class C^n if the nth derivative $D^n P$ exists and is continuous. If P is of class C^n, then $D^n P(f)\{h_1, h_2, \dots, h_n\}$ is a multilinear symmetric mapping $D^n P: (U \subset F) \times F \times \cdots \times F \to G$. A mapping P is of class C^∞ if it is of class C^n for all $n \in \mathbb{N}$.

Let $P: U \subset F \to V \subset G$ be a continuous mapping between open subsets of Fréchet subspaces. Its tangent $TP: (U \subset F) \times F \to (V \subset G) \times G$ is the mapping defined by the formula $TP(f, h) = (P(f), DP(f)h)$. We note that TP is defined and continuous if DP is defined and continuous.

We recall the definition of the Riemann integral for functions with values in a Fréchet space. Let $C([a, b], F)$ be the Fréchet space of all continuous functions on the closed interval $[a, b]$ taking values in a Fréchet space F. Define the seminorms on $C([a, b], F)$ by $\|f\| = \sup_t \|f(t)\|_n$, where $\|f(t)\|_n$ are seminorms on F. A function $f(t)$ is said to be linear if $f(t) = tf_1 + f_2$ for some f_1 and f_2 in F, and piecewise linear if it is continuous and there exists a partition $a = t_0 \le t_1 \le \cdots \le t_k = b$ such that $f(t)$ is linear on each closed interval $[t_{i-1}, t_i]$ for all $1 \le i \le k$. The vector subspace $PL([a, b], F)$ of piecewise linear functions on the closed interval $[a, b]$ with values in F is evidently an everywhere dense subspace of $C([a, b], F)$. For a piecewise linear function the integral is defined by the formula

$$\int_a^b f(t)\,dt = \sum_{i=1}^{k} \frac{1}{2}[f(t_{i-1}) + f(t_i)](t_i - t_{i-1}).$$

Since the integral is a continuous linear functional on the everywhere dense subspace $PL([a, b], F)$, it can be extended by continuity to a continuous linear functional on the entire space $C([a, b], F)$. The Hahn-Banach theorem ensures the uniqueness of the extension. The definition of the integral immediately implies the following statements:

(1) $l(\int_a^b f(t)\,dt) = \int_a^b l(f(t))\,dt$ for every continuous linear functional $l: F \to \mathbb{R}$;

(2) $\|\int_a^b f(t)\,dt\| \le \int_a^b \|f(t)\|\,dt$ for every seminorm $\|\ \|: F \to \mathbb{R}$;

(3) $\int_a^b f(t)\,dt + \int_b^c f(t)\,dt = \int_a^c f(t)\,dt$;

(4) $\int_a^b [f(t) + g(t)]\,dt = \int_a^b f(t)\,dt + \int_a^b g(t)\,dt$;

(5) $\int_a^b cf(t)\,dt = c\int_a^b f(t)\,dt$;

(6) if $f(t)$ is a C^1 curve on the closed interval $[a, b]$ with values in the Fréchet space F, then the Newton-Leibnitz formula

$$f(b) - f(a) = \int_a^b f'(t)\,dt$$

holds; and

(7) if the condition of the preceding statement is satisfied and, in addition, $\|f'(t)\| \le K$, then $\|f(b) - f(a)\| \le K(b - a)$.

DEFINITION 1.2. A Hausdorff topological space $\mathfrak{M}$ is called a Fréchet manifold if

(1) a covering of the space $\mathfrak{M}$ by open sets $\{U_\alpha\}_{\alpha \in A}$ is given;

(2) for each index $\alpha \in A$ there is a homeomorphism $h_\alpha: U_\alpha \to V_\alpha$, where

V_α is an open subset of a Fréchet space F (the homeomorphisms h_α are called charts on $\mathfrak{M}$); and

(3) for each pair of indices $\alpha, \beta \in A$ the mapping $h_\alpha \circ h_\beta^{-1}$ is smooth in its domain of definition.

We now list some Fréchet manifolds that are most essential for what follows.

As noted above, if M^n is a compact finite-dimensional manifold, then the space of infinitely differentiable functions on M^n forms a Fréchet space and is therefore a Fréchet manifold.

Let M^n and N^k be smooth manifolds, where M^n is compact. Then the space $C^\infty(M^n, N^k)$ of infinitely differentiable mappings is a Fréchet manifold. Indeed, we can easily check that $C^\infty(M^n, N^k)$ is a Hausdorff topological space. Let $f \in C^\infty(M^n, N^k)$. Find an open neighborhood U_f of the point $f \in C^\infty(M^n, N^k)$ homeomorphic to an open subset V_f of some Fréchet space. For that choose a complete Riemannian metric of class C^∞ on N^k and define the exponential mapping $\exp: TN^k \to N^k$ corresponding to this Riemannian metric. Here $p: TN^k \to N^k$ is the tangent bundle of the manifold N^k. As we know, this mapping is a diffeomorphism of a sufficiently small neighborhood of the origin in the tangent space $T_x(N^k)$ onto a neighborhood of the point x of the manifold N^k. Let $e: TN^k \to N^k \times N^k$ be a mapping defined by $e = \exp \times p$. Clearly, e is a diffeomorphism of a neighborhood V of the zero section of the bundle $p: TN^k \to N^k$ onto a neighborhood W of the diagonal in $N^k \times N^k$. We can assume that the vectors in V are bounded in absolute value. Let $f^*(TN^k)$ be the inverse image of the bundle TN^k under the mapping $f: M^n \to N^k$. Recall that $f^*(TN^k)$ is defined as the subset of the direct product $TN^k \times M^n: \{(\xi, m) \in TN^k \times M^n | p(\xi) = f(m)\}$ with projection $p^*(\xi, m) = m$. Set $\overline{f}(\xi, m) = \xi$. The diagram

$$\begin{array}{ccc} f^*(TN^k) & \xrightarrow{\overline{f}} & TN^k \\ {\scriptstyle p^*}\downarrow & & \downarrow{\scriptstyle p} \\ M^n & \xrightarrow[f]{} & N^k \end{array}$$

commutes. Here the mapping $\overline{f}$ is an isomorphism on the fibers. Denote by $\Gamma^\infty(M^n, TN^k)$ the space of sections of the bundle $f^*(TN^k)$. These sections can be identified with C^∞-sections $\gamma: M^n \to TN^k$ such that $p \circ \gamma = f$ (sections along the mapping f). For an open neighborhood U_f of the point $f: M^n \to N^k$ we can take the open set consisting of the mappings $g: M^n \to N^k$ such that $(f(x), g(x)) \in W$ $(g \in C^\infty(M^n, N^k))$. For V_f we can take the open set in $C^\infty(M^n, TN^k)$ consisting of sections γ such that $\gamma(x) \in$

V $(x \in M^n)$. Clearly, the mapping $h_f: U_f \to V_f$ defined by the formula $h_f(g)(x) = g(x) - f(x)$ $(g(x) \in U_f)$ is a coordinate homeomorphism with the inverse $h_f^{-1}(\gamma)(x) = f(x) + \gamma(x)$ $(\gamma(x) \in V_f)$. It can be shown that the coordinate homeomorphisms h_f constructed in this way are smoothly compatible. This procedure defines the structure of a Fréchet manifold on the space $C^\infty(M^n, N^k)$; this structure does not depend on the choice of the Riemannian metric.

Let M^n be a smooth compact manifold. Then the set $\mathrm{Diff}(M^n)$ of diffeomorphisms of M^n is open in $C^\infty(M^n, M^n)$ and is, therefore, a Fréchet manifold. The openness of $\mathrm{Diff}(M^n)$ is essentially the consequence of the fact that the group $GL(n, \mathbb{R})$ is open in the space of all matrices.

As in the case of a finite-dimensional manifold, one can define the tangent space to a Fréchet manifold $\mathfrak{M}$ at a point $\mathfrak{m} \in \mathfrak{M}$. Denote by $S_\mathfrak{m}(\mathfrak{M})$ the set of C^∞ curves $f: \mathbb{R} \to \mathfrak{M}$ satisfying the condition $f(0) = \mathfrak{m}$. If f_1 is tangent to f_2, then for any chart h_α the equality $(d/dt)(h_\alpha \circ f_1(0)) = (d/dt)(h_\alpha \circ f_2(0))$ holds. It follows from the chain rule that 'tangency at a point $\mathfrak{m}$' is a well-defined equivalence relation on the set $S_\mathfrak{m}(\mathfrak{M})$. Denote by $T_\mathfrak{m}(\mathfrak{M})$ the set of classes of equivalence in $S_\mathfrak{m}(\mathfrak{M})$. It is evident that $T_\mathfrak{m}(\mathfrak{M})$ is a Fréchet space. Let us consider the tangent spaces on the Fréchet manifolds $C^\infty(M^n, N^k)$ and $\mathrm{Diff}(M^n)$. It can be shown without difficulty that if $t \to f_t$ and $t \to g_t$ are smooth curves in $C^\infty(M^n, N^k)$ satisfying the condition $f_0 = g_0$, then f_t and g_t are tangent at $t = 0$ if and only if for each point $x \in M^n$ the curves $t \to f_t(x)$, $t \to g_t(x)$ (belonging to the manifold N^k) are tangent at $t = 0$ [55]. Let $\xi \in T_f C^\infty(M^n, N^k)$ and let $t \to f_t$ be a curve representing the vector ξ. Define the section $\xi': M^n \to TN^k$ by the formula $\xi'(x) = (d/dt)(f_t(x))|_{t=0}$. Clearly, this definition does not depend on the choice of f_t. Therefore the curve $t \to f_t$ generates a section $\bar{\xi}$ in $T_f C^\infty(M^n, N^k)$ and thus induces a mapping $T_f^\infty(M^n, TN^k) \to T_f C^\infty(M^n, N^k)$. It can be shown that there exists an isomorphism $T_f C^\infty(M^n, N^k) \approx \Gamma_f^\infty(M^n, TN^k)$. Hence it follows that $T_{\mathrm{id}\, M^n} \mathrm{Diff}(M^n) \approx \Gamma^\infty(M^n, T(M^n))$.

The notion of a vector bundle is naturally extended to Fréchet manifolds. Let $\mathfrak{M}$ and $\mathfrak{B}$ be Fréchet manifolds, and $\pi: \mathfrak{B} \to \mathfrak{M}$ a projection mapping such that for each point $f \in \mathfrak{M}$ the fiber $\pi^{-1} f$ has the structure of a vector space. If the condition of local triviality is satisfied, then $\pi: \mathfrak{B} \to \mathfrak{M}$ is called a Fréchet vector bundle. Such is, for example, the tangent bundle $T\mathfrak{M}$ of the Fréchet manifold $\mathfrak{M}$.

Let $\mathfrak{M}$ be a Fréchet manifold and $\mathfrak{N}$ a closed subset of $\mathfrak{M}$. The subset $\mathfrak{N}$ is a submanifold of $\mathfrak{M}$ if for each point in $\mathfrak{N}$ there exists a coordinate chart in $\mathfrak{M}$ whose range in the product $F \times G$ of Fréchet spaces is such that only the points of $\mathfrak{N}$ are mapped into the set $F \times 0$. Consider, by way of example, the following situation. Let M^n be a compact smooth manifold and P^s a submanifold of a smooth manifold N^k. Then the Fréchet manifold $C^\infty(M^n, P^s)$ of smooth mappings from M^n into P^s is a smooth Fréchet submanifold of the Fréchet manifold of smooth mappings from M^n into N^k.

Let $\mathfrak{M}$ and $\mathfrak{N}$ be smooth Fréchet manifolds. A mapping $P: \mathfrak{M} \to \mathfrak{N}$ is said to be smooth if for each point $f \in \mathfrak{M}$ and its image $P(f) \in \mathfrak{N}$ one can find local charts around them such that the local representative of P in these charts is a smooth mapping. A smooth mapping P induces the tangent mapping $DP(f): T\mathfrak{M}_f \to T\mathfrak{N}_{P(f)}$ of the corresponding tangent bundles that takes the fiber over $f \in \mathfrak{M}$ into the fiber over $P(f) \in \mathfrak{N}$ and is linear on each fiber. The local representatives for the tangent mapping coincides with the derivatives of the mapping P.

A Fréchet Lie group is a Fréchet manifold $\mathfrak{G}$ equipped with a group structure such that the multiplication mapping and the inverse mapping

$$C: \mathfrak{G} \times \mathfrak{G} \to \mathfrak{G}, \qquad C(g, h) = gh,$$
$$v: \mathfrak{G} \to \mathfrak{G}, \qquad V(g) = g^{-1}$$

are smooth. For example, let M^n be a compact smooth manifold. Then the group of diffeomorphisms $\mathrm{Diff}(M^n)$ is a Fréchet Lie group.

We say that a Fréchet Lie group $\mathfrak{G}$ acts on a Fréchet manifold $\mathfrak{M}$ if there is a smooth mapping

$$A: \mathfrak{G} \times \mathfrak{M} \to \mathfrak{M}, \qquad A(g, f) = gf,$$

such that $1f = f$ and $(g_1 g_2)f = g_1(g_2 f)$.

Consider, as an example of this kind of action, the following situation. Let M^n and N^k be compact smooth manifolds without boundary, and $C^\infty(M^n, N^k)$ the Fréchet manifold of smooth mappings from M^n into N^k. The action of the Fréchet Lie group $\mathfrak{G} = \mathrm{Diff}\, M^n \times \mathrm{Diff}\, N^k$ on $C^\infty(M^n, N^k)$ is given by the rule

$$\mathrm{Diff}\, M^n \times \mathrm{Diff}\, N^k \times C^\infty(M^n, N^k) \to C^\infty(M^n, N^k), \quad (\varphi, \psi)f = \psi \circ f \circ \varphi^{-1}.$$

We conclude this section with the well-known (for Banach spaces) inverse function theorem in the infinite-dimensional situation.

INVERSE FUNCTION THEOREM. *Let $P: U \subset F \to V \subset G$ be a smooth mapping between open subsets U and V of Banach spaces F and G, respectively. Suppose that for some $f_0 \in U$ the derivative $DP(f_0): F \to G$ is an invertible linear mapping. Then we can find neighborhoods $\tilde{U}$ of f_0 and $\tilde{V}$ of $g_0 = P(f_0)$ such that the mapping P gives a one-to-one mapping of $\tilde{U}$ onto $\tilde{V}$, and the inverse mapping $P^{-1}: \tilde{V} \subset G \to \tilde{U} \subset F$ is smooth.* □

For Fréchet spaces the inverse function theorem does not hold in this form. We give a counterexample following Hamilton [58].

Consider the Fréchet space of smooth functions defined on the closed interval $[-1, 1]$. Let $P: C^\infty[-1, 1] \to C^\infty[-1, 1]$ be the differential operator acting according to the formula $Pf = f - x \cdot f(df/dx)$. Clearly, P is of class C^∞, and its derivative is given by the formula

$$DP(f)g = g - xg\frac{df}{dx} - xf\frac{dg}{dx}.$$

The derivative of the operator P at $f = 0$ is the identity operator. Since $P(0) = 0$, if the inverse function theorem were true, then the image of P would contain a neighborhood of zero. We will show that this is not the case. Consider the sequence of functions $g_n = 1/n + b_n x^n$, where we assume that $b_n \neq 0$ and the sequence of numbers b_n tends to zero. Then evidently $g_n \to 0$ in the space $C^\infty[-1, 1]$. Let us prove that this sequence does not belong to the image of the operator P. This can be seen by examining power series. Every function in $C^\infty[-1, 1]$ has a formal power expansion at 0 (which does not necessarily converge to it). It is not difficult to show that if $f = a_0 + a_1 x + a_2 x^2 + a_3 x^3 + \cdots$, then

$$P(f) = a_0 + (1 - a_0)a_1 x + (a_2 - a_1^2 - 2a_0 a_2)x^2$$
$$+ (a_3 - 3a_1 a_2 - 3a_0 a_3)x^2 + \cdots.$$

Suppose that $P(f) = 1/n + b_n x^n$. First we have $a_0 = 1/n$. If $n > 1$, then $a_0 \neq 1$ and $(1 - a_0)a_1 = 0$, whence $a_1 = 0$. The next term of the series is then equal to $(1 - 2a_0)a_2 x^2$. If $n = 2$, then $a_0 = 1/2$ and this term is zero, which contradicts $Pf = 1/2 + b_2 x^2$, $b_2 \neq 0$. If $n > 2$, then $a_0 \neq 1/2$ and we conclude that $a_2 = 0$. Suppose that $a_1 = a_2 = \cdots = a_{k-1} = 0$, and consider the term $(1 - ka_0)a_k x^k$. If $k < n$, then $a_0 = 1/n$ and again this term of the series vanishes. Thus no term in the series

$$P(f) = a_0 + (1 - a_0)a_1 x + (a_2 - a_1^2 - 2a_0 a_2)x^2 + \cdots$$

is equal to $b_n x_n$ under the condition that $b_n \neq 0$, $n > 1$. Hence it follows that the sequence of functions $1/n + b_n x^n$ does not belong to the image of the operator $P = f - xf(df/dx)$.

Note that the derivative of the operator P at points close to zero is not an invertible operator. This is easy to see by evaluating the derivative at $f = 1/n$ applied to $h = x^k$:

$$DP\left(\frac{1}{n}\right)x^k = \left(1 - \frac{k}{n}\right)x^k.$$

This example may imply that in order for the inverse function theorem for Fréchet spaces to be true it is necessary to require that the derivative must be invertible in a whole neighborhood of the point of the image. The following example shows that this condition is not sufficient [59].

Let H be the vector space of entire holomorphic functions. Let

$$\|f\|_n = \sup\{|f(z)|: |z| \leq n\}.$$

Then H is a Fréchet space. Consider the nonlinear mapping $P: H \to H$, $P(f) = e^f$. Clearly, P is a smooth mapping, and its derivative

$$DP: H \times H \to H, \qquad k = DP(f)h = e^f h,$$

is an invertible linear operator for all f. It is easy to see that the mapping inverse to the derivative

$$h = DP(f)^{-1}k = e^{-f}k$$

is smooth. However, the following simple argument demonstrates that the mapping P is not locally invertible.

Let S be the subset of H consisting of nonvanishing functions. Then S is a relatively closed subset of $H \setminus 0$ and contains no open set. Indeed, it is not hard to show, using Rouché's theorem, that if $f(z) \in H\setminus\{0\}\setminus S$, then any function that is sufficiently close to $f(z)$ belongs to $H \setminus \{0\} \setminus S$. Therefore the set S is relatively closed in $H\setminus\{1\}$. Since the space of polynomials is an everywhere dense set in H, the set S cannot contain an open set. Evidently, $P(f) = e^f$ does not vanish, and therefore the image of P is not contained in S. Now it is clear that P is not a locally invertible mapping.

The above examples illustrate the difficulties arising in the treatment of Fréchet spaces.

§2. The Nash-Moser-Sergereart-Hamilton category

The basic idea, which paved the way for the proof of the inverse function theorem for Fréchet spaces, was suggested by Nash [104]. Then Moser fashioned it into an abstract theorem of functional analysis and demonstrated its wide applicability [103]. These ideas are close to those contained in the earlier works by A. N. Kolmogorov and I .V. Arnol′d on small denominators in the problems of mechanics [4, 73]. There are by now many versions of this theorem, which is usually called the Nash-Moser inverse function theorem, and the choice of the exposition depends on the applications one has in view. Both the theorem and its applications were considered in the works of N. N. Bogolyubov, Yu. A. Mitropol′skiĭ, A. M. Samoĭlenko, A. L. Gromov, L. V. Ovsyannikov, Hamilton, Hörmander, Jacobowitz, Kuranishi, and Sergereart [10, 56, 110, 58, 64, 67, 74, 124].

In our exposition we follow the line of Sergereart and Hamilton. They have picked out a category of Fréchet spaces for which the inverse function theorem is satisfied. We call it the Nash-Moser-Sergereart-Hamilton category. Following Hamilton and Sergereart, we say that the objects and morphisms of this category are *tame* [58, 124].

DEFINITION 1.3. A Fréchet space whose seminorms $\{\| \ \|_n : n \to \mathbb{N}\}$ satisfy the inequalities

$$\|f\|_0 \leq \|f\|_1 \leq \|f\|_2 \leq \cdots$$

is said to be a *graded Fréchet space*.

Now we give several examples of graded Fréchet spaces.

Let B be a Banach space with norm $\| \ \|_B$. Denote by $\Sigma(B)$ the space of all sequences $\{f_k\}$, where $f_k \in B$, such that

$$\|\{f_k\}\|_n = \sum_{k=0}^{\infty} e^{nk}\|f_k\|_B < \infty$$

for all $n \geq 0$. Then $\Sigma(B)$ is a graded Fréchet space.

Let M^n be a compact smooth manifold. Then $C^\infty(M^n, V)$ with the collection of norms

$$\|g\|_n = \sum_\alpha \|g\|_n^{U_\alpha}$$

is a graded Fréchet space. If V is a vector bundle over M^n, then the space C^∞ of smooth sections of V is also a graded Fréchet space.

The cartesian product $F \times G$ of two graded Fréchet spaces F and G is a graded Fréchet space with the grading

$$\|(f, g)\|_n = \|f\|_n + \|g\|_n.$$

A closed subspace of a graded Fréchet space is a graded Fréchet space.

DEFINITION 1.4. A linear mapping $L: F \to G$ of one graded Fréchet space into another is said to be *tame* if there exist positive integers m and r such that the estimates $\|Lf\|_n \leq C\|f\|_{n+r}$ are satisfied for all $n \geq m$, where C is a constant which may depend on n.

A tame mapping is evidently continuous. A composition of tame linear mappings is a tame linear mapping. An isomorphism L is said to be tame if both L and L^{-1} are tame. Consider two examples of tame linear mappings.

Let $F = G = \Sigma(B)$. Define the mapping $L: \Sigma(B) \to \Sigma(B)$ by the formula $(Lf)_k = e^{rk} f_k$. Then $\|Lf\|_n \leq \|f\|_{n+r}$, so L is tame.

Let $C^\infty[a, b]$ denote the graded Fréchet space of smooth functions on the closed interval $[a, b]$ with the grading

$$\|f\|_n = \sup_{k\leq n} \sup_{a\leq x\leq b} |D^k(f(x))|.$$

Define a linear mapping $L: C^\infty[0, 1] \to C^\infty[-1, 1]$ by the formula $Lf(x) = f(x^2)$. Then L is tame and $\|Lf\|_n \leq C\|f\|_n$. The image of L is the closed subspace $C_s^\infty[-1, 1]$ of symmetric functions with $f(-x) = f(x)$. Since L is one-to-one, the inverse linear continuous mapping

$$L^{-1}: C_s^\infty[-1, 1] \to C^\infty[0, 1]$$

exists but is not tame. This follows from the fact that the Taylor series of a symmetric function contains only even powers of x, so the inverse mapping satisfies the estimate

$$\|L^{-1}g\|_n \leq C\|g\|_{2n}.$$

Define $L: \Sigma(B) \to \Sigma(B)$ by the formula $(Lf)_k = f_{2k}$. Then $\|Lf\|_n \leq \|f\|_{2n}$, and L is not tame.

DEFINITION 1.5. A continuous (nonlinear) mapping $P: F \to G$ of graded Fréchet spaces is said to be tame if there exist positive integers m and r such that the estimates

$$\|P(f)\|_n \le C(1 + \|f\|_{n+r})$$

hold for all $n \ge m$, where C is a constant which may depend on n.

It is not difficult to show that a composition of tame mappings is tame. A mapping L is tame if and only if it is both linear and tame. Any continuous mapping of a graded Fréchet space into a Banach space is tame.

DEFINITION 1.6. Let F and G be two graded Fréchet spaces and $P: U \subset F \to G$ a nonlinear continuous mapping. We say that P is a *smooth tame* mapping if P is smooth and all its derivatives $D^k P$ are tame.

For example, the mapping $P: C^\infty[a, b] \to C^\infty[a, b]$ defined by the formula $P(f) = e^f$ is a smooth tame mapping.

Now we give the following important definition.

DEFINITION 1.7. A graded Fréchet space is said to be tame if there is a one-parameter family of linear mappings $S(t): F \to F$ such that

$$\|S(t)f\|_{n+r} \le Ct^r \|f\|_n, \quad \|S(t)f - f\|_n \le \overline{C}t^{-r}\|f\|_{n+r} \qquad (t > 0),$$

where the constants C and $\overline{C}$ may depend on n and r.

By definition, the Nash-Moser-Sergereart-Hamilton category consists of tame Fréchet spaces and tame mappings between them.

It is not difficult to show that the Cartesian product of tame Fréchet spaces is a tame Fréchet space.

The following theorem holds.

THEOREM 1.1. *Let M^n be a compact manifold. Then $C^\infty(M^n, \mathbb{R})$ is a tame Fréchet space.* □

We do not give a proof of this theorem which can be found in [58, 124].

Now we state the inverse function theorem.

THEOREM 1.2 (Nash-Moser in the form of Sergereart-Hamilton). *Let F and G be tame Fréchet spaces and $P: U \subset F \to G$ a smooth tame mapping. Suppose that the equation for the derivatives $DP(f)h = k$ has a unique solution $h = VP(f)k$ for all f in U and all k, and that the family of inverses $VP: U \times G \to F$ is a smooth tame mapping. Then P is locally invertible, and each local inverse P^{-1} is a smooth tame mapping.* □

The proof of this theorem is given in [58].

DEFINITION 1.8. A tame Fréchet manifold is a Fréchet manifold with coordinate charts in tame Fréchet spaces, and the transition functions are smooth tame mappings.

As shown by a number of authors (see, for example, [124]), if M^n is a compact smooth manifold and V a fiber bundle over M^n, then the space

$C^\infty(M^n, V)$ of smooth sections of V is a tame Fréchet manifold. In particular, if M^n and N^k are smooth manifolds and M^n is compact, then the space $C^\infty(M^n, N^k)$ of smooth mappings is a tame Fréchet manifold. The space $C^\infty(N^n, N^k)$ can be considered as the space of sections of the product bundle $M^n \times N^k$ over M^n [124].

By definition, a smooth tame Lie group is a tame Fréchet manifold $\mathfrak{G}$ equipped with a group structure such that the multiplication mapping $\mathfrak{G} \times \mathfrak{G} \to \mathfrak{G}$ and the inverse mapping are tame mappings. The following theorem holds.

THEOREM 1.3. *Let M^n be a compact smooth manifold, and* $\operatorname{Diff} M^n$ *the group of diffeomorphisms on M^n. Then* $\operatorname{Diff} M^n$ *is a tame Lie group.*

The proof of this theorem is given in [58].

The action of a tame Lie group on a tame Fréchet manifold is defined in an obvious manner. As a model example, consider the action of the tame Lie group $\mathfrak{G} = \operatorname{Diff} M^n \times \operatorname{Diff} N^k$, where M^n and N^k are smooth (compact) manifolds, on the tame Fréchet manifold $C^\infty(M^n, N^k)$ given by the formula

$$\langle(\varphi, \psi)f\rangle = \psi \circ f \circ \varphi^{-1}. \qquad \square$$

In each particular case the proof of the fact that a manifold is a tame Fréchet manifold is, as a rule, far from trivial.

In conclusion we note that Hamilton defines a tame Fréchet space as a direct summand of a graded Fréchet space $\Sigma(B)$, where B is a Banach space. He defines the smoothing operators $S(t)\colon \Sigma(B) \to \Sigma B$ by the formula $(S(t)f)_k = s(t-k)f_k$, where $s(u)$ is a smooth function such that $s(u) = 0$, $u \le 0$, $s(u) = 1$, $u \ge 1$, $0 \le s(u) \le 1$, and $f = \{f_k\}$ is the sequence from $\Sigma(B)$. The smoothing operators satisfy the following estimates:

$$\|S(t)f\|_{n+r} \le Ce^{rt}\|f\|_n, \qquad \|(\mathrm{Id} - S(t))f\|_n \le \overline{C}e^{-rt}\|f\|_{n+r},$$

where the constants C and $\overline{C}$ may depend on n and r. Tame mappings are defined in the same way. For tame spaces in the sense of Hamilton there is the inverse function theorem. It is stated in the same way. In defining tame Fréchet manifolds in the sense of Hamilton we must consider tame Fréchet spaces in the sense of Hamilton; see [58] for a detailed exposition.

§3. Stratification of smooth functions on a manifold

Let M^n be a closed smooth manifold and $C_K^\infty(M^n, \mathbb{R})$ a tame Fréchet space of smooth functions $f\colon M^n \to \mathbb{R}$, $K = f(M^n)$. The group $\operatorname{Diff} M^n \times \operatorname{Diff}_K \mathbb{R}$ acts on $C_K^\infty(M^n, \mathbb{R})$ by the formula

$$(\varphi, \psi)f = \psi \circ f \circ \varphi^{-1} \qquad (\psi|_{\mathbb{R}\setminus K} = \mathrm{id}).$$

Fix a function $f \in C_K^\infty(M^n, \mathbb{R})$ and consider the mapping

$$\Phi\colon \operatorname{Diff} M^n \times \operatorname{Diff}_K \mathbb{R} \to C^\infty(M^n, \mathbb{R}).$$

It follows from the preceding section that Φ is a tame C^∞ mapping. Consider the differential of Φ at the point $(\operatorname{id} M^n, \operatorname{id}\mathbb{R})$. It is known that

$$D = D\Phi(\operatorname{id} M^n, \operatorname{id}\mathbb{R})(\xi_1, \xi_2) = df(\xi_1) + \xi_2 \circ f,$$

where

$$\xi_1 \in \Gamma^\infty(M^n, TM^n), \qquad \xi_2 \in \Gamma_K^\infty(\mathbb{R}, T\mathbb{R}),$$
$$df(\xi_1) + \xi_2 \circ f \in \Gamma^\infty(M^n, f^*T\mathbb{R}),$$

and $\Gamma^\infty(M^n, TM^n) \oplus \Gamma_K^\infty(\mathbb{R}, T\mathbb{R})$ are considered as tangent spaces to the manifold $\operatorname{Diff} M^n \times \operatorname{Diff}_K \mathbb{R}$ at the point $(\operatorname{id} M^n, \operatorname{id}\mathbb{R})$.

DEFINITION 1.9. The codimension $c(f)$ of the function f is the codimension of the image of D in $\Gamma^\infty(M^n, f^*T\mathbb{R})$.

Following Sergereart, we compute the codimension $c(f)$ of the function f by considering its critical points. Making use of the canonical identification of the space $\Gamma_K^\infty(\mathbb{R}, T\mathbb{R})$ of smooth sections with the space $C_K^\infty(\mathbb{R}, \mathbb{R})$ on $\mathbb{R}$ and, similarly, of $\Gamma^\infty(M^n, f^*T\mathbb{R})$ with $C^\infty(M^n, \mathbb{R})$, we can rewrite the mapping D in the form

$$D: \Gamma^\infty(M^n, TM^n) \times C_K^\infty(\mathbb{R}, \mathbb{R}) \to C^\infty(M^n, \mathbb{R}).$$

Let $x \in M^n$. Denote by $C_x^\infty(M^n)$ the space of germs of functions belonging to $C^\infty(M^n, \mathbb{R})$ at the point x, and by $\Gamma_x^\infty(TM^n)$ the space of germs of sections at x. Suppose that x_0 is a critical point and $a = f(x_0)$. The mapping induced by D can be written in the form

$$Dx_0: \Gamma_{x_0}^\infty(TM^n) \times C_a^\infty(\mathbb{R}) \to C_{x_0}^\infty(M^n).$$

Clearly, the codimension d of the image D_{x_0} satisfies the inequality $d \le c(f)$. Set $A = df\Gamma_{x_0}^\infty(x_0)(TM^n)$. The space A is evidently an ideal of $C_{x_0}^\infty(M^n)$ generated by the partial derivatives of the function f. Let $A_n = A + (f - f(x_0))^n$. If $\xi_2 \in C_a^\infty(\mathbb{R})$, then, by the Taylor formula,

$$(\xi_2 \cdot f)(x) = \sum_{i=0}^{\infty} \frac{\xi_2^{(i)}(a)}{i!}(f(x) - f(x_0))^i$$
$$+ \frac{(f(x) - f(x_0))^n}{(n-1)!} \int_0^1 (1-t)^{n-1} \xi_2^{(n)}[tf(x) + (1-t)f(x_0)]\, dt.$$

Hence it follows that the image of D_{x_0} belongs to the subspace

$$A_n + \sum_{i=0}^{n-1} (f - f(x_0))^i \mathbb{R}.$$

Let k be the least positive integer such that

$$\operatorname{Im} D_{x_0} = A \oplus \sum_{i=0}^{k-1} \mathbb{R}(f - f(x_0))^i.$$

DEFINITION 1.10. Let $x_0 \in M^n$, $f \in C^\infty(M^n, \mathbb{R})$, and $a = f(x)$. Then

(1) the codimension of the function $c(f, x_0)$ at the point x_0 is the dimension of $C^\infty_{x_0}(M^n)/A$ as a vector space over $\mathbb{R}$;

(2) the dimension of the function f for the pair (x_0, a) and $d(f, a, x_0)$ is the least positive integer k such that $(f - f(x_0))^k \in A$; and

(3) the dimension of f at the point a is the number

$$d(f, a) = \sup_{f(x)=a} d(f, a, x);$$

if $a \notin f(M^n)$, we set $d(f, a) = 0$.

PROPOSITION 1.1. *Let $f: M^n \to \mathbb{R}$ be a function of finite codimension $c(f)$. Then the numbers $c(f, a)$, $d(f, a, x_0)$, and $d(f, a)$ are finite and*

$$c(f) = \sum_{x_i \in M^n} c(f, x_i) - \sum_{a_i = f(x_i)} d(f, a_i)$$

(x_i are critical points of the function f). □

Suppose that the function $f: M^n \to \mathbb{R}$ is of codimension n. Choose the functions $f_1, \dots, f_n$ forming the basis of the quotient space

$$C^\infty(M^n, \mathbb{R})/D(\Gamma^\infty(M^n, TM^n) \times C^\infty_K(\mathbb{R}, \mathbb{R})) \approx \mathbb{R}^n.$$

Consider the mapping

$$D: \Gamma^\infty(M^n, TM^n) \times C^\infty_K(\mathbb{R}, \mathbb{R}) \times \mathbb{R}^n \to C^\infty(M^n, \mathbb{R}),$$

defined by the formula

$$D(\xi_1, \xi_2, \lambda_1, \lambda_2, \dots, \lambda_n) = df\xi_1 + \xi_2 \circ f + \sum_{i=1}^n \lambda_i f_i.$$

The following two statements are due to Sergereart [124].

PROPOSITION 1.2. *There exists a tame mapping*

$$L: C^\infty(M^n, \mathbb{R}) \to \Gamma^\infty(M^n, TM^n) \times C^\infty_K(\mathbb{R}, \mathbb{R}) \times \mathbb{R}^n,$$

such that the composition mapping $D \circ L$ is the identity mapping of $C^\infty(M^n, \mathbb{R})$. □

THEOREM 1.4. *Let $f: M^n \to \mathbb{R}$ be of codimension n. There exist a neighborhood of the function $\mathfrak{V} \ni f$ in $C^\infty(M^n, \mathbb{R})$, n functions $f_1, \dots, f_n$ in this neighborhood, and two tame C^∞-mappings*

$$S_1, S_2: \mathfrak{V} \to \operatorname{Diff} M^n \times \operatorname{Diff} \mathbb{R} \times \mathbb{R}^n$$

such that if

$$S_1(g) = (\varphi_1, \varphi_2, \lambda_1, \dots, \lambda_n), \qquad S_2(g) = (\psi_1, \psi_2, \mu_1, \dots, \mu_n),$$

then

$$g = \varphi_2 \circ f \circ \varphi_1 \circ \sum_{i=1}^{n} \lambda_i f_i = \psi_2 \circ \left(f + \sum_{i=1}^{n} \mu_i f_i \right) \circ \psi_1. \qquad \square$$

This theorem shows that the action of the group $\operatorname{Diff} M^n \times \operatorname{Diff}_K \mathbb{R}$ on the space $C^\infty(M^n, \mathbb{R})$ is locally trivial. We note that the proof of this theorem is based on the inverse function theorem discussed in the preceding section.

Denote by $\mathscr{F}^i$ the subspace of $C^\infty(M^n, \mathbb{R})$ consisting of functions of codimension i. Following Cerf, the space $\mathscr{F}^i$ can be used for the stratification of the space $C^\infty(M^n, \mathbb{R})$.

DEFINITION 1.11. Let X be a topological space. A stratification ΣX of X is a collection of subspaces $(X^0, X^1, \dots, X^i, \dots, X^\infty)$ in X satisfying the following conditions:

(1) $X^i \cap X^j = \varnothing$;

(2) $\bigcup_i X^i = X$; and

(3) $\bigcup_{0 \le i \le n} X^i$ is an open set in X for each n.

DEFINITION 1.12. A stratification ΣX of the space X is said to be locally trivial if for any point $x \in X$ there is a stratified set E with the point stratum $\{0\}$, a topological space Y with the trivial stratification and a chosen point y, and a morphism $\psi: E \times Y \to X$ of stratified spaces such that

(1) $\psi(0, y) = x$;

(2) the set $\psi(E \times Y)$ is open in X; and

(3) the mapping ψ is a homeomorphism onto its image.

In the next definition we follow Cerf [17].

DEFINITION 1.13. The "natural" stratification $\Sigma C^\infty(M^n, \mathbb{R})$ is the collection of subspaces

$$\mathscr{F}^0, \mathscr{F}^1, \dots, \mathscr{F}^i, \dots, \mathscr{F}^\infty.$$

It is known that the "natural" stratification of $C^\infty(M^n, \mathbb{R})$ has a number of remarkable properties:

(1) $\Sigma C^\infty(M^n, \mathbb{R})$ is locally trivial (this is essentially a consequence of Theorem 1.4);

(2) the stratification is invariant under the action of the group $\operatorname{Diff} M^n \times \operatorname{Diff} \mathbb{R} = \mathfrak{G}$;

(3) if $f \in \mathscr{F}^i$ $(i < \infty)$, then the mapping Φ_f defines a locally trivial bundle $\mathfrak{G} \to$ orbit of f;

(4) if $i \le 5$, then $\mathscr{F}^i$ coincides in a neighborhood of the function f with its orbit (note that in codimension greater than 5 the latter condition does not necessarily hold; a counterexample is provided by the function

$$f = x^5 + x^2 y^2 + y^4$$

constructed by Hendrics [61]); and

(5) for each i the subspace $\mathscr{F}^i$ is of codimension i in $C^\infty(M^n, \mathbb{R})$.

It follows from condition (5) that any mapping $S_i \to C^\infty(M^n, \mathbb{R})$ can be deformed in $\mathcal{F}^0 \cup \mathcal{F}^1 \cup \cdots \cup \mathcal{F}^{i+1}$.

Let us describe the strata in codimensions 0, 1, and 2. The stratum in codimension 0 consists of Morse functions in general position. We remind the reader that a Morse function is a smooth function $f: M^n \to \mathbb{R}$ whose critical points are all nondegenerate. A critical point x_0 of the function f is said to be nondegenerate if there is a coordinate chart in a neighborhood of x_0, where f can be represented in the form

$$f = f(x_0) - \sum_{i=1}^{\lambda} x_i^2 + \sum_{i=\lambda+1}^{n} x_i^2.$$

The number λ is called the index of the critical point. A Morse function is in general position if it assumes distinct values at its critical points. Morse functions are considered in the subsequent sections.

A stratum in codimension 1 consists of two components: $\mathcal{F}_\alpha^1 \cup \mathcal{F}_\beta^1$. The component $\mathcal{F}_\alpha^1$ includes "birth" functions, i.e., those functions for which all but one critical points are nondegenerate, and in a neighborhood of the exceptional point the function can be represented in the form

$$f(x) = f(x_0) - \sum_{i=1}^{\lambda} x_i^2 + \sum_{i=\lambda+1}^{n-1} x_i^2 + x_n^3.$$

Furthermore, the functions in $\mathcal{F}_\alpha^1$ assume distinct values at their critical points. The component $\mathcal{F}_\beta^1$ consists of Morse functions whose values coincide at two critical points and are distinct at the others. In what follows, we call a critical point at which the function has such a singularity a "birth-death" point. In order to describe the stratum in codimension 2 we shall need the notion of a swallow tail singularity. We remind the reader that a function $f: M^n \to \mathbb{R}$ has a swallow tail singularity at the point x_0 if there exists a coordinate chart containing x_0, where f can be represented in the form

$$f(x) = f(x_0) - \sum_{i=1}^{\lambda} x_i^2 + \sum_{i=\lambda+1}^{n-1} x_i^2 + x_n^4.$$

The stratum in codimension 2 has six components. The component $\mathcal{F}_\alpha^2$ consists of functions whose all but one critical points are nondegenerate. The exceptional point has a swallow tail singularity, while at the critical points any function from $\mathcal{F}_\alpha^2$ assumes distinct values. The component $\mathcal{F}_\beta^2$ contains functions with exactly two critical points being birth-death points, other critical points being nondegenerate. Functions in this component assume distinct values at critical points. The component $\mathcal{F}_\gamma^2$ contains only such functions for which only one critical point is a birth-death point, and all the others are nondegenerate. Moreover, each function f in this component has two nondegenerate critical points, where f assumes the same value. At all other critical points f assumes distinct values. The component $\mathcal{F}_\delta^2$ consists of functions having one critical birth-death point, other critical points being nondegenerate. For each function in $\mathcal{F}_\delta^2$ the level surface containing the critical birth or death point contains just one nondegenerate critical point. The values of the function at other critical points are distinct. The component $\mathcal{F}_\varepsilon^2$ contains Morse functions that assume the same value at three critical points and distinct values at all the others. Finally, the component $\mathcal{F}_\xi^2$ includes Morse functions having two pairs of critical points at which they assume the same values. At all other critical points the values of the functions are distinct.

Let $\gamma: [0, 1] \to C^\infty(M^n, \mathbb{R})$ be a curve in the space of functions.

DEFINITION 1.14. A path γ is said to be in general position if γ belongs to C^∞, $\gamma(0), \gamma(1) \in \mathcal{F}_0$, $\gamma([0, 1]) \cap \mathcal{F}_1$ consists of finitely many points the intersection at which is transversal, and $\gamma([0, 1])$ does not intersect strata in codimension 2 or greater.

Since $C^\infty(M^n, \mathbb{R})$ is a smooth Fréchet manifold, we can speak of differentiability of a path and of a transversal intersection. The connectivity of $C^\infty(M^n, \mathbb{R})$ and the transversality theory implies the following proposition whose proof can be found in [17, 19, 59].

PROPOSITION 1.3. *Let f_1 and f_0 be Morse functions in general position defined on the manifold M^n. There exists a path $\gamma: [0, 1] \to C^\infty(M^n, \mathbb{R})$ joining f_0 and f_1.* $\square$

It can be shown that if the path intersects no strata in codimension one or greater, then there exist diffeomorphisms $h: M^n \to M^n$, $\varphi: \mathbb{R} \to \mathbb{R}$ isotopic to the identity and such that $\gamma(0) = \varphi_0 \gamma(1) \circ h$.

If the path γ intersects $\mathcal{F}^1$ at the point $\gamma(t_0) \in \mathcal{F}_\beta^1$, then the number of critical points is the same for all the functions $\gamma(\tau)$ (where τ belongs to an ε-neighborhood of the point t_0). If the path γ intersects $\mathcal{F}^1$ at the point $\gamma(t_1) \in \mathcal{F}_\alpha^1$, then, in a neighborhood of the point t_1, the number of critical points of index λ and $\lambda+1$ differ by one. In other words, when γ intersects $\mathcal{F}_\alpha^1$, the critical points of index λ and $\lambda + 1$ either are born or die for the corresponding Morse functions. This is considered in more detail in [17, 19, 59]. In the following chapters we consider obstacles arising if in the space $C^\infty(M^n, \mathbb{R})$ we attempt to join two Morse functions lying in the subspace of Morse functions.

CHAPTER II

Minimal Morse Functions on Simply Connected Manifolds

Morse functions on smooth manifolds play an important role in many areas of mathematics. Classical topics of Morse theory are well covered in textbooks, and we do not devote much space to them in this book. Our main focus is on modern aspects of Morse theory.

In the first section we give a short proof of Smale's theorem on the existence of a minimal Morse function on a simply connected manifold in dimension greater than 5. In the second section we study epimorphisms of free abelian groups into a finitely generated abelian group. The results obtained there are applied to the classification of minimal Morse functions. In the subsequent sections we introduce three important equivalence relations for Morse functions—homotopy, isotopy, and conjugation—and give necessary and sufficient conditions for them to be satisfied for minimal Morse functions. Let us state one of these conditions.

Two Morse functions f_0 and f_1 are isotopic if there is a path $\gamma: I \to C^\infty(M^n, \mathbb{R})$ in the space $C^\infty(M^n, \mathbb{R})$ such that $\gamma(0) = f_0$, $\gamma(1) = f_1$, and $\gamma(t)$ is a Morse function for all $t \in I$. The following result holds. Two minimal Morse functions on a simply connected closed manifold M^n $(n > 5)$ are isotopic if and only if their homology invariants coincide.

It turns out that on a simply connected manifold there exist finitely many nonisotopic minimal Morse functions expressed in explicit form.

§1. Ordered minimal functions

Let $f: M^n \to [0, 1]$ be a Morse function in general position. We say that f is an ordered function if $f(x_i^\lambda) < f < f(y_j^{\lambda+1})$, where x_i^λ $(y_j^{\lambda+1})$ are critical points of index λ $(\lambda + 1)$. A proof that such functions exist can be found in [96]. Morse functions are associated with a gradient-like vector field. Let us recall its definition. Suppose that a Riemannian metric ν is fixed on a smooth manifold M^n. A vector field ξ is gradient-like for a Morse function f if $\xi > 0$ on the complement of the set of critical points of f and $\xi = \operatorname{grad}_\nu f$ in a neighborhood U of each critical point. It is well known that for any Morse function on the manifold M^n there exists a gradient-like vector field ξ with respect to a Riemannian metric ν [96]. In what follows, as a rule, we will not specify the Riemannian metric with respect to which the gradient-like vector field is given.

Suppose that a Morse function f and a gradient-like vector field ξ are fixed. Let x_0 be a critical point of the function f. Choose $\varepsilon > 0$ such that the manifold

$$f^{-1}[f(x_0), f(x_0+\varepsilon)] \cup f^{-1}[f(x_0), f(x_0-\varepsilon)]$$

contains no critical points other than x_0. We suppose that f is an ordered Morse function. The integral curve of the field ξ entering x_0 issues from a uniquely defined point on the manifold $f^{-1}(f(x_0)-\varepsilon)$. The set of segments of integral curves issuing from the manifold $f^{-1}(f(x_0)-\varepsilon)$ and entering the point x_0 is called the middle disk $D_L(x_0)$ of the critical point x_0. The manifold

$$D_L(x_0) \cap f^{-1}(f(x_0)-\varepsilon)$$

is called the middle sphere $S_L(x_0)$ of the point x_0. The submanifold $D_L(x_0)$ is a smooth image of a Euclidean disk and with boundary $S_L(x_0)$. The comiddle disk is defined analogously. Consider the segments of integral curves beginning at the point x_0 and ending on $f^{-1}(f(x_0)+\varepsilon)$. The resulting set is called the comiddle disk of the point x_0, and its boundary the comiddle sphere $S_R(x_0)$. Clearly, both middle and comiddle disks depend on the choice of the vector field ξ. By varying ξ, we can select middle and comiddle disks with desired properties. It is known that the middle and comiddle disks are parts of the stable and nonstable manifolds of the critical point x_0. Denoting by φ_t the one-parameter group of diffeomorphisms generated by the vector field, define the stable and nonstable manifolds of the point x_0 by the relations

$$W(x_0) = \left\{x \in M^n \mid \lim_{t\to\infty} \varphi_t(x) = x_0\right\},$$

$$W^*(x_0) = \left\{x \in M^n \mid \lim_{t\to-\infty} \varphi_t(x) = x_0\right\},$$

whence $D_L(x_0) = f^{-1}(f(x_0)-\varepsilon)\cap W(x_0)$, $D_R(x_0) = f^{-1}(f(x_0)+\varepsilon)\cap W^*(x_0)$.

Denote by $\mathscr{F}_\lambda(M^n) \subset C^\infty(M^n, \mathbb{R})$ the set of Morse functions with the minimal possible number of critical points of index λ.

DEFINITION 2.1. A function $f \in \bigcap_{\lambda=0}^{n} \mathscr{F}_\lambda(M^n)$ is called a minimal (exact) Morse function on the manifold M^n.

As we shall see in what follows, minimal Morse functions do not necessarily exist on every smooth manifold, and one of the central questions we are discussing in this book is the condition of existence of minimal Morse functions.

Suppose that W^n is a smooth compact manifold with boundary $\partial W^n = V_0 \cup V_1$. In the sequel by a Morse function on W^n we mean any function

$f: W^n \to [0, 1]$ such that $f^{-1}(0) = V_0$, $f^{-1}(1) = V_1$, and all critical points of f are nondegenerate and lie inside W_n. The sets V_0 and V_1 may be empty.

The following theorem was proved by Smale [143, 144].

THEOREM 2.1. *Let W^n be a smooth compact manifold with boundary $\partial W^n = V_0 \cup V_1$, $\pi_1(W^n, x) \approx \pi_1(V_0, x) \approx \pi_1(V_1, \overline{x}) = 0$, $n \geq 6$. Then there exists a minimal Morse function on W^n whose number of critical points of index λ is*

$$N_\lambda = \mu(H_\lambda(W^n, V_0, \mathbb{Z})) + \mu(\operatorname{Tors} H_{\lambda-1}(W^n, V_0, \mathbb{Z})),$$

where $\mu(H)$ is the minimal number of generators of the group H. □

The theorem has a long history beginning with the work of Morse [102] who laid the foundations of the theory which now bears his name. Many mathematicians took part in the development of Morse theory. The most important results here belong to Bott, Lyusternik, Shnirel′man, Chogoshvili, Elsholz, Pitcher, and Reeb [78, 20, 21, 44, 113, 121].

Now we give a proof of Theorem 2.1 different from the original proof of Smale and more appropriate for our purposes. We will also prove a generalization of this theorem illustrating the impact of the fact that the boundary is non-simply-connected. We follow mainly the works [126, 128, 130]. For technical reasons, we will use, instead of Morse functions, decomposition of a manifold into handles. Let us recall the facts necessary for what follows (see [47, 28] for more details).

Suppose that a smooth embedding

$$\varphi: S^{\lambda-1} \times D^{n-\lambda} \to \partial W^n$$

is given, where ∂W^n is the boundary of the smooth manifold W^n. Let $\overline{W}^n_\varphi$ be the manifold obtained from the disjoint union of W^n and $D^\lambda \times D^{n-\lambda}$ by identifying their points by means of the mapping φ. One can introduce a differentiable structure on the manifold $\overline{W}^n_\varphi$ [27]. We say that the manifold $\overline{W}^n_\varphi$ is obtained from W^n by attaching a handle of index λ with respect to φ. Let $h: D^\lambda \times D^{n-\lambda} \to \overline{W}^n_\varphi$ be the natural embedding. Then $h(\partial D^\lambda \times 0)$ is called the middle sphere, $h(0 \times \partial D^{n-\lambda})$ the comiddle sphere, $h(D^\lambda \times 0)$ the middle disk, and $h(0 \times Dn - \lambda)$ the comiddle disk of the handle.

If $\overline{\varphi}: S^{\lambda-1} \times D^{n-\lambda} \to \partial W^n$ is an embedding isotopic to the embedding φ in ∂W^n, then $\overline{W}^n_\varphi$ is diffeomorphic to $\overline{W}^n_{\overline{\varphi}}$. In what follows, if the manifold $\overline{W}^n$ is obtained from W^n by attaching handles of index λ, we have

$$\overline{W}^n = W^n + h_1^\lambda + h_2^\lambda + \cdots + h_k^\lambda.$$

It is known that a smooth compact manifold without boundary can be obtained from a disk by consecutively attaching handles of increasing index.

The images of the mappings corresponding to handles of the same index do not intersect. Such a decomposition into handles is said to be ordered. A similar statement holds for a manifold with boundary, the only difference being that the disk is replaced by the collar of the boundary [47].

There is a connection, established by Smale, between ordered Morse functions on a manifold W^n and the ordered decomposition of this manifold into handles. Let us now make a precise statement. Let $f: W^n \to [0, 1]$ be an ordered Morse function. Define a Riemannian metric ν on W^n and construct a gradient-like vector field ξ for which no integral curve joins two critical points of the same index. The proof of the existence of such a vector field can be found in the book [94].

Choose a sequence of numbers $c_1, \dots, c_{n-1}$ satisfying the inequalities $f(x_i^\lambda) < c^\lambda < f(y_j^{\lambda+1})$, where x_i^λ $(y_j^{\lambda+1})$ are the critical points of index λ $(\lambda + 1)$. Using the vector field ξ, we construct middle and comiddle disks of the critical points of index λ in such a way that the middle and comiddle spheres lie on the manifolds $f^{-1}(c_{\lambda-1})$ and $f^{-1}(c_\lambda)$, respectively. Consider the tubular neighborhoods of sufficiently small radius for each middle (comiddle) disk on the submanifold $\Omega = f^{-1}[c_\lambda, c_{\lambda-1}]$. After smoothing the angles they can be interpreted as handles of index λ. It can be shown that the manifold $W_\lambda = f^{-1}[0, c_\lambda]$ is diffeomorphic to the manifold $W_{\lambda-1} = f^{-1}[0, c_{\lambda-1}]\bigcup_{i=1}^k h_i^\lambda$, where k is the number of critical points of index λ and the h_i^λ are the corresponding handles of index λ.

The manifolds $W_\lambda = f^{-1}[0, c_\lambda]$ define a filtration

$$W_0 \subset W_1 \subset \dots \subset W_i \subset \dots \subset W_n = W^n.$$

It is known that

$$H_i(W_\lambda, W_{\lambda-1}, \mathbb{Z}) \approx \begin{cases} 0 & \text{for } i \neq \lambda, \\ \underbrace{\mathbb{Z} \oplus \dots \oplus \mathbb{Z}}_{k} & \text{for } i = \lambda. \end{cases}$$

The basis of the homology group $H_\lambda(W_\lambda, W_{\lambda-1}, \mathbb{Z})$ is given by the middle disks of critical points (handles) of index λ. Making use of the exact homology sequence for the triple $W_{\lambda-1} \subset W_\lambda \subset W_{\lambda+1}$, we can write out the chain complex of free abelian groups

$$\{C, \partial\}: C_0 \leftarrow \dots \leftarrow C_{\lambda-1} \xleftarrow{\partial_\lambda} C_\lambda \xleftarrow{\partial_{\lambda+1}} C_{\lambda+1} \leftarrow \dots \leftarrow C_n,$$

whose homologies coincide with those of the manifold W^n. Here we have $C_\lambda = H_\lambda(W_\lambda, W_{\lambda-1}, \mathbb{Z})$.

Suppose that W^n is an oriented manifold. Given an orientation of the middle disk D_L^λ of the critical point (handle) of index λ, the orientation of the comiddle disk $D_R^{n-\lambda}$ of this point (handle) is defined uniquely from the condition that the index of intersection $\varepsilon(D_L^\lambda, D_R^{n-\lambda})$ be equal to $+1$. The orientation of the handle h_i^λ is obtained as the product of the orientations of D_L^λ and $D_R^{n-\lambda}$. The orientations of the middle and comiddle spheres are induced by the orientations of the middle and comiddle disks. Therefore the choice of orientations makes it possible to compute the index of intersection of the middle and comiddle spheres of the critical points (handles) of index λ and $\lambda+1$ in the manifold $f^{-1}(c_\lambda)$ (under the assumption that they intersect transversally).

The matrix of the homomorphism ∂_λ is given by the matrix of intersection indices of the middle and comiddle spheres of critical points (handles) of index $\lambda+1$ and λ in the manifold $f^{-1}(c_\lambda)$. The matrix of ∂_λ can be modified algebraically by any sequence of the following elementary operations:

(1) permutation of rows (columns);
(2) addition of a row to another row; and
(3) multiplication of a row by -1.

These operations can be realized geometrically. Operation (1) corresponds to the renumbering of the critical points (handles) of the index $\lambda+1$ (λ), operation (2) to the addition of handles, and operation (3) to the reversal of the orientation of the middle disk of the critical point (handle) of index $\lambda+1$ (λ). Note that operations (1) and (2) change neither the Morse function f nor the gradient-like vector field ξ, while operation (3) alters both f and ξ in some neighborhoods of critical points of index $\lambda+1$ and λ. This process is described in detail in [28].

If the comiddle sphere of a critical point (handle) of index $\lambda+1$ intersects the middle sphere of a critical point (handle) of index λ in a single point, then these critical points (handles) can be eliminated. In other words, there is a smooth path $\gamma: [0, 1] \to C^\infty(M^n, \mathbb{R})$ such that the function $\gamma(t_0)$ ($t_0 \in [0, 1]$) belongs to the component $\mathscr{F}_\alpha^1$ of the stratum of codimension $+1$. For $t < t_0$ the Morse function $\gamma(t)$ has the same critical points as f. For $t > t_0$ the number of the critical points of index λ and $\lambda+1$ of the Morse function $\gamma(t)$ is one less than that of f, while the remaining critical points coincide with the critical points of f. Both f and ξ are essentially altered in a neighborhood containing just these two critical points. For more details see [96].

Along with the Morse functions f, we can consider the functions $-f$. The critical points of index λ for f become critical points of index $n-\lambda$ for $-f$. In the language of handles, we obtain a dual handle decomposition of the manifold W^n in the sense that a handle of index λ is replaced by the handle of index $n-\lambda$. Similarly, we can construct the dual chain complex

$$\{C^*, \partial^*\}: C_0^* \leftarrow \cdots \leftarrow C_\lambda^* \xleftarrow{\partial_{\lambda+1}^*} C_{\lambda+1}^* \xleftarrow{\partial_{\lambda+2}^*} C_{\lambda+2}^* \leftarrow \cdots \leftarrow C_n^*,$$

whose matrix ∂_λ^* is equal to $(-1)^{-\lambda+1}\overline{\partial}_{n-\lambda}$ ($\overline{\partial}_{n-\lambda}$ is the transpose of $\partial_{n-\lambda}$).

Proof of Smale's theorem. Fix an ordered Morse function f on the manifold W^n. The homotopic clauses of the theorem imply that there exists a Morse function on W^n having no critical points of index $0, 1, n-1, n$. The proof of this assertion can be found in [123]. Suppose that the function f satisfies this condition. Consider the filtration

$$V_0 \times I = W_0 \subset W_2 \subset \cdots \subset W_{n-2} \subset W_n = W^n.$$

Here $W_\lambda = f^{-1}[0, c_\lambda]$, and c_λ satisfies the inequalities $f(x_\lambda) < c_\lambda < f(y_{\lambda+1})$, where x_λ $(y_{\lambda+1})$ is a critical point of index λ $(\lambda+1)$. This filtration makes it possible to construct the chain complex

$$\{C, \partial\}: C_2 \xleftarrow{\partial_3} C_3 \longleftarrow \cdots \xleftarrow{\partial_{n-2}} C_{n-2}.$$

By using the handle addition procedure (i.e., by modifying the function f) the matrices ∂_λ can be reduced to the diagonal form

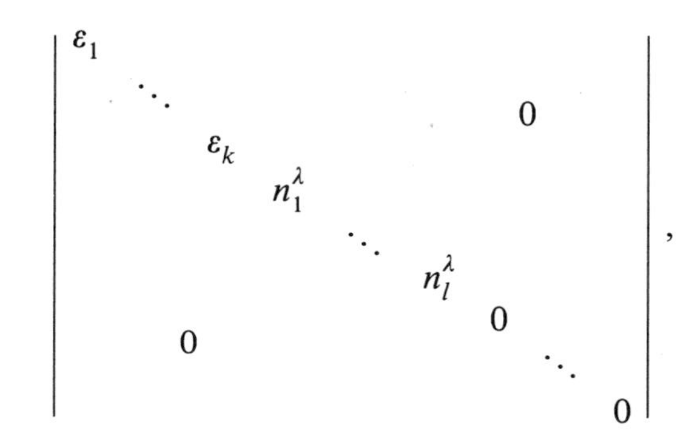

where $\varepsilon_i = \pm 1$ and n_j divides n_{j+1}, $n_j \in \mathbb{Z}$, $n_j \neq \pm 1, 0$. The handles of adjacent indices with the index of intersection of middle and comiddle spheres equal to ε_i can be cancelled, so they make no contribution to the homology of the manifold. The handles of index λ that do not intersect those of index $\lambda - 1$ and $\lambda + 1$ and correspond to the free generators of the group $H_\lambda(W^n, V_0, \mathbb{Z})$, remain. No cancellation occurs for the handles of index λ and $\lambda + 1$ whose middle and comiddle spheres have intersection index equal to n_j. These handles define a torsion in the group $H_\lambda(W^n, V_0, \mathbb{Z})$. Thus, we can construct a Morse function on W^n without "redundant" critical points. It is evidently a minimal Morse function. The proof of the theorem is similar if either or both of the manifolds V_i $(i = 0, 1)$ are empty. In these cases one or two critical points of index 0 and n are added. □

In what follows, a minimal Morse function whose associated chain complex is of diagonal form will be called a distinguished minimal Morse function.

Let us now prove a more general theorem.

THEOREM 2.2. *Let W^n be a smooth compact manifold with the boundary $\partial W^n = V_0 \cup V_1$, $\pi_1(W^n, x) = 0$ $(n \geq 6)$. Then there exists a minimal Morse function on W^n having no critical points of index* 0, 1, $n-1$, n, *and with*

$$\mu(\pi_2(W^n, V_0))$$

critical points of index 2,

$$\mu(\pi_2(W^n, V_0)) + \mu(H_n(W^3, V_0, \mathbb{Z})) + \mu(H_2(W^n, V_0, Q))$$

critical points of index 3,

$$\mu(\pi_2(W^n, V_1)) + \mu(H_{n-3}(W^n, V_0, Q)) \\ + \mu(\operatorname{Tors} H_{n-4}(W^n, V_0, \mathbb{Z})) - \mu(H_{n-2}(W^n, V_0, \mathbb{Z}))$$

critical points of index $n-3$,

$$\mu(\pi_2(W^n, V_1))$$

critical points of index $n-2$, *and*

$$\mu(H_\lambda(W^n, V_0, \mathbb{Z})) + \mu(\operatorname{Tors} H_{\lambda-1}(W^n, V_0, \mathbb{Z}))$$

critical points of index λ $(4 \leq \lambda \leq n-4)$.

Here $\pi_2(W^n, V_i)$ $(i = 0, 1)$ is considered as a crossed $\pi_1(V_i)$-module. Such modules are discussed in detail in Chapter VI.

To prove the theorem we shall need the following lemma.

LEMMA 2.1. *Let $\pi_1(W^n, x) \approx \pi_1(V_1, x) = 0$. Then there exists a minimal Morse function on the manifold having no critical points of index* 0, 1, $n-1$, *and n and having $\mu(\pi_2(W^n, V_0))$ critical points of index* 2,

$$\mu(\pi_2(W^n, V_0)) + \mu(H_3(W^n, V_0, \mathbb{Z})) - \mu(H_2(W^n, V_0, Q))$$

critical points of index 3, *and*

$$\mu(H_\lambda(W^n, V_0, \mathbb{Z})) + \mu(\operatorname{Tors} H_{\lambda-1}(W^n, V_0, \mathbb{Z}))$$

critical points of index λ $(4 \leq \lambda \leq n-2)$.

PROOF OF LEMMA 2.1. Instead of Morse functions, we shall use the handlebody decomposition of the manifold, starting with $V_0 \times I$. It is known that there is a decomposition having no handles of index 0 and 1, because $\pi_1(W_0^n, V_0) = 0$ [123]. Let $a_1, a_2, \dots, a_{\mu_1}$ be a minimal system of generators of $\pi_2(W^n, V_0)$ as a $\pi_1(V_0, x)$-module. Let us realize them as smooth nonintersecting embeddings of the disks $\varphi_i: (D_i^2, S_i^1) \to (W^n, V_0 \times 1)$, such that S_i^1 is the transversal intersection of $\varphi_i(D_i^2)$ with $V_0 \times 1$. Let Γ_i be tubular neighborhoods of $\varphi(D_i^2)$ in $\overline{W^n \setminus V_0 \times I}$. Clearly, Γ_i is diffeomorphic to $D^2 \times D^{n-2}$. Let

$$Y = (V_0 \times I) \bigcup_i (\Gamma_i), \qquad X = \overline{W^n \setminus Y}, \qquad \overline{V} = \partial Y \setminus V_0.$$

A standard application of the van Kampen-Seifert theorem shows that $\pi_1(Y, y) = \pi_1(X, y) = \pi_1(\overline{V}, y) = 0$. Consider the following commutative diagram:

$$\begin{array}{ccccccccc} \pi_2(V_2) & \longrightarrow & \pi_2(Y) & \longrightarrow & \pi_2(Y, V_0) & \longrightarrow & \pi_1(V_0, x) & \longrightarrow & (e) \\ & & \downarrow{\scriptstyle\alpha} & & \downarrow{\scriptstyle\beta} & & & & \\ \pi_2(V_0) & \longrightarrow & \pi_2(W^n) & \longrightarrow & \pi_2(W^n, V_0) & \longrightarrow & \pi_1(V_0, x) & \longrightarrow & (e) \\ & & \downarrow & & & & & & \\ & & \pi_2(W^n, Y) & & & & & & \end{array}$$

By construction, β is an epimorphism. The lemma on four homomorphisms [81] imply that α is an epimorphism. Therefore $\pi_2(W^n, Y) = 0$ and by the Hurewicz theorem $H_2(W^n, Y) = 0$. Making use of the excision theorem and the Poincaré duality, we obtain

$$H_2(W^n, Y, \mathbb{Z}) \approx H_2(X, \overline{V}, \mathbb{Z}) \approx H_{n-2}(X, V_1, \mathbb{Z}) \approx \operatorname{Tors}(X, V_1, \mathbb{Z}) = 0.$$

Consider the pair W^n, X. By construction, the manifold W^n is obtained from X by attaching handles of index $n-2$. Therefore

$$H_{n-2}(W^n, X, \mathbb{Z}) = \mu_1\mathbb{Z}, \qquad H_i(W^n, X, \mathbb{Z}) = 0 \quad (i \neq n-2).$$

Let

$$0 \to H_{n-2}(W^n, V_1, \mathbb{Z}) \to H_{n-2}(W^n, X, \mathbb{Z}) \to H_{n-3}(X, V_1, \mathbb{Z}) \to H_{n-3}(W^n, V_1, \mathbb{Z}) \to 0$$

be a segment of the exact homology sequence for the triple (W^n, X, Y). It is not difficult to compute that

$$H_{n-3}(X, V_1, \mathbb{Z}) = (\mu_1 + \mu_2 - \mu_3 - \mu_4)\mathbb{Z},$$

where $\mu_2 = \mu(H_{n-3}(W^n, V_1, \mathbb{Z}))$, $\mu_3 = \mu(H_{n-2}(W^n, V_1, \mathbb{Z}))$, and $\mu_4 = \mu(\operatorname{Tors} H_{n-3}(W^n, V_1, \mathbb{Z}))$. Using the Smale theorem, let us construct the following minimal Morse function on the manifold X:

$$f_1: X \to [1/2, 1], \qquad f_1^{-1}(1/2) = V_1, \qquad f_1^{-1}(1) = \overline{V}.$$

For this function the number of critical points of index λ is equal to

$$N_\lambda = \mu(H_\lambda(W^n, V_1, \mathbb{Z})) + \mu(\operatorname{Tors} H_{\lambda-1}(W^n, V_1, \mathbb{Z})) \qquad (2 \le \lambda \le n-4),$$
$$N_{n-3} = \mu_1 + \mu_2 - \mu_3 - \mu_4 + \mu(\operatorname{Tors} H_{\lambda-1}(W^n, V_1, \mathbb{Z})).$$

An easy computation based on the Poincaré duality yields

$$N_{n-3} = \mu_1 + \mu(H_3(W^n, V_0, \mathbb{Z})) - \mu(H_2(W^n, V_0, Q)).$$

On the manifold Y, there exists a Morse function $f_2: Y \to [0, 1/2]$ such that $f_2^{-1}(0) = V_0$, $f_2^{-1}(1/2) = \overline{V}$, and f_2 has μ_1 critical points of index 2.

Consider the function $f = f_2 \cup (1 - f_1)$. We claim that f is a minimal Morse function on the manifold W^n. Indeed, let g be an arbitrary ordered Morse function on W^n and ξ a gradient-like vector field for g no integral curve of which connects two critical points of the same index. Let $W_2 = g^{-1}[0, 1/2]$ be a submanifold containing all critical points of index 2 and only these critical points. We will show that the number of critical points of index 2 for the function g is not less than that for the function f. The mapping

$$\pi_2(W_2, V_0) \to \pi_2(W^n, V_0)$$

is an epimorphism, because each element of $\pi_2(W^n, V_0)$ can be realized as an embedding of a two-dimensional disk and then, by the general position argument, transformed by isotopy into the manifold W_2. The middle disks of critical points of g of index 2 provide generators for $\pi_2(W^n, V_0)$ as a $\pi_1(V_0, x)$-module, and therefore their number is not less than μ_1. Here we make use of the Whitehead theorem on the structure of the second relative homotopy group; this is shown in more detail in Chapter VI. Let $\mu(\pi(W_2, V_0)) = k$. Set $\widehat{V} = \partial W_2 \setminus V_0$ and $\widehat{X} = \overline{W^n \setminus W_2}$. It is easy to prove that $\pi_1(W_2) \approx \pi_1(\widehat{V}) \approx \pi_1(\widehat{X}) = 0$. Making use of the argument above, we can show that

$$H_i(\widehat{X}, V_1, \mathbb{Z}) \approx H_i(W^n, V_1, \mathbb{Z}) \qquad (i < n-3),$$
$$H_{n-3}(\widehat{X}, V_1, \mathbb{Z}) = (\mu_1 + k - \mu_3 - \mu_4)\mathbb{Z}.$$

It suffices to take $\widehat{X}$ instead of X. The Smale theorem implies that g has at least as many critical points as f. This concludes the proof of the lemma. □

Proof of Theorem 2.2. Let $a_1, \dots, a_r$ be a minimal system of generators of $\pi_2(W^n, V_1)$ as a $\pi_1(V_1, v)$-module. As in the preceding lemma, let us realize them by handles of index 2 attached to $(V_1 \times I, V_1 \times 1)$. Denote the resulting manifold by Z. Let $X = \overline{W^n \setminus Z}$. It easy to show that $\pi_1(X, x) \approx \pi_1(\partial X \setminus V_0, x) = 0$. It follows from the exact homology sequence for the triple (W^n, X, V_0) and the Poincaré duality that

$$H_{n-3}(X, V_0, \mathbb{Z}) = (\mu(H_{n-3}(W^n, V_0, Q)) + \mu(\pi_2(W^n, V_1)) - \mu(H_{n-2}(W^n, V_0)))\mathbb{Z}.$$

Making use of the preceding lemma, we can construct a minimal Morse function on the submanifold X and extend it in an obvious manner to the manifold W^n. The fact that this Morse function is a minimal one is proved as in Lemma 2.1. □

Corollary 2.1. *Let W^n be a compact, contractible manifold, $n \geq 6$. Then there exists a minimal Morse function*

$$f: W^n \to [0, 1], \qquad f^{-1}(0) = \partial W^n$$

on W^n *with critical points only of index* 2, 3, n. *The number of these points is* $N_2 = N_3 = \pi_2(W^n, \partial W^n)$, $N_n = 1$. $\square$

Smale proved that if $\pi_1(\partial W^n, w) = 0$, then W^n is diffeomorphic to a standard disk D^n and, therefore, there exists a Morse function on W^n with only one critical point of index n [143]. Corollary 2.1 illustrates how the fact that the boundary is non-simply-connected affects the number of critical points of a minimal Morse function.

By an n-dimensional homology sphere (not simply connected) we usually mean a smooth connected closed manifold whose integral homologies in dimensions other than 0 or n are trivial. A homotopy sphere is a smooth manifold homotopy-equivalent to a standard sphere. As shown by Kervaire, for each homology sphere Σ^n, $n \geq 4$, there exists a homotopy sphere S^n (unique up to a diffeomorphism) such that the connected sum of Σ^n and S^n bounds a contractible manifold [42, 71].

In setting forth the results of this section we mainly followed [125, 130].

§2. Equivalent epimorphisms

In what follows we shall need several assertions on epimorphisms of free abelian groups into a finitely-generated abelian group.

Let G be a finitely generated abelian group and $\mu(G) = k + l$ the minimal number of its generators.

DEFINITION 2.2. Let $\varphi_i: F_i \to G$ be epimorphisms, where F_i are free abelian groups of rank $k + l$ $(i = 1, 2)$. The epimorphism φ_1 is said to be equivalent to φ_2 if there exists an isomorphism $f: F_1 \to F_2$ such that $\varphi_1 = \varphi_2 \circ f$.

Let $g_1, \dots, g_k, \overline{g}_1, \dots, \overline{g}_l$ be a minimal system of generators of the group G, where g_i are free generators $(1 \leq i \leq k)$ and $\overline{g}_j$ are torsion generators such that the order n_j divides n_{j+1} $(1 \leq j \leq l)$. Let $\varphi: F \to G$ be an epimorphism, where F is a free group of rank $k + l$. Choose a basis $f_1, \dots, f_k, f_{k+1}, \dots, f_{k+l}$ in the group F. Then it is evident that $\varphi(f_s)$ can be uniquely represented in the form

$$\varphi(f_s) = \sum_{i=1}^{k} a_{is} g_i + \sum_{j=1}^{l} a_{k+js} \overline{g}_j,$$

where $a_{js} \in \mathbb{Z}$, $a_{k+js} \in \mathbb{Z}_{n_j}$ $(1 \leq s \leq k + l)$.

Consider the square matrix

$$A_\varphi = \begin{vmatrix} a_{11} & \cdots & a_{1k} & a_{1\,k+1} & \cdots & a_{1\,k+l} \\ \vdots & & & & & \vdots \\ a_{k+l\,1} & \cdots & a_{k+l\,k} & a_{k+l\,k+1} & \cdots & a_{k+l\,k+l} \end{vmatrix},$$

whose entries in the first k columns are integers, and those in columns numbered $(k + j)$ are elements of the group $\mathbb{Z}_{n_j}$. Elementary transformations of the basis F correspond to elementary transformations applied to rows of the matrix. Note that these transformations are performed over $\mathbb{Z}$ in the first k columns of the matrix, and over $\mathbb{Z}_{m_j}$ in the other columns.

LEMMA 2.2. *Let $\varphi: F \to G$ be an epimorphism, where G is a free abelian group of rank $k+l$. Then there exists a basis $f_1, \dots, f_k, f_{k+1}, \dots, f_{k+l}$ such that φ can be represented in the form*

$$\begin{aligned} \varphi(f_i) &= g_i \qquad (1 \le i \le k), \\ \varphi(f_{k+1}) &= \alpha \overline{g}_1, \\ \varphi(f_{k+j}) &= \overline{g}_j \qquad (2 \le j \le l), \end{aligned}$$

where $(\alpha, n_1) \equiv 1$, $|\alpha| < 1/2$.

We call α the determinant of the epimorphism φ.

THEOREM 2.3. *Let $\varphi_i: F_i \to G$ be the epimorphisms of free abelian groups of rank $k+l$ $(i = 1, 2)$. The epimorphism φ_1 is equivalent to the epimorphism φ_2 if and only if their determinants coincide.*

PROOF. *Necessity.* Assume the contrary, i.e, that $\alpha^1 \ne \alpha^2$. The preceding lemma makes it possible to choose bases $f_1^i, \dots, f_k^i, f_{k+1}^i, \dots, f_{k+l}^i$ in the groups F_i such that φ_i can be represented in the form

$$\begin{aligned} \varphi_i(f_s^i) &= g_s \qquad (1 \le s \le k), \\ \varphi_i(f_{k+1}^i) &= \alpha^i \overline{g}_{k+1}, \\ \varphi_i(f_{k+j}^i) &= \overline{g}_{k+j} \qquad (2 \le j \le l). \end{aligned}$$

Since φ_1 is equivalent to φ_2, there exists an isomorphism $f: F_1 \to F_2$ such that $\varphi_1 = \varphi_2 \circ f$. In these bases f is of the form

$$\begin{aligned} f(f_i^1) &= f_i^2 + \sum_{j=1}^{l} a_{ij} f_j^2 \qquad (1 \le i \le k), \\ f(f_{k+1}^1) &= (\beta + a_{k+1\,k+1}) f_{k+1}^2 + \sum_{j=2}^{l} a_{k+1\,k+j} f_{k+j}^2, \\ f(f_{k+j}^1) &= (1 + a_{k+j\,k+j}) f_{k+j}^2 + \sum_{s=\{1,2,\dots,l\}\setminus j} a_{k+s\,k+s} f_{k+s}^2, \end{aligned}$$

where $\beta\alpha^2 = \alpha^1$, $a_{ij} \equiv a_{k+1\,k+1} \equiv a_{k+j\,k+j} \equiv 0 \pmod{m_j}$, and $\alpha^1, \alpha^2 \in \mathbb{Z}_{m_1}$. Denote by A the matrix of the isomorphism f expressed in these bases:

$$A = \begin{vmatrix} 1 & 0 & \cdots & 0 & a_{1\,k+1} & a_{1\,k+2} & \cdots & a_{1\,k+l} \\ 0 & 1 & \cdots & 0 & a_{2\,k+1} & a_{2\,k+2} & \cdots & a_{2\,k+l} \\ \vdots & \vdots & & \vdots & \vdots & \vdots & & \vdots \\ 0 & 0 & \cdots & 1 & a_{k\,k+1} & a_{k\,k+2} & \cdots & a_{k\,k+l} \\ 0 & 0 & \cdots & 0 & \beta + a_{k+1\,k+1} & \cdots & \cdots & a_{k+1\,k+l} \\ 0 & 0 & \cdots & 0 & a_{k+2\,k+1} & 1 + a_{k+2\,k+2} & \cdots & a_{k+2\,k+l} \\ \vdots & \vdots & & \vdots & \vdots & \vdots & & \vdots \\ 0 & 0 & \cdots & 0 & a_{k+l\,k+1} & 1 + a_{k+l\,k+2} & \cdots & 1 + a_{k+l\,k+l} \end{vmatrix}.$$

We can assume without any loss of generality that $\det(A) = 1$. Otherwise we replace the element f_1^1 of the basis by $-f_1^1$. Since $a_{i\,k+j} \equiv 0 \pmod{n_j}$ $(1 \le i \le k+l,\quad 1 \le j \le l)$, then $a_{i\,k+j} \equiv 0 \pmod{n_1}$ because $n_j \equiv 0 \pmod{n_1}$. The natural epimorphism $p: \mathbb{Z} \to \mathbb{Z}/n\mathbb{Z}$ extends to the group homomorphism $\overline{p}: GL(m, \mathbb{Z}) \to GL(m, \mathbb{Z}_n)$. For any $\tilde{A} \in GL(m, \mathbb{Z})$ denote by $\overline{p}(\tilde{A})$ its reduction modulo n. Clearly $\det(\overline{p}(\tilde{A})) = \det \tilde{A} \pmod{n}$. Consider the reduction of the matrix A modulo n_1:

$$\overline{p}(A) = \begin{vmatrix} 1 & 0 & \cdots & 0 & 0 & \cdots & \cdots & 0 \\ 0 & 1 & \cdots & 0 & 0 & \cdots & \cdots & 0 \\ \vdots & \vdots & & \vdots & \vdots & \vdots & & \vdots \\ 0 & 0 & \cdots & 1 & 0 & \cdots & \cdots & 0 \\ 0 & 0 & \cdots & 0 & p(\beta) & \cdots & \cdots & 0 \\ 0 & 0 & \cdots & 0 & 0 & 1 & \cdots & 0 \\ \vdots & \vdots & & \vdots & \vdots & \vdots & & \vdots \\ 0 & 0 & \cdots & 0 & 0 & 0 & \cdots & 1 \end{vmatrix}$$

Since $\det(A) = 1$, the preceding formula implies that $p(\beta) \equiv 1 \pmod{n_1}$. Therefore $\beta\alpha^2 = \beta \pmod{n_1}\alpha^2 = \alpha^2 = \alpha^1$. The necessity is proved.

Sufficiency. If the epimorphisms φ_1 and φ_2 have the same determinant, then they are evidently equivalent. The proof of the theorem is complete. □

We now prove Lemma 2.2.

Let $f_1, \ldots, f_k, f_{k+1}, \ldots, f_{k+l}$ be an arbitrary basis in the group F. Then

$$\varphi(f_s) = \sum_{i=1}^{k} a_{is} g_i + \sum_{j=1}^{l} a_{k+j} \overline{g}_j \qquad (1 \le s \le k+l), \quad a_{is} \in \mathbb{Z},\ a_{k+j} \in \mathbb{Z}_{n_j}.$$

We form the square matrix

$$\hat{A} = \begin{vmatrix} a_{1\,1} & \cdots & a_{1\,k+l} \\ \vdots & & \vdots \\ a_{k+l\,1} & \cdots & a_{k+l\,k+l} \end{vmatrix}.$$

We will reduce the matrix to the required form by transforming its rows. Let a_{s1} be the minimal in absolute value element in the first column. By adding or subtracting multiples of the sth row to or from all other rows of the matrix with nonzero first element we can make the first element of each row either zero or smaller than a_{s1}. Again take the minimal element in the first column and repeat the same operation. After finitely many steps we obtain a basis $\overline{f}_1, \dots, \overline{f}_k, \overline{f}_{k+1}, \dots, \overline{f}_{k+l}$ in which the matrix $\hat{A}$ is of the form

$$\hat{A}_1 = \begin{Vmatrix} \overline{a}_{11} & \overline{a}_{12} & \cdots & \overline{a}_{1\,k+l} \\ 0 & \overline{a}_{22} & \cdots & \overline{a}_{2\,k+l} \\ \vdots & \vdots & & \vdots \\ 0 & \overline{a}_{k+l\,2} & \cdots & \overline{a}_{k+l\,k+l} \end{Vmatrix}.$$

Since f is an epimorphism, there exists an element $\overline{\overline{f}}_1 = \pm\overline{f}_1 + \sum_{j=2}^{k+l} x_j \overline{f}_j$ such that $f(\overline{\overline{f}}_1) = g_1$. Take $\overline{\overline{f}}_1, \overline{f}_2, \dots, \overline{f}_{k+l}$ for the elements of the basis. The matrix $\hat{A}_1$ takes the form

$$\hat{A}_2 = \begin{Vmatrix} 1 & 0 & \cdots & 0 \\ 0 & \overline{a}_{22} & \cdots & \overline{a}_{2\,k+l} \\ \vdots & \vdots & & \vdots \\ 0 & \overline{a}_{k+l\,2} & \cdots & \overline{a}_{k+l\,k+l} \end{Vmatrix}.$$

Proceeding in the same manner, find a basis $\overline{\overline{f}}_1, \dots, \overline{\overline{f}}_k, \overline{f}_{k+1}, \dots, \overline{f}_{k+l}$ in which the matrix of the epimorphism f is written in the form

$$\hat{A}_3 = \begin{Vmatrix} 1 & \cdots & 0 & 0 & \cdots & 0 \\ \vdots & & \vdots & \vdots & & \vdots \\ 0 & \cdots & 1 & 0 & \cdots & 0 \\ 0 & \cdots & 0 & \overline{\overline{a}}_{k+1\,k+1} & \cdots & \overline{\overline{a}}_{k+1\,k+l} \\ \vdots & & \vdots & \vdots & & \vdots \\ 0 & \cdots & 0 & \overline{\overline{a}}_{k+l\,k+1} & \cdots & \overline{\overline{a}}_{k+l\,k+l} \end{Vmatrix}.$$

For the following argument we shall need only the basis elements $\overline{f}_{k+1}, \dots, \overline{f}_{k+l}$ and the lower block of the matrix $\hat{A}_3$. Denoting $f_j = \overline{f}_{k+j}$ and $a_{kj} = \overline{\overline{a}}_{k+j\,k+j}$, we obtain the matrix

$$\overline{A} = \begin{Vmatrix} a_{11} & \cdots & a_{1l} \\ \vdots & & \vdots \\ a_{l1} & \cdots & a_{ll} \end{Vmatrix}$$

in the basis $f_1, \dots, f_l$, where the elements of the jth column belong to the group $\mathbb{Z}_{n_j}$. Making use of the preceding argument, we find a basis $\overline{f}_1, \dots, \overline{f}_l$

in which the matrix $\overline{A}$ takes the triangular form

$$\overline{\overline{A}} = \begin{vmatrix} \overline{a}_{11} & 0 & \cdots & 0 \\ \overline{a}_{21} & \overline{a}_{22} & \cdots & 0 \\ \vdots & \vdots & & \vdots \\ \overline{a}_{l1} & \cdots & \cdots & \overline{a}_{ll} \end{vmatrix}.$$

Clearly, $\overline{a}_{ll}$ is a generator of $\mathbb{Z}_{n_l}$, since otherwise the mapping φ takes no linear combination of the elements of the basis $\overline{f}_1, \dots, \overline{f}_l$ into the element g_{ll}. If $\overline{a}_{l-1\,l-1}$ is a generator in the group $\mathbb{Z}_{n_{l-1}}$, then we can obtain zero in this position. Suppose that $\overline{a}_{l-1\,l-1}$ is not a generator of $\mathbb{Z}_{n_{l-1}}$ and let us prove that the converse statement is then true. Since φ is an epimorphism, there exists an element $\gamma = x_1\overline{f}_1 + \cdots + x_{l-1}\overline{f}_{l-1} + x_l\overline{f}_l$ which goes into g_{l-1}. Clearly, $x_l \equiv 0 \pmod{n_{l-1}}$, since otherwise $\varphi(\gamma) \neq g_{l-1}$. Let us prove that $x_l\overline{f}_l = 0$. Consider $x_l\overline{a}_{l\,l-1}$. In the group $\mathbb{Z}_{n_{l-1}}$ the order of $\overline{a}_{l\,l-1}$ divides n_{l-1} and therefore divides n_l. Thus $n_{l-1} \equiv 0 \pmod{\overline{a}_{l-1\,l-1}}$, while $n_l \equiv 0 \pmod{n_{l-1}}$ by assumption. We have $x_l \equiv 0 \pmod{n_l}$, whence it follows that $x_l \equiv 0 \pmod{n_{l-1}} \equiv 0$ (mod order of $\overline{a}_{l-1\,l-1}$). Thus $x_l\overline{a}_{l\,l-1} = 0$. Similarly, we can prove that $x_l \cdot \overline{a}_{lj} \equiv 0 \pmod{n_j}$ for all $1 \leq j \leq l-2$. It is now clear that $x_l f_l = 0$. Therefore the element γ is of the form $\gamma = x_1\overline{f}_1 + \cdots + x_{l-1}\overline{f}_{l-1}$. The element $\overline{a}_{l-1\,l-1}$ is evidently a generator of the group $\mathbb{Z}_{n_{l-1}}$ (otherwise no linear combination of the basis elements $\overline{f}_1, \dots, \overline{f}_{l-1}$ would go into g_{l-2}). We can find a basis in which the matrix of the epimorphism φ is of the form

$$\overline{\overline{A}}_1 = \begin{vmatrix} \overline{a}_{11} & \cdots & 0 \\ \vdots & & \vdots \\ \overline{a}_{l-1\,1} & \cdots & \overline{a}_{l-1\,l-1} \\ \overline{a}_{l\,1} & \cdots & \overline{a}_{l\,l} \end{vmatrix}.$$

A similar argument proves that the elements $\overline{a}_{jj}$ are generators of $\mathbb{Z}_{n_j}$ $(1 \leq j \leq l-2)$. Therefore there is a basis in which the epimorphism φ is expressed by the matrix

$$\overline{\overline{A}}_2 = \begin{vmatrix} \overline{a}_{11} & & 0 \\ & \ddots & \\ 0 & & \overline{a}_{ll} \end{vmatrix}.$$

Consider the block

$$\begin{vmatrix} \overline{a}_{l-1\,l-1} & 0 \\ 0 & \overline{a}_{ll} \end{vmatrix}$$

and transform it into the form

$$\begin{vmatrix} \overline{a}_{l-1\,l-1} & 1 \\ 0 & \overline{a}_{ll} \end{vmatrix},$$

where 1 is the unit element of the ring $\mathbb{Z}_{n_j}$. The following transformations are evident:

$$\begin{vmatrix} \overline{a}_{l-1\,l-1} & 1 \\ 0 & \overline{a}_{ll} \end{vmatrix} \to \begin{vmatrix} \tilde{a}_{l-1\,l-1} & 1 \\ \tilde{a}_{ll-1} & 1 \end{vmatrix} \to \begin{vmatrix} \tilde{a}_{l-1\,l-1} & 0 \\ \tilde{a}_{ll-1} & 1 \end{vmatrix}.$$

An argument similar to the one given above shows that $\tilde{a}_{l-1\,l-1}$ is a generator of $\mathbb{Z}_{n_{l-1}}$. Therefore our block can be transformed into the form

$$\begin{vmatrix} a_{l-1\,l-1} & 0 \\ 0 & 1 \end{vmatrix}.$$

Working from bottom to top, we can reduce the matrix of the epimorphism φ to the form

$$\begin{vmatrix} \alpha_1 & & & 0 \\ & 1 & & \\ & & \ddots & \\ 0 & & & 1 \end{vmatrix}.$$

If $|\alpha| > n_1/2$, reverse the sign of the corresponding element of the basis in F. The lemma is proved. □

Let $\varphi(N)$ be the Euler function, i.e., the number of positive integers smaller than N that are coprime with N. The following statement holds.

PROPOSITION 2.1. *There exist $\varphi(n_1)/2$ nonequivalent epimorphisms of free abelian groups of rank $k+l$ into the group G.*

The proof follows from Theorem 2.3 and Lemma 2.2. □

The results of this section were published by the author in [125, 126, 128, 139, 140]. Lemma 2.2 was proved in the work of Webb [158].

§3. Homotopy equivalent functions

Let $f: M^n \to [0, 1]$ be an ordered Morse function. Each such function defines a filtration

$$M_0^f \subset M_1^f \subset \cdots \subset M_\lambda^f \subset \cdots \subset M_n^f = M^n,$$

which is constructed as follows. Choose a sequence of numbers $c_1, \dots, c_{n-1} \in [0, 1]$ such that $f(x_i^\lambda) < c_\lambda < f(y_j^{\lambda+1})$, where x_i^λ $(y_j^{\lambda+1})$ are critical points of index λ $(\lambda+1)$. Set $M_\lambda^f = f^{-1}[0, c_\lambda]$. Thus the submanifolds M_λ^f contain all critical points of index $0, 1, \dots, \lambda$ and none other.

DEFINITION 2.3. Two ordered Morse functions f and g on a manifold M^n are homotopy equivalent if there exists a diffeomorphism $h: M^n \to M^n$ isotopic to the identity one such that

$$M_\lambda^f = h(M_\lambda^g).$$

Agoston and Matsumoto constructed manifolds on which there exist minimal Morse functions that are not homotopy equivalent [1, 86]. It follows from the definition that if the function f is modified on the manifold

$\operatorname{Int} f^{-1}[c_\lambda, c_{\lambda-1}]$, then the new function is also homotopy equivalent to f. For example, one can modify the function in such a way that its middle disks will form the required basis in the homology groups.

We shall now give a homotopy criterion for minimal Morse functions on a simply connected manifold of dimension greater than five.

Fix a minimal system of generators $h_1, \dots, h_k, \overline{h}_1, \dots, \overline{h}_l$ of the homology group $H_\lambda(M^n, \mathbb{Z})$, where h_j are free generators, $\overline{h}_j$ are torsion generators, and the order of $\overline{h}_j$ is equal to n_j and divides the order n_{j+1}.

Let $f: M^n \to [0, 1]$ be a minimal Morse function on the manifold M^n and let $M_0^f \subset M_1^f \subset \cdots \subset M_n^f = M^n$ be the filtration associated with f. Consider the exact homology sequence for the pair (M^n, M_λ^f):

$$\to H_\lambda(M_\lambda^f, \mathbb{Z}) \xrightarrow{i_*} H_\lambda(M^n, \mathbb{Z}) \to H_\lambda(M^n, M_\lambda^f, \mathbb{Z}) \to .$$

Clearly $H_\lambda(M^n, M_\lambda^f, \mathbb{Z}) = 0$, because on the manifold $\overline{M^n \setminus M_\lambda^f}$ there are no critical points of index λ since they are responsible for the homology in dimension λ. Therefore the homomorphism i_* is an epimorphism. It is not difficult to show that $H_\lambda(M_\lambda^f, \mathbb{Z})$ is a free abelian group of rank $\mu(H_\lambda(M^n, \mathbb{Z}))$. By Lemma 2.2 there is a basis in the group $H_\lambda(M_\lambda^f, \mathbb{Z})$ in which the epimorphism i_* can be expressed in the form

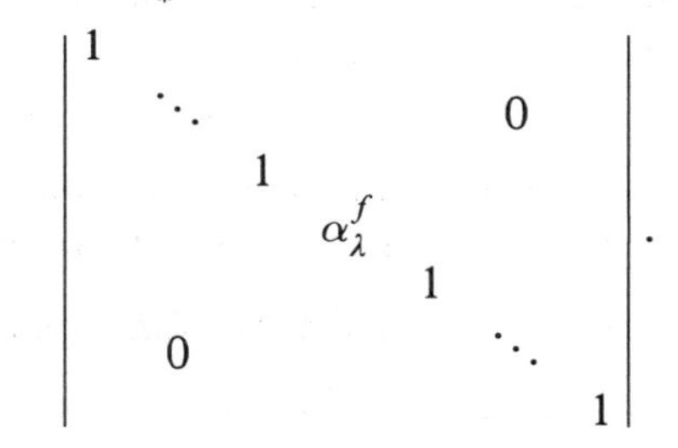

Thus to each Morse function f there corresponds an element $\alpha_\lambda^f \in \mathbb{Z}_{n_1}$ which we will call the homology invariant of f in dimension λ. If $\operatorname{Tors} H_\lambda(M^n, \mathbb{Z}) = 0$, then we set $\alpha_\lambda^f = 0$. Each minimal Morse function on the manifold M^n defines the set of homological invariants $\alpha_2^f, \alpha_3^f, \dots, \alpha_{n-3}^f$. The following theorem holds.

THEOREM 2.4. *Two minimal Morse functions f and g on a closed simply connected manifold M^n $(n \geq 6)$ are homotopy equivalent if and only if their homological invariants coincide.*

To prove this theorem we shall need the following lemma.

LEMMA 2.3. *Let f and g be minimal Morse functions on a manifold M^n, $\pi_1(M^n, m) = 0$ $(n \geq 6)$. Suppose that $M_{\lambda-1}^f = M_{\lambda-1}^g = M_{\lambda-1}$ and $\alpha_\lambda^f = \alpha_\lambda^g$ $(0 \leq \lambda < n-3)$. Then there exists a diffeomorphism $h: M^n \to M^n$ isotopic to the identity such that $h|M_{\lambda-1} = \mathrm{id}$, $M_\lambda^f = h(M_\lambda^g)$.*

PROOF OF THE LEMMA. Let R_f (R_g) be the handle decomposition of the manifold M^n constructed by the Morse function f (g) and the gradient-like vector field ξ (η) whose integral curves connect no two critical points of the same index. By a general position argument we can find a diffeomorphism $\overline{h}: M^n \to M^n$ isotopic to the identity and such that $\overline{h}|M_{\lambda-1} = \mathrm{id}$, $\overline{h}(M^g_\lambda) \subset \mathrm{Int}(M^f_\lambda)$. Indeed, the submanifold M^g_λ admits a representation

$$M^g_\lambda = M_{\lambda-1} \cup D^\lambda_1 \times D^{n-\lambda}_1 \cup \cdots \cup D^\lambda_k \times D^{n-\lambda}_k .$$

Denote by $\overline{D}^i_s \times \overline{D}^{n-i}_s$ the handles from the decomposition R_f. Clearly, there exists $\tilde{\lambda}$ such that $M^g_\lambda \subset \mathrm{Int}\, M^f_{\tilde{\lambda}}$. If $(D^\lambda_\tau \times 0) \cap (0 \times \overline{D}^{n-\tilde{\lambda}}_s) = \varnothing$, then we can easily construct a diffeomorphism $h_1: M^n \to M^n$, $h_1|_{M_{\lambda-1}} = \mathrm{id}$, isotopic to the identity such that $h_1(M^g_\lambda) \subset \mathrm{Int}\, M^f_{\tilde{\lambda}-1}$. This can be done by making use of integral curves of the vector field ξ on $M^f_{\tilde{\lambda}} \setminus M^f_{\tilde{\lambda}-1}$. If $(D^\lambda_\tau \times 0) \cap (0 \times \overline{D}^{n-\tilde{\lambda}}_s) \neq \varnothing$, then, taking into account that $\lambda + n - \tilde{\lambda} < n$, the middle and comiddle disks D^λ_i and $D^{n-\tilde{\lambda}}_s$ can be separated and the problem reduces to the first case. The desired diffeomorphism is constructed by gradually retracting the manifold M^g_λ from the manifolds $M^f_{\tilde{\lambda}}$ $(\tilde{\lambda} > \lambda)$. Consider the following embeddings:

$$\begin{array}{ccc} (M^f_\lambda, M_{\lambda-1}) & \xleftarrow{\overline{h}} & (M^g_\lambda, M_{\lambda-1}) \\ {}_{i_1}\searrow & & \swarrow {}_{i_2} \\ & (M^n, M_{\lambda-1}) & \end{array}$$

Write the commutative diagram:

$$0 \to H_\lambda(M^g_\lambda, \mathbb{Z}) \xrightarrow{\psi_2} H_\lambda(M^g_\lambda, M_{\lambda-1}, \mathbb{Z}) \to H_{\lambda-1}(M_{\lambda-1}, \mathbb{Z}) \to H_{\lambda-1}(M^g_\lambda, \mathbb{Z}) \to 0$$

$$j^2_* \searrow$$

$$\overline{h}_* \downarrow 0 \to H_\lambda(M^n, \mathbb{Z}) \to H_\lambda(M^n, M_{\lambda-1}, \mathbb{Z}) \to H_{\lambda-1} \to (M_{\lambda-1}, \mathbb{Z}) \to$$

$$j^1_* \nearrow$$

$$0 \to H_\lambda(M^f_\lambda, \mathbb{Z}) \xrightarrow{\psi_1} H_\lambda(M^f_\lambda, M_{\lambda-1}, \mathbb{Z}) \to H_{\lambda-1}(M_{\lambda-1}, \mathbb{Z}) \to H_{\lambda-1}(M^f_\lambda, \mathbb{Z}) \to 0,$$

$$j^1: M^f_\lambda \to M^n, \quad j^2: M^g_\lambda \to M^n.$$

Since j^1_* and j^2_* are epimorphisms, $\overline{h}_*$ is either an isomorphism or a monomorphism. If $\overline{h}_*$ is an isomorphism then, by virtue of the Whitehead theorem, $\overline{h}$ is a homotopy equivalence since both M^f_λ and M^g_λ are simply connected manifolds. Set $Y = \overline{M^f_\lambda \setminus h(M^g_\lambda)}$. It is easy to see that $\pi_1(Y, y) \approx \pi_1(\partial Y, y) = 0$. It follows from the exact homology sequence for the pair $(M^f_\lambda, \overline{h}(M^g_\lambda))$ that $H_i(M^f_\lambda, \overline{h}(M^g, \lambda), \mathbb{Z}) = 0$. Therefore Y is an h-cobordism. By the h-cobordism theorem the manifold Y is diffeomorphic to $\partial M^f_\lambda \times [0, 1]$. Making use of the $t \in [0, 1]$-coordinate, it is easy to find a diffeomorphism $h_1\colon M^n \to M^n$ that is isotopic to the identity and $h_1(\overline{h}(M_\lambda^g)) = M_\lambda^f$, $h|_{M_{\lambda-1}} = \mathrm{id}$. Take for h the diffeomorphism $h = h_1 \circ \overline{h}$.

Suppose that $\overline{h}_*$ is a monomorphism. Using the homology torsion, we can adjust $\overline{h}_*$, i.e., find a new diffeomorphism $\tilde{h}\colon M^n \to M^n$ that is isotopic to the identity, $\tilde{h}|_{M_{\lambda-1}} = \mathrm{id}$, $\tilde{h}(M_\lambda^g) \subset \mathrm{Int}\, M_\lambda^f$, and $\tilde{h}_*\colon H_\lambda(M_\lambda^g, \mathbb{Z}) \to H_\lambda(M_\lambda^f, \mathbb{Z})$ is an isomorphism. This reduces the problem to the first case.

By the hypothesis, $\alpha_\lambda^f = \alpha_\lambda^g$. Choose the basis elements $\{a_i\}$ in $H_\lambda(M_\lambda^f, \mathbb{Z})$ and $\{b_i\}$ in $H_\lambda(M_\lambda^g, \mathbb{Z})$ such that the matrices of the epimorphisms j_*^1 and j_*^2 are diagonal, as in Lemma 2.2. Then it is evident that

$$\overline{h}_*(a_i) = b_i + \sum_{j=k_\lambda+1}^{k_\lambda+l_\lambda} \theta_{ij} b_{ij} \qquad (1 \le i \le k_\lambda),$$

$$\overline{h}_*(a_i) = \sum_{j=k_\lambda+1}^{k_\lambda+l_\lambda} \hat{\theta}_{ij} b_{ij} \qquad (1 + k_\lambda \le i \le k_\lambda + l_\lambda).$$

Middle disks of critical points of index λ of the function f (g) define generators in the group $H_\lambda(M_\lambda^f, M_{\lambda-1}, \mathbb{Z})$ $(H_\lambda(M_\lambda^g, M_{\lambda-1}, \mathbb{Z}))$ containing the subgroup $\psi_1(H_\lambda(M_\lambda^f, \mathbb{Z}))$ as a direct summand. Let us modify f and ξ on $M_\lambda^f \setminus M_{\lambda-1}$, and, respectively, g and η on $M_\lambda^g \setminus M_{\lambda-1}$ in such a way that the new middle disks realize the elements $\psi_1(a_i) \in H_\lambda(M_\lambda^f, M_{\lambda-1}, \mathbb{Z})$ for the function f, and $\psi_2(b_i) \in H_\lambda(M_\lambda^g, M_{\lambda-1}, \mathbb{Z})$ for g. The modified Morse functions will be denoted by the same letters f and g. We can assume without any loss of generality that the intersection $(0 \times \partial \overline{D}_j^{n-\lambda}) \cap (\partial \overline{D}_i^{\lambda+1} \times 0)$ $(1 \le j \le l)$ is transversal and consists of n_j points. Here $\overline{D}_j^{\lambda+1} \times 0$ is the middle disk of a critical point of index $\lambda+1$ of the function f, and $0 \times \overline{D}_j^{n-\lambda}$ is the comiddle disk of a critical point of index λ of the same function. By a general position argument and the Whitney lemma [96] we can make the intersection $(\overline{D}_i^\lambda \times 0) \cap (0 \times \overline{D}_j^{n-\lambda})$ transversal and consisting of θ_{ij} points. The conditions of the Whitney lemma are satisfied, because the manifolds are simply connected. We will show that with the help of the connected sum of $\overline{D}_i^\lambda \times 0$ with $\partial(\overline{D}_j^{\lambda+1} \times 0)$ we can reduce or increase the number of points of intersection of middle and comiddle disks $\overline{D}_i^\lambda \times 0$ and $\partial(\overline{D}_j^{\lambda+1} \times 0)$ by n_j.

Let $\zeta\colon (\partial \overline{D}_j^{\lambda+1} \times 0) \times [0, \varepsilon] \to \overline{D}_j^\lambda \times \overline{D}_j^{n-\lambda}$, and ε be an isotopy such that

$$\zeta[(\partial \overline{D}_j^{\lambda+1} \times 0) \times \varepsilon] \subset \mathrm{Int}\, \overline{D}_j^\lambda \times \overline{D}_j^{n-\lambda},$$

$$\zeta[(\partial \overline{D}_j^{\lambda+1} \times 0) \times \varepsilon] \cap \overline{D}_i^\lambda \times \overline{D}_i^{n-\lambda} = \varnothing, \qquad i \ne j.$$

Fix an appropriate orientation of $\zeta[(\partial\overline{D}_j^{\lambda+1} \times 0) \times \varepsilon]$ and consider the connected sum $A_i = (D_i \times 0) \coprod \zeta[(\partial\overline{D}_j^{\lambda+1} \times 0) \times \varepsilon]$. Clearly, A_i is a disk attached to $\partial M_{\lambda-1}$. Take a tubular neighborhood of A_i in $\operatorname{Int} M_\lambda^f$ of sufficiently small radius. It is easy to see that we have constructed a new handle $\hat{D}_i^\lambda \times \hat{D}_i^{n-\lambda}$ of index λ attached to $\partial M_{\lambda-1}$ via the old mapping. By construction, A_i and $\partial\overline{D}_i^{\lambda+1} \times 0$ are isotopic in $M^n \setminus M_{\lambda-1}$, because $\partial\overline{D}_i^{\lambda+1} \times 0$ bounds the disk $\overline{D}_i^{\lambda+1} \times 0$. This isotopy can be extended to the entire manifold in such a way that it is the identity on $M_{\lambda-1}$ and takes $D_i^\lambda \times D_i^{n-\lambda}$ into the tubular neighborhood A_i. The number of points of intersection of $A_i \cap 0 \times D_j^{n-\lambda}$ becomes equal to $\theta_{ij} \pm n_j$. By repeating this operation finitely many times we can make $\theta_{ij} = \delta_{ij}$. Now it is easy to construct a diffeomorphism $\overline{h}: M^n \to M^n$ satisfying the above conditions. Clearly, $\overline{h}_*$ is given by the identity matrix, and therefore h_* is an isomorphism. The proof of the lemma is complete. □

PROOF OF THEOREM 2.4. *Necessity*. If the Morse functions f and g are equivalent, then their homological invariants evidently coincide. This follows from the definition of equivalent epimorphisms and Theorem 2.3.

Sufficiency. Using the Palais theorem on the disk [96], find a diffeomorphism $h_0: M^n \to M^n$ isotopic to the identity and such that $h_0(M_0^g) = M_0^f$. Suppose that we have found the diffeomorphisms

$$h_i: M^n \to M^n, \qquad i = 2, \dots, \lambda - 1 < n - 3,$$

satisfying the following conditions:

(1) h_i is isotopic to the identity;

(2) $h_i/h_{i-1}(M_{i-1}^g)$ is the identity; and

(3) $h_i(M_i^g) = M_i^f$.

Let $\overline{g} = g \circ h_{\lambda-1}^{-1}$. Then $M_{\lambda-1}^f = M_{\lambda-1}^{\overline{g}}$. Applying Lemma 2.3 to the functions f and $\overline{g}$, we find a diffeomorphism $h_\lambda: M^n \to M^n$ isotopic to the identity and satisfying the conditions

$$h/M_{\lambda-1}^{\overline{g}} = \mathrm{id}, \qquad h_\lambda(M_{\lambda-1}^{\overline{g}}) = M_\lambda^f.$$

Clearly, the diffeomorphism h_λ satisfies conditions (1)–(3). If $\lambda = n - 2$, then a similar argument can be applied to the functions $1 - f$ and $1 - \overline{g}$. □

We note that Agoston considered the case where the homology of the manifold is torsion-free, and proved that then all minimal Morse functions are homotopy equivalent [1].

The general case, where arbitrary ordered Morse functions on a simply connected manifold of dimension greater than six are homotopy equivalent, was considered in the work of Mikhaĭlyuk [98].

§4. The realization theorem

Consider the homology group of the manifold $H_\lambda(M^n, \mathbb{Z})$. Let

$$g_1^\lambda, \dots, g_{k_\lambda}^\lambda, \overline{g}_1^\lambda, \dots, \overline{g}_{l_\lambda}^\lambda$$

be its minimal system of generators, where g_i^λ are free generators, $\overline{g}_j^\lambda$ torsion generators, and the order of $\overline{g}_j^\lambda$ divides the order of $\overline{g}_{j+1}^\lambda$. Let us call the generators $\overline{g}_1^\lambda$ the distinguished elements in $H_\lambda(M^n, \mathbb{Z})$. Denote by $G(\overline{g}_1^\lambda)$ the subgroup generated by the element $\overline{g}_1^\lambda$. In the present section we shall prove that for each set of generators of the form $\alpha_2, \dots, \alpha_{n-3}$ in the groups $G(\overline{g}_1^\lambda)$ one can construct an ordered minimal Morse function on the manifold M^n whose homological invariants coincide with $\alpha_2, \dots, \alpha_{n-3}$. In what follows we will say that $\alpha_2, \dots, \alpha_{n-3}$ is an admissible set of generators.

THEOREM 2.5. *For any admissible set of generators $\{\alpha_\lambda\}$ on a simply connected closed manifold M^n, $n \geq 6$, there exists a minimal ordered Morse function whose homological invariants coincide with $\{\alpha_\lambda\}$.*

PROOF. Let $f: M^n \to [0, 1]$ be an ordered minimal Morse function on the manifold M^n whose homological invariants are all equal to 1. We can assume without any loss of generality that the gradient-like vector field ξ of the function f satisfies the condition $f(x_\lambda^b) < f(x_\lambda^z)$, where x_λ^b is a critical point of index λ whose middle sphere intersects the comiddle sphere of a critical point of index $\lambda - 1$, and x_λ^z is a critical point of index λ whose middle sphere intersects no comiddle sphere of critical points of index $\lambda - 1$. Choose two sequences of numbers $\{c_\lambda\}$ and $\{d_\lambda\}$ in the closed interval $[0, 1]$ satisfying the relations

$$f(x_\lambda^b) < d_\lambda < f(x_\lambda^z), \qquad f(x_\lambda) < c_\lambda < f(y_{\lambda+1}),$$

where x_λ $(y_{\lambda+1})$ are critical points of index λ $(\lambda + 1)$.

Consider the submanifolds $M_\lambda = f^{-1}[0, c_\lambda]$, $N_\lambda = f^{-1}[0, d_\lambda]$. By the hypothesis, there exists a basis $m_1, \dots, m_{k_\lambda}, \overline{m}_1, \dots, \overline{m}_{l_\lambda}$ in the homology group $H_\lambda(M_\lambda, \mathbb{Z})$ such that the embedding $i: M_\lambda \to M^n$ induces the mapping $i_*(m_{i_\lambda}) = g_{i_\lambda}$, $i_*(\overline{m}_{j_\lambda}) = \overline{g}_{j_\lambda}$. As noted above, the homomorphism of the homology groups

$$\psi: H_\lambda(M_\lambda, \mathbb{Z}) \to H_\lambda(M_\lambda, M_{\lambda-1}, \mathbb{Z})$$

appearing in the exact sequence of the pair $(M_\lambda, M_{\lambda-1})$ is a monomorphism onto the direct summand. Without loss of generality we can assume that the function f is chosen in such a way that the middle disks of critical points of index λ realize a system of generators $a_1, \dots, a_{k_\lambda}, \overline{a}_1, \dots, \overline{a}_{l_\lambda}$ in $H_\lambda(M_\lambda, M_{\lambda-1})$ such that $\psi(m_i^\lambda) = a_i$ and $\psi(\overline{m}_j^\lambda) = \overline{a}_j$.

A critical point corresponding to the generator m_1^λ will be said to be distinguished. It is not difficult to prove that

$$H_i(N_\lambda, \mathbb{Z}) \approx H_i(M^n, \mathbb{Z}), \quad 0 \le i \le \lambda - 1, \qquad H_j(N_\lambda, \mathbb{Z}) = 0, \quad i > \lambda.$$

We construct an ordered minimal Morse function g on the manifold M^n such that

$$g^{-1}[0, d_\lambda] = \overline{N}_\lambda = N_\lambda, \qquad g^{-1}[0, c_\lambda] = \overline{M}_\lambda \subset \operatorname{Int} M_\lambda,$$

and the homological invariants of g coincide with α_λ. The numbers α_λ and c_λ satisfy the inequalities for the function g. The construction is achieved by induction.

Let $M_2 = f^{-1}[0, c_2]$. Set $f = g$ on M_2. Suppose that we have the sequence of submanifolds

$$M_2 = \overline{N}_2 \subset \overline{N}_3 \subset \cdots \subset \overline{N}_{\lambda-1}, \qquad \overline{N}_2 = \overline{M}_2 \subset \overline{M}_3 \subset \cdots \subset \overline{M}_{\lambda-1},$$

and Morse functions on them with the required properties. Let us show how to construct the submanifolds $\overline{N}_\lambda$ and $\overline{M}_\lambda$.

By construction, $\overline{N}_\lambda = N_\lambda \subset M_\lambda$ and the submanifold M_λ is of the form

$$M_\lambda = N_\lambda \cup D_1^\lambda \times D_1^{n-\lambda} \cup \cdots \cup D_{k_\lambda + l_\lambda}^\lambda \times D_{k_\lambda + l_\lambda}^{n-\lambda}.$$

Let $D_1^\lambda \times D_1^{n-\lambda}$ be the handle corresponding to the distinguished critical point of index λ. It is attached to the manifold N_λ via an embedding

$$\varphi: \partial D_1^\lambda \times D_1^{n-\lambda} \to \partial N.$$

Choose s_λ distinct points $x_1, x_2, \ldots, x_{s_\lambda}$ in $\varphi(D_1^\lambda \times D_1^{n-\lambda})$, where the number s_λ satisfies the relation $s_\lambda \cdot 1 = \alpha_\lambda$ in the group $G(\overline{g}_1^\lambda)$. Let $\partial D_{s_i}^\lambda$ be concentric spheres in $\varphi(D_1^\lambda \times D_1^{n-\lambda})$ containing the points x_i, and $D_{x_i}^\lambda$ the disks in $D_1^\lambda \times D_1^{n-\lambda}$ bounding them. Consider the connected sum

$$\Gamma = D_{x_1}^\lambda \coprod \cdots \coprod D_{s_\lambda}^\lambda.$$

Clearly, if $D_1^\lambda \times 0$ defines the element m_1^λ, then Γ defines the element $\alpha_\lambda m_1^\lambda$. Let U be a tubular neighborhood of Γ in $D_1^\lambda \times D_1^{n-\lambda}$ such that

$$U \cap \varphi(D_1^\lambda \times D_1^{n-\lambda}) \approx S^{\lambda-1} \times D_1^{n-\lambda}.$$

The tubular neighborhood U can be regarded as a handle of index λ attached to $\partial N_{\lambda-1}$. Denote $\overline{M}_\lambda = \overline{N}_\lambda \cup U \cup D_2^\lambda \times D_2^{n-\lambda} \cup$ (other handles of index λ). By construction $\overline{M}_\lambda \subset M_\lambda$ and $H_i(\overline{M}_\lambda, \mathbb{Z}) \approx H_i(M_\lambda, \mathbb{Z})$. Write the commutative diagram:

$$\begin{array}{ccccccccc}
0 & \longrightarrow & H_\lambda(\overline{M}_\lambda, \mathbb{Z}) & \longrightarrow & H_\lambda(\overline{M}_\lambda, \overline{N}_{\lambda-1}, \mathbb{Z}) & \xrightarrow{\overline{\partial}} & H_{\lambda-1}(\overline{N}_{\lambda-1}, \mathbb{Z}) & \longrightarrow \\
 & & \downarrow{\scriptstyle j_*} & & \downarrow{\scriptstyle i_*} & & \downarrow & \\
0 & \longrightarrow & H_\lambda(M_\lambda, \mathbb{Z}) & \longrightarrow & H_\lambda(M_\lambda, N_{\lambda-1}, \mathbb{Z}) & \xrightarrow{\partial} & H_{\lambda-1}(N_{\lambda-1}, \mathbb{Z}) & \longrightarrow
\end{array}$$

By construction i_* is a monomorphism, and therefore j_* is also a monomorphism. In the chosen bases, i_* is given by the matrix

$$A(i_*) = \begin{vmatrix} 1 & & & & & & & \\ & 1 & & & & & 0 & \\ & & \ddots & & & & & \\ & & & 1 & & & & \\ & & & & \alpha_\lambda & & & \\ & & & & & 1 & & \\ & 0 & & & & & \ddots & \\ & & & & & & & 1 \end{vmatrix}.$$

Recall that the bases in the groups $H_\lambda(\overline{M}_\lambda, \overline{N}_{\lambda-1}, \mathbb{Z})$ and $H_\lambda(M_\lambda, N_{\lambda-1}, \mathbb{Z})$ are given by the middle disks of the handles. Thus there is a basis $r_1, \dots, r_k$, $\bar{r}_1, \dots, \bar{r}_{l_\lambda}$ in the group $H_\lambda(\overline{M}_\lambda, \mathbb{Z})$ in which j_* is given by the matrix $A(i_*)$. Clearly,

$$H_\lambda(\overline{M}_\lambda, \mathbb{Z}) \to H_\lambda(M^n, \mathbb{Z})$$

is an epimorphism. Consider the pair $(N_\lambda, \overline{M}_\lambda)$ and write for it the exact sequence

$$\begin{aligned} 0 \to H_{\lambda+1}(N_\lambda, \overline{M}_\lambda, \mathbb{Z}) &\to H_\lambda(\overline{M}_\lambda, \mathbb{Z}) \xrightarrow{t_*} H_\lambda(N_\lambda, \mathbb{Z}) \\ &\to H_\lambda(N_\lambda, \overline{M}_\lambda, \mathbb{Z}) \to H_{\lambda-1}(\overline{M}_{\lambda-1}, \mathbb{Z}) \to H_{\lambda-1}(N_\lambda, \mathbb{Z}) \to . \end{aligned}$$

We claim that t_* is an epimorphism. Indeed, we have the sequence of embeddings $\overline{M}_\lambda \subset N_\lambda \subset M^n$ and the corresponding chain of mappings

$$H_\lambda(\overline{M}_\lambda, \mathbb{Z}) \xrightarrow{t_*} H_\lambda(N_\lambda, \mathbb{Z}) \xrightarrow{p_*} H_\lambda(M^n, \mathbb{Z}).$$

By construction, p_* is an isomorphism, $p_* \circ t_*$ is an epimorphism, and therefore t_* is an epimorphism. The exact homology sequence implies that

$$H_i(N_\lambda, \overline{M}_\lambda, \mathbb{Z}) = 0 \quad \text{for } i \le \lambda$$

and that $H_{\lambda+1}(N_\lambda, \overline{M}_\lambda, \mathbb{Z})$ is a free abelian group of rank $\mu(\operatorname{Tors} H_\lambda(M^n, \mathbb{Z}))$ (where $\mu(H)$ is the minimal number of generators of the group H). Consider the manifold $L = \overline{N_\lambda \setminus \overline{M}_\lambda}$. By the Smale theorem we can construct a minimal Morse function $\bar{g}$ with just the critical points of index $\lambda + 1$ whose number equals precisely $\mu(\operatorname{Tors} H_\lambda(M^n, \mathbb{Z}))$. Thus, if we have a function $\tilde{g}$ on $\overline{N}_{\lambda-1}$, we extend it to $\overline{M}_\lambda$ using the handles constructed, and then to $\overline{M}_\lambda \cup L$ using $\bar{g}$. It is clear that we have to take N_λ for $\overline{N}_\lambda$. This completes the inductive step. A consecutive application of the above argument yields a Morse function with the desired properties. The theorem is proved. □

Using the Poincaré duality one can easily prove the following corollary.

COROLLARY 2.2. *On a closed simply connected manifold* M^n $(n \geq 6)$ *there exist*

$$P(M^n) = \prod_{\lambda=2}^{k-1} \left(\left[\frac{\varphi(n_\lambda) + 1}{2} \right] \right)^2 \qquad (n = 2k),$$

$$P(M^n) = \varphi(n_k) \prod_{\lambda=2}^{k-1} \left(\left[\frac{\varphi(n_\lambda) + 1}{2} \right] \right)^2 \qquad (n = 2k + 1)$$

minimal Morse functions that are not homotopy equivalent. Here n_λ *is the order of the group* $G(\overline{g}_1^\lambda) \in H_\lambda(M^n, \mathbb{Z})$, $\varphi(n_\lambda)$ *is the Euler function, and* $[n]$ *is the integer part of* n. □

The above statements yield a practical algorithm for finding out whether two minimal Morse functions on a simply connected closed manifold M^n $(n \geq 6)$ are homotopy equivalent.

Starting with Morse functions f and g, construct the handle decomposition and, using the diffeomorphism isotopic to the identity, transform the manifold M_λ^f into $\operatorname{Int} M_\lambda^g$. Let A_λ be the integer matrix of the indices of intersection of the oriented middle disks of index λ of the function f with the comiddle disks of the handles of the same index of the function g. If $\det A_\lambda \equiv 1 \pmod{n_1^\lambda}$ for all λ such that $\operatorname{Tors} H_\lambda(M^n, \mathbb{Z}) \neq 0$, then the functions f and g are homotopy equivalent. Otherwise they are not homotopy equivalent.

The question of whether minimal Morse functions are homotopy equivalent can be approached differently. Let f be an arbitrary minimal Morse function on a closed simply connected manifold M^n, and let ξ be its gradient-like vector field. Starting with f and ξ, construct the chain complex of free abelian groups

$$\{C, \partial, f\}: C_2 \xleftarrow{\partial_3} C_3 \longleftarrow \cdots \xleftarrow{\partial_{n-2}} C_{n-2}.$$

Without loss of generality we can assume that f and g are chosen in such a way that the matrices of the boundary homomorphisms ∂_λ are diagonal. If g is another ordered minimal Morse function on M^n and η is its gradient-like vector field, then, as mentioned above, there exists a diffeomorphism $h: M^n \to M^n$ isotopic to the identity and taking the submanifolds M_λ^g into $\operatorname{Int} M_\lambda^f$. Using g and η, construct the chain complex

$$\{C, d, g\}: D_2 \xleftarrow{d_3} D_3 \longleftarrow \cdots \xleftarrow{d_{n-2}} D_{n-2}.$$

The mapping h induces the chain mapping of the following chain complexes:

$$\begin{array}{ccccccc}
D_2 & \xleftarrow{d_3} & D_3 & \longleftarrow & \cdots & \xleftarrow{d_{n-3}} & D_{n-3} \\
{\scriptstyle h_*}\downarrow & & \downarrow{\scriptstyle h_*} & & & & \downarrow{\scriptstyle h_*} \\
C_2 & \xleftarrow{\partial_3} & C_3 & \longleftarrow & \cdots & \xleftarrow{\partial_{n-3}} & C_{n-3}
\end{array}$$

This chain mapping is homotopy equivalent to the chain isomorphism if and only if the Morse functions f and g are homotopy equivalent. Next, the subgroups of cycles are included into these chain complexes as direct summands and there is a canonical minimal epimorphism

$$0 \longleftarrow H_\lambda(M^n, \mathbb{Z}) \xleftarrow{\ p\ } Z_\lambda$$

(Z_λ is the group of cycles of the chain complex $\{C, \partial, f\}$). Therefore, the question of when the chain mapping $h_*: \{D, d, g\} \to \{C, \partial, f\}$ is homotopy equivalent to the chain isomorphism reduces to the question of equivalence of minimal epimorphisms

$$C_\lambda \supset Z_\lambda \xrightarrow{\ p\ } H_\lambda(M^n, \mathbb{Z}) \xleftarrow{\ \overline{p}\ } \overline{Z}_\lambda \subset D_\lambda .$$

If the epimorphisms p and $\overline{p}$ are equivalent, we can, if necessary, modify the chain mapping using some chain homotopy $\varphi_i: D_i \to C_{i+1}$. This is what was actually done in the proof of Lemma 2.2.

§5. Isotopic functions

Let f_0 and f_1 be Morse functions on a manifold M^n.

DEFINITION 2.4. The Morse function f_0 is isotopic to the Morse function f_1 if there exists a path $\gamma: [0, 1] \to C^\infty(M^n, \mathbb{R})$ such that $\gamma(0) = f_0$, $\gamma(1) = f_1$, and $\gamma(t)$ is a Morse function for all $t \in [0, 1]$.

This definition does not exclude the possibility of the Morse functions $\gamma(t)$ having large codimension.

The following easy proposition will be used below.

PROPOSITION 2.2. *If $h: M^n \to M^n$ is a diffeomorphism isotopic to the identity and $f: M^n \to [0, 1]$ is a Morse function, then there exists a path $\gamma: [0, 1] \to C^\infty(M^n, \mathbb{R})$ joining the functions f and $f \circ h$ and such that $\gamma(t)$ are Morse functions for all $t \in [0, 1]$.*

In this section we will give a necessary and sufficient condition of isotopic equivalence for ordered minimal Morse functions on a simply connected manifold of dimension greater than 5.

We now give a construction belonging to Cerf [17].

Let $\gamma: [0, 1] \to C^\infty(M^n, \mathbb{R})$ be a smooth path joining the Morse functions $f_0 = \gamma(0)$ and $f_1 = \gamma(1)$ and satisfying the condition that $\gamma(t)$ are Morse functions for all $t \in [0, 1]$. Consider the mapping

$$A: M^n \times [0, 1] \to [0, 1] \times [0, 1], \qquad (m, t) = (\gamma(t)(m), t), \quad m \in M^n.$$

Let $x_0 \in M^n$ be a critical point of the function $f_0 = \gamma(0)$ and $a_0 = f_0(x_0)$ the critical value. Under variation of the parameter $t \in [0, 1]$ the critical point x_0 describes a path γ_{x_0} on $M^n \times [0, 1]$. Let $(\tilde{x}_0, 1)$ be the endpoint of this path, $f_1(\tilde{x}_0) = b_0$. Consider the image $A(\gamma_{x_0})$. We obtain a path on $[0, 1]\times[0, 1]$ joining the points a_0 and b_0. Consider all critical points of the function f_0 and plot the corresponding curves on $[0, 1] \times [0, 1]$. In general, these curves may either intersect or coincide (partially or completely). In what follows we will call this picture the development of the path γ.

The one-parameter ordering theorem due to Cerf asserts that a path γ: $[0, 1] \to C^\infty(M^n, \mathbb{R})$ can be deformed into a path $\overline{\gamma}$ whose development satisfies the following condition: the paths corresponding to critical points of index λ do not intersect the paths corresponding to critical points of index $\overline{\lambda}$ [17].

Let W^n be a compact smooth manifold with boundary $\partial W^n = V_0 \cup V_1$. Suppose that $\pi_1(W^n w) \approx \pi_1(V_0, v) \approx \pi_1(V_1, \overline{v}) = 0$ $(n \geq 6)$ and

$$H_i(W^n, V_0, \mathbb{Z}) = \begin{cases} 0 & \text{if } i \neq \lambda, \\ \underbrace{\mathbb{Z} \oplus \cdots \oplus \mathbb{Z}}_{k} & \text{if } i = \lambda \quad (2 \leq \lambda \leq n-3). \end{cases}$$

Fix a basis $a_1, \ldots, a_k$ in the homology group $H_\lambda(W^n, V_0, \mathbb{Z})$. Choose a minimal Morse function in general position $f: W^n \to [0, 1]$ on the manifold W^n such that $f_0^{-1} = V_0$, $f^{-1}(1) = V_1$. By the Smale theorem such a function on W^n always exists and has k critical points of index λ. Let ξ be a gradient-like vector field for the function f. The middle disks of the critical points of the function define a basis $b_1, \ldots, b_k$ in the group $H_\lambda(W^n, V_0, \mathbb{Z})$. However, it is not defined uniquely and is modified by a variation of the gradient-like vector field ξ. Let us describe this process.

Let A be the transition matrix from the basis $b_1, \ldots, b_k$ to the basis $a_1, \ldots, a_k$. Denote by $T(k, \mathbb{Z})$ the subgroup of upper triangular matrices of the form

$$\begin{vmatrix} 1 & a_{12} & \cdots & a_{1n} \\ & 1 & & \\ & & \ddots & \\ 0 & & & 1 \end{vmatrix}$$

from the group $GL(n, \mathbb{Z})$. As shown by Cerf [17], for any Morse function f and any matrix B from $T(k, \mathbb{Z})$ one can always choose a gradient-like vector field $\overline{\xi}$ such that the new middle disks of critical points of index λ of the function f define a basis $c_1, \ldots, c_k$ in the group $H_\lambda(W^n, V_0, \mathbb{Z})$ for which the transition matrix from the basis $c_1, \ldots, c_k$ to the basis $a_1, \ldots, a_k$ is of the form AB. This statement means the following. Suppose that the critical points $x_1, \ldots, x_k$ of the function f are ordered in such a way that $f(x_i) > f(x_{i+1})$. Denote by $D_1^\lambda \times 0, \ldots, D_k^\lambda \times 0$ their middle disks constructed by means of the field ξ. The geometrical situation enables one, by varying the vector field ξ for a fixed function f, to attach to the middle disk $D_i^\lambda \times 0$ only a linear combination of the middle disks $D_j^\lambda \times 0$, $i > j$.

This construction makes it possible to associate with each minimal Morse function in general position on the manifold W^n a coset $\omega(f)$ from $GL(n, \mathbb{Z})/T(n, \mathbb{Z})$ which we will call the Cerf invariant of the function f.

The following statement also belongs to Cerf.

THEOREM 2.6. *Let* W^n *be a compact smooth manifold with boundary* $\partial W^n = V_0 \cup V_1$ $(n \geq 6)$. *The groups* $\pi_1(W^n, w) \approx \pi_1(V_0, w) \approx \pi_1(V_1, v) = 0$, *and* $H_i(W^n, V_0, \mathbb{Z}) = 0$ *for* $i \neq \lambda$, $H_\lambda(W^n, V_0, \mathbb{Z}) \approx \underbrace{\mathbb{Z} \oplus \cdots \oplus \mathbb{Z}}_{k}$. *Two minimal Morse functions* f *and* g *in general position on the manifold* W^n *belong to the same connected component of the stratum of codimension* 0 *if and only if they have the same Cerf invariant* $\omega(f) = \omega(g)$. □

Consider $\mathscr{F}_\mu(W^n)$, the space of all minimal Morse functions on the manifold W^n. The following proposition belongs to Matsumoto [86].

PROPOSITION 2.3. *Suppose that* W^n *is a smooth compact manifold with boundary* $\partial W^n = V_0 \cup V_1$ *and* $\pi_1(W^n, w) \approx \pi_1(V_0, w) \approx \pi_1(V_1, v) = 0$ $(n \geq 6)$, $H_i(W^n, V_0, \mathbb{Z})$ *are free abelian groups. Then the space of minimal Morse functions on* W^n *is path-connected.* □

The proof of this statement is based on the fact that by increasing the value of the function at a critical point (i.e., if we are able to intersect the component $\mathscr{F}_\beta^1(M^n)$ of the stratum in codimension 1) and varying the gradient-like vector field, we can realize any basis in the homology group formed by the middle disks of critical points.

We shall now prove the following theorem.

THEOREM 2.7. *Two ordered minimal Morse functions* f_0 *and* f_1 *on a simply connected closed manifold* M^n, $\pi_1(M^n, x) = 0$ $(n \geq 6)$, *are isotopic if and only if the homological invariants of* f_0 *and* f_1 *coincide, i.e., if and only if* f_0 *and* f_1 *are homotopy equivalent.*

PROOF. *Necessity.* Let $\gamma: [0, 1] \to C^\infty(M^n, \mathbb{R})$ be a path joining the functions f_0 and f_1 such that $\gamma(t)$ is a Morse function for each $t \in [0, 1]$. Let us draw the development of γ. By a repeated application of the one-parameter ordering theorem, we can deform the path γ into a path $\overline{\gamma}$ whose development is such that no path corresponding to the critical points of index λ intersects a path corresponding to the critical points of index $\overline{\lambda}$ $(\overline{\lambda} > \lambda)$. Let c_λ $(\overline{c}_\lambda)$ be a regular value of the function f_0 (f_1) lying between the values of f_0 (f_1) at critical points of index λ and $\lambda + 1$. Take a smooth path $\theta(t)$ on the development joining these points. We can assume without any loss of generality that the path $\theta(t)$ intersects each line $t \times [0, 1]$ at precisely one point and intersects no path corresponding to the critical points. Denote by S_λ the part of $[0, 1] \times [0, 1]$ bounded by the segments $[0, 1] \times 0$, $[0, c_\lambda] \subset 0 \times [0, 1]$, $[1, c_\lambda] \subset 1 \times [0, 1]$, and the curve $\theta(t)$.

Consider $E_\lambda = A^{-1}[S_\lambda]$. Clearly, E_λ is diffeomorphic to $f^{-1}(c_\lambda) \times [0, 1]$. For the proof one must use the relative Meyer-Vietoris theorem and the relative theorem on h-cobordism. The projection of E_λ onto M^n yields the isotopy between $f^{-1}[0, c_\lambda]$ and $f^{-1}[0, \bar{c}_\lambda]$. Therefore the epimorphisms

$$i_{1*}: H_\lambda(f_0^{-1}[0, c_\lambda], \mathbb{Z}) \to H_\lambda(M^n, \mathbb{Z}),$$
$$i_{2*}: H_\lambda(f_1^{-1}[0, \bar{c}_\lambda], \mathbb{Z}) \to H_\lambda(M^n, \mathbb{Z})$$

are equivalent and the functions f_0 and f_1 have the same homology invariants in dimension λ.

Sufficiency. Let c_λ $(\bar{c}_\lambda)$ be regular values of the functions f_0 (f_1) lying between the values of the functions f_0 (f_1) at critical points of index λ and $\lambda+1$. Since the functions f_0 and f_1 are homotopy equivalent, there exists a diffeomorphism isotopic to the identity such that $f_0^{-1}[0, c_\lambda] = h(f_1^{-1}[0, \bar{c}_\lambda])$ for all λ. By Proposition 2.3, there is a path in the space of Morse functions joining f_0 to the function $\bar{f}_0^{-1} = f_1 \circ h$ such that

$$\bar{f}_0^{-1}[\bar{c}_\lambda, \bar{c}_{\lambda-1}] = f_1^{-1}[\bar{c}_\lambda, \bar{c}_{\lambda-1}] = N_\lambda.$$

Now, using Proposition 2.3, join the restrictions of the functions f_0 and $\bar{f}_0$ to the submanifold N_λ by a path $\tilde{\gamma}(t)$ such that $\tilde{\gamma}(t)$ is a Morse function for each $t \in [0, 1]$. By repeating this argument, we find a path in the space of Morse functions joining the functions f_0 and f_1. □

COROLLARY 2.3. *On a manifold M^n there exist $P(M^n)$ nonisotopic minimal Morse functions.*

PROOF. Making use of the theorem on regrouping critical points [17, 60], join an arbitrary minimal Morse function to an ordered Morse function, and then apply Theorem 2.7 and Corollary 2.2. □

Note that on non-simply-connected manifolds there exist homotopy equivalent but nonisotopic Morse functions [60].

In conclusion let us discuss the question: When do two minimal ordered Morse functions on a closed manifold M^n (≥ 6), $\pi_1(M^n) = 0$, belong to the same connected component of the stratum in codimension 0?

DEFINITION 2.5. Two Morse functions f and g on the manifold M^n are said to be conjugate if there exist diffeomorphisms $h: M^n \to M^n$, $k: [0, 1] \to [0, 1]$, both isotopic to the (respective) identity and satisfying the condition $f = k \circ g \circ h$.

If the functions f and g are conjugate, then they belong to the same connected component of the stratum containing them. Making use of a gradient-like vector field depending on a parameter, we can show that the converse statement is also true.

Let $\{C, \partial\}$ be a free chain complex of abelian groups. We say that $\{C, \partial\}$ is an ordered basis chain complex (OBC complex) if ordered basis elements are fixed in the chain groups C_i. The following change of bases operations are allowed:

(1) if $c_1 < c_2 < \cdots < c_k$ is a basis in C_i, then $c_1 < c_2 < \cdots < \hat{c}_j < \cdots < c_k$ is a new basis in C_i, where either $\hat{c}_j = -c_j$ or

$$\hat{c}_j = \sum_{n=1}^{j-1} a_n c_n, \qquad a_n \in \mathbb{Z};$$

(2) iterated application of the operation (1).

Two OBC complexes $\{C, \partial\}$ and $\{\overline{C}, \overline{\partial}\}$ are said to be equivalent if there exist sequences of admissible operations over the bases in C_i and $\overline{C}_i$ such that in the new basis the matrices of differentials for both complexes coincide.

Let there be given chain mappings of the OBC complexes $\{C, \partial\}$ and $\{\overline{C}, \overline{\partial}\}$ into an OBC complex $\{D, d\}$:

$$p: \{C, \partial\} \longrightarrow \{D, d\} \longleftarrow \{\overline{C}, \overline{\partial}\} : q.$$

Suppose that $\{C, \partial\}$ and $\{\overline{C}, \overline{\partial}\}$ are equivalent OBC complexes. Choose bases in the complexes $\{C, \partial\}$ and $\{\overline{C}, \overline{\partial}\}$ in which the matrices of their differentials coincide and consider the chain mappings not as mappings of different complexes but as mappings of the same chain complex into the complex $\{D, d\}$. In this case one can speak about the homotopy equivalence of the chain mappings of OBC complexes. Note that such bases in the complexes $\{C, \partial\}$ and $\{\overline{C}, \overline{\partial}\}$ are not chosen uniquely. These bases can be modified by means of admissible operations in such a way that the matrices of the boundary homomorphisms remain unchanged. In this case the chain mappings into the chain complex $\{D, \partial\}$ also vary. If they were not homotopy equivalent, then they may become homotopy equivalent and vice versa.

Suppose that the chain isomorphisms of minimal OBC complexes

$$\{C, \partial\} \xrightarrow{p} \{D, d\} \xleftarrow{q} \{C, \partial\}$$

are homotopy equivalent. Then this diagram can be closed by an isomorphism $\varphi: \{C, \partial\} \to \{C, \partial\}$ that is homotopy equivalent to the identity. We will use this fact below.

Fix an arbitrary minimal ordered Morse function f and a gradient-like vector field ξ on the manifold M^n. The function f and the field ξ define an OBC complex up to admissible transformations of the bases. If $f(x) > f(y)$, then $c_x > c_y$, where x, y are critical points of index λ, and c_x, c_y are the basis elements in c_λ corresponding to x and y. If $g: M^n \to [0, 1]$ is another ordered minimal Morse function and η is its gradient-like vector field, then there always exists a chain mapping

$$h_*: \{C(g, \eta, \partial^g)\} \to \{C(f, \xi, \partial^f)\},$$

where $\{C(g, \eta, \partial^g)\}$ $(\{C(f, \xi, \partial^f)\})$ is the chain complex constructed from the pair g, η (f, ξ). The mapping h_* is induced by the diffeomorphism $h: M^n \to M^n$ that is isotopic to the identity. This mapping is not defined uniquely and may be modified step by step as in the proof of Lemma 2.3.

THEOREM 2.8. *Let M^n be a closed simply connected manifold, $n \geq 6$. Minimal ordered Morse functions g_1 and g_2 on the manifold M^n are conjugate if and only if the associated OBC complexes $\{C(g_1, \eta_1, \partial^{g_1})\}$ and $\{C(g_2, \eta_2, \partial^{g_2})\}$ are equivalent and the induced chain mappings*

$$h_{1*}: C(g_1, \eta_1, \partial^{g_1}) \longrightarrow C(f, \xi, \partial^f) \longleftarrow C(g_2, \eta_2, \partial^{g_2}) : h_{2*}$$

are homotopy equivalent.

PROOF. *Necessity.* Obvious.

Sufficiency. We can assume without any loss of generality that the critical values of the functions g_1 and g_2 coincide. Otherwise we can make them coincide by using a diffeomorphism $k: [0, 1] \to [0, 1]$, $k(0) = 0$. We can also assume that the function f is a distinguished minimal ordered Morse function (the matrices of the boundary homomorphisms of the chain complex $C(f, \xi, \partial^\xi)$ are diagonal). Since the chain complexes $C(g_1, \eta_1, \partial^{\eta_1})$ and $C(g_2, \eta_2, \partial^{\eta_2})$ are equivalent and the chain mappings h_{1*} and h_{2*} are homotopy equivalent, we can, making use of the chain homotopy, modify the mapping h_{2*} in such a way that it will coincide with h_{1*}. In geometric terms, this chain homotopy can be realized as a sequence of isotopies of the submanifolds $M_\lambda^{g_2}$ in the manifold M^n by means of diffeomorphisms $h_j: M^n \to M^n$ isotopic to the identity (similarly to the proof of Lemma 2.3). Set $h = h_n \circ \cdots \circ h_2 \circ h_0$. Since the chain complexes $C(g_1, \eta_1, \partial^{\eta_1})$, $C(g_2, \eta_2, \partial^{\eta_2})$, and $C(f, \xi, \partial^\xi)$ are minimal, the chain mappings h_{1*} and h_{2*} can only be monomorphisms on the groups of chains. Hence there exists a chain mapping $\varphi: C(g_1, \eta_1, \partial^{\eta_1}) \to C(g_2, \eta_2, \partial^{\eta_2})$ expressed in the chosen basis by the identity matrices. In geometrical terms, this means that the submanifolds $M_\lambda^{g_1}$ and $M_\lambda^{g_2 \circ h^{-1}}$ coincide and the restrictions of the functions g_1 and $g_2 \circ h^{-1}$ to $M_\lambda^{g_1}$ have the same Cerf invariant. Thus, the functions g_1 and $g_2 \circ h^{-1}$ can be joined by a path lying in the same connected component of the stratum of codimension 0. □

Note that all the assertions of this chapter concerning closed simply connected manifolds of codimension greater than 5 are valid for simply connected manifolds W^n with simply connected boundaries $\partial W^n = V_0 \cup V_1$ $(n \geq 6)$ and the Morse functions $f: W^n \to [0, 1]$, $f^{-1}(0) = V_0$, $f^{-1}(1) = V_1$.

CHAPTER III

Stable Algebra

Substantial difficulties arising in the analysis of non-simply-connected manifolds are due, in the first place, to various algebraic reasons. Here questions of combinatorial group theory, general and homological algebra, and K-theory are closely interrelated. Unfortunately, our knowledge of many of them is unsatisfactory, which impedes the progress in this area of topology.

In this chapter we develop the algebraic apparatus needed for the analysis of Morse functions on non-simply-connected manifolds. All rings are assumed to be associative with unity. Unless otherwise stated, all modules will be taken to be left finitely-generated ones.

In the first section we study the behavior of a minimal number of generators of a finitely-generated Λ-module over an IBN-ring. In the second section we define the numerical invariant of a $\mathbb{Z}[\pi]$-module M

$$S(M) = \lim_{n\to\infty} (\mu(M \oplus n\Lambda) - n) - \mu(\mathbb{Z} \otimes_{\mathbb{Z}[\pi]} M),$$

which plays an important role in the sequel.

The third section is devoted to the relation of the stable rank of a ring Λ and the minimal number of generators of a Λ-module M. In the fourth and fifth sections we discuss the properties of epimorphisms of modules. In the sixth section the existence of minimal resolutions of modules over s-rings is proved. In the seventh section we consider extensions of modules.

§1. Numerical invariants of modules

Let Λ be a ring and M a Λ-module. Recall that a family of elements $\{m_i\} \in M$ generating the module M is said to be the system of generators of M. If the direct sum of embeddings $m_i\Lambda \to M$ defines an isomorphism

$$f\colon m_1\Lambda \oplus \cdots \oplus m_k\Lambda \to M,$$

then $\{m_i\}$ is called a basis for the module M and k the basis number for M. The basis number depends in general on the choice of basis, and therefore cannot be taken as an invariant of the module $M \approx m_1\Lambda\oplus\cdots\oplus m_k\Lambda$. A ring Λ such that the basis number of any free module over it is determined uniquely is called an IBN-ring (ring with Invariant Basis Number). It is well known that if there exists a nontrivial homomorphism of Λ into a commutative ring, then Λ is an IBN-ring [45]. Therefore, every integral group ring $\mathbb{Z}[G]$ is an IBN-ring, since it has the augmentation homomorphism

$$\varepsilon\colon \mathbb{Z}[G] \to \mathbb{Z} \qquad \left(\varepsilon\left(\sum_i n_i g_i\right) = \sum_i n_i, \quad g_i \in G\right).$$

Noetherian rings are also IBN-rings.

Throughout this book we denote by $\mu(M)$ the minimal number of generators of a module M. We set $\mu(M) = 0$ if and only if M is the zero module. In what follows, we will, as a rule, denote the free module of rank k by

$$k\Lambda = \underbrace{\Lambda \oplus \cdots \oplus \Lambda}_{k}.$$

DEFINITION 3.1. For any Λ-module M define the number $d(M)$ by

$$\lim_{k\to\infty} (\mu(M \oplus k\Lambda) - k) = d(M).$$

DEFINITION 3.2. Let M be a Λ-module. The minimal number of generators $\mu(M)$ is additive if $d(M) = \mu(M)$, i.e., if $\mu(M \oplus k\Lambda) = \mu(M) + k$ for all positive integers k.

The next lemma shows that for sufficiently "large" modules M the number $\mu(M)$ is always additive.

LEMMA 3.1. *For an arbitrary Λ-module M there exists a number k such that $\mu(M \oplus k\Lambda)$ is additive.*

PROOF. Consider the graph of the function

$$y = \mu(M \oplus [x]\Lambda) + (\mu(M \oplus [x+1]\Lambda) - \mu(M \oplus [x]\Lambda) - \mu(M \oplus [x]\Lambda))(x - [x])$$

on the plane. At integer points this function is equal to $\mu(M \oplus n\Lambda)$. The graph consists of straight line segments of two kinds: those parallel to the straight line $y = x$, and those parallel to the x-axis. Since $\mu(M \oplus k\Lambda) \geq k$, there is a value of k beyond which the graph is parallel to the straight line $y = x$. Hence is follows that the module $\overline{M} = M \oplus k\Lambda$ has the property $\mu(\overline{M} \oplus n\Lambda) = \mu(\overline{M}) + n$. □

Two Λ-modules M and N are said to be stably isomorphic if $M \oplus k\Lambda \approx N \oplus k\Lambda$ for some positive integer k.

COROLLARY 3.1. *If Λ-modules M and N are stably isomorphic, then $d(M) = d(N)$. For any Λ-module M the equality $d(M \oplus k\Lambda) = d(M) + k$ holds for all positive integers k.* □

In the sequel we will provide an estimate for the number k starting with which $\mu(M \oplus l\Lambda)$ is additive. If $\Lambda = \mathbb{Z}[G]$, then it is easy to see that

$$\mu(\mathbb{Z} \otimes_{\mathbb{Z}[G]} M) = \mu(\mathbb{Z} \otimes_{\mathbb{Z}[G]} N)$$

for any stably isomorphic $\mathbb{Z}[G]$-modules M and N (here $\mathbb{Z}$ is considered as a trivial $\mathbb{Z}[G]$-module).

The next definition plays a key role in what follows.

DEFINITION 3.3. Let M be a $\mathbb{Z}[G]$-module. Set

$$S(M) = d(M) - \mu(\mathbb{Z} \otimes_{\mathbb{Z}[G]} M).$$

The above statements and Corollary 3.1 imply the following corollary.

COROLLARY 3.2. *If the $\mathbb{Z}[G]$-modules $M \oplus k\mathbb{Z}[G]$ and $N \oplus n\mathbb{Z}[G]$ are isomorphic, then*

$$S(M) = S(N) = S(M \oplus k\mathbb{Z}[G]) = S(N \oplus n\mathbb{Z}[G]). \quad \square$$

It is not difficult to show that $S(M) \geq 0$, because

$$\lim_{n\to\infty} (\mu(M \oplus k\mathbb{Z}[G]) - k) \geq \lim_{k\to\infty} (\mu(\mathbb{Z} \otimes_{\mathbb{Z}[G]} (M \oplus k\mathbb{Z}[G])) - k) = \mu(\mathbb{Z} \otimes_{\mathbb{Z}[G]} M).$$

Recall that a ring Λ is called a Hopf ring if any epimorphism of a free Λ-module onto itself is an isomorphism. Similarly, in a free module of rank n over Λ, any n generators are free, i.e., form a basis. A theorem due to Kaplansky implies that every integral group ring is a Hopf ring [100]. Evidently, for any nonzero module M over a Hopf ring one has $d(M) > 0$. Let us give an example of a module, for which $S(M) > 0$. Let $I(G)$ be the augmentation ideal of a finitely generated perfect* group G (i.e., $I(G)$ is the kernel of the augmentation homomorphism $\varepsilon: \mathbb{Z}[G] \to \mathbb{Z}$). Recall that a group is said to be perfect if it coincides with its commutator. It is known that $I[G]$ is a finitely-generated $\mathbb{Z}[G]$-module in which one can take for generators the elements $g_1 - 1, \dots, g_l - 1$ (where $g_1, \dots, g_l$ is the system of generators of the group G). It is also known [62] that

$$\mathbb{Z} \otimes_{\mathbb{Z}[G]} I[G] = I[G]/(I[G])^2 \approx G/[G, G] = 0.$$

Therefore, $S(I[G]) > 0$.

§2. Stably free modules

We recall that a Λ-module M is called stably free if the direct sum of M with some free module is free, i.e., $M \oplus n\Lambda \approx m\Lambda$. The number $m - n$ is said to be the rank of the stably free module M. A ring Λ is called an s-ring if every stably free Λ-module is free. We note that the question of whether a given ring Λ is an s-ring is in general a complicated one [147, 148, 115]. We list here some groups for which it is known that their integral group ring is an s-ring. In what follows such groups will be called s-groups.

A theorem of Jacobinski asserts that if a finite group has no epimorphic mapping onto the generalized quaternion group

$$Q^n = \{a, b; a^{2^{n-1}} = 1, b^2 = a^{2^{n-2}}, ba = a^{-1}b, n > 1\},$$

*Editor's note. The term *complete* group is used in English literature.

the binary tetrahedral group

$$T = \{a, b; a^3 = b^2 = (ab)^2\},$$

the octahedral group

$$O = \{a, b; a^3 = b^4 = (ab)^2\},$$

or the icosahedral group

$$I = \{a, b; a^3 = b^5 = (ab)^2\},$$

then it is an s-group. Therefore, among the finite groups the abelian, the simple, and those of odd order are s-groups. Free groups and free abelian groups are also s-groups [3, 98, 104].

LEMMA 3.2. *If the minimal number of generators $\mu(M)$ of a stably free module M is additive, then M is a free module.*

PROOF. Choose a positive integer k such that $M \oplus k\Lambda$ is a free module. Select a minimal number of generators $m_1, \dots, m_l$ in the module M, and let $e_1, \dots, e_k$ be a basis in $k\Lambda$. Then $m_1, \dots, m_l, e_1, \dots, e_k$ is a minimal system of generators for the module $M \oplus k\Lambda = F$. Suppose that $f_1, \dots, f_s$ is a basis in $M \oplus k\Lambda = F$. Clearly, $s \geq k + l$. We construct an epimorphism $g: F \to F$ by putting $g(f_i) = m_i$ $(1 \leq i \leq l)$, $g(f_{l+j}) = e_j$ $(1 \leq j \leq k)$, and $g(f_n) = 0$ $(l + k < n \leq s)$. Let $K = \ker g$. We have $K \oplus F = F$, and therefore $\mu(K \oplus F) = \mu(K) + \mu(F) = \mu(F)$, whence $K = 0$ and g is an isomorphism. Let $\overline{F}$ be the free submodule in F generated by the elements $f_1, \dots, f_l$. Since $g(\overline{F}) = M$, we conclude that M is a free module. □

It turns out that for all modules over s-rings the minimal number of generators is additive.

LEMMA 3.3. *Let Λ be an s-ring and suppose that M is a Λ-module. Then the number $\mu(M)$ is additive.*

PROOF. It is sufficient to prove that the equality $\mu(M \oplus \Lambda) = \mu(M) + 1$ holds for any Λ-module M. We claim that the following inequalities are true simultaneously:

$$\mu(M \oplus \Lambda) \leq \mu(M) + 1, \qquad \mu(M \oplus \Lambda) \geq \mu(M) + 1.$$

The first is obvious. We prove the second. Let $m_1, \dots, m_k$ be a minimal system of generators of the module $N = M \oplus \Lambda$, F a free module of rank k, and $f: F \to N$ an epimorphism. Denote by $p: \Lambda \oplus M \to M$ the projection onto the first summand. Using the composition of homomorphisms

$$F \xrightarrow{f} \Lambda \oplus M \xrightarrow{p} \Lambda,$$

find the decomposition into the direct sum

$$\overline{F} \oplus \Lambda \xrightarrow{\overline{f} + i} M \oplus \Lambda,$$

where $\overline{F} \oplus \Lambda = F$, $\overline{f} \oplus i = f$. Since Λ is an s-ring, the module $\overline{F}$ is free; consequently, $\mu(N) = \mu(\overline{F} \oplus \Lambda) = \mu(\overline{F}) + 1 = k$. Since $\overline{f}: \overline{F} \to M$ is an epimorphism, we have $\mu(\overline{F}) \geq \mu(M)$, whence it follows that $\mu(N) = \mu(\overline{F}) + 1 \geq \mu(M) + 1$. □

These lemmas yield the following characterization of s-rings.

COROLLARY 3.3. *A ring Λ is an s-ring if and only if for any stably free module P over Λ the following equality is satisfied*:

$$\mu(M \oplus \Lambda) = \mu(M) + 1. \qquad \square$$

Therefore, in order to find the number $S(M)$ for modules over s-rings, we can use the formula

$$S(M) = \mu(M) - \mu(\mathbb{Z} \otimes_{\mathbb{Z}[G]} M).$$

We conclude with some examples of nonfree but stably free modules.

Let $\Lambda = \mathbb{R}[X, Y, Z]/X^2 + Y^2 + Z^2 = 1$ be the coordinate ring of the two-dimensional sphere. Define a Λ-module M by $M = 3\Lambda/(x, y, z)\Lambda$ (where x, y, z are images of X, Y, Z in Λ). Then $M \oplus \Lambda \approx 2\Lambda \oplus \Lambda$. Therefore M is a nonfree but stably free module corresponding to the tangent bundle over the two-dimensional sphere [151].

Now let k be an arbitrary field and n a positive integer. Set

$$\Lambda = k[X_1, \dots, X_{2n}]/\left(\sum_{i=1}^{n} X_i X_{n+i} - 1\right)$$

and denote by x_i the images of the variables X_i in the ring Λ. Define a homomorphism of free Λ-modules $f: n\Lambda \to \Lambda$ by the formula

$$(\lambda_1, \dots, \lambda_n) \xrightarrow{f} \sum_{i=1}^{n} \lambda_i x_i.$$

It was proved in [148, 151] that f is an epimorphism. Let M be the kernel of the homomorphism f; then $M \oplus \Lambda \approx n\Lambda$ and M is a nonfree module.

Let G be a finite group of order n. Let $N = \sum_i g_i$ be a norm in the ring $\mathbb{Z}[G]$. If the number p is relatively prime with n, then the ideal (p, N) generated by the elements p and N is a projective module over the ring $\mathbb{Z}[G]$. It is known that if (p, N) is a stably free module, then $(p, N) \oplus \mathbb{Z}[G] \approx \mathbb{Z}[G] \oplus \mathbb{Z}[G]$.

Many nonfree but stably free modules over integral group rings for non-abelian infinite groups are constructed in [148, 151].

§3. Stable rank of a ring and additivity of $\mu(M)$

Consider a free module of rank k over a ring Λ. Recall that an element $a = (\lambda_1, \dots, \lambda_k) \in k\Lambda$ is said to be unimodular in the module $k\Lambda$ if there exists a homomorphism of free modules $f: k\Lambda \to \Lambda$ such that $f(a) = 1 \in \Lambda$. Similarly, $\lambda_1\Lambda \oplus \cdots \oplus \lambda_k\Lambda = \Lambda$. Let $k \geq 1$. Following Bass, we say that k is the stable rank of the ring Λ if k is the least positive integer satisfying the condition: for any unimodular element $a = (\lambda_1, \dots, \lambda_k) \in k\Lambda$ there exist $\gamma_1, \dots, \gamma_{k-1}$ in Λ such that the element $b = (\lambda_1 + \gamma_1\lambda_k, \dots, \lambda_{k-1} + \gamma_{k-1}\lambda_k)$ is unimodular in $(k-1)\Lambda$. Denote the stable rank of the ring Λ by $SR(\Lambda)$. In other words, the condition $SR(\Lambda) = k$ means that the unit left ideal in the ring Λ is generated by k elements "in general position". The following statement is due to Bass and characterizes the stable rank [8].

If Λ is a commutative ring whose space of maximal ideals is Noetherian and has Krull dimension equal to d, while Λ is an algebra finitely generated as a module A, then $SR(\Lambda) \leq d + 1$.

It is known that if Λ is a semilocal ring or the ring of integers in a numerical field, then $SR(\Lambda) = 1$; if Λ is a Dedekind ring, then $SR(\Lambda) = 2$. The Bass proposition implies that the stable rank of the ring of polynomials in n variables over a field equals $n + 1$ [8].

Denote by $E(n, \Lambda)$ the subgroup of the group of invertible matrices over the ring Λ generated by elementary matrices $e_{ij}(\lambda)$ $(1 \leq i \neq j \leq n, \lambda \in \Lambda)$. Here $e_{ij}(\lambda)$ denotes the matrix whose diagonal elements are all equal to one, while the only nonzero nondiagonal element appears in the ith row and the jth column and is equal to λ. Bass proved that if $n = SR(\Lambda)$, then $E(r, \Lambda)$ acts transitively on the set of unimodular elements in $r\Lambda$ if $r > n$ [8]. We shall use this fact in the proof of the following lemma.

LEMMA 3.4. *Let $f: n\Lambda \to m\Lambda$ be an epimorphism of free modules. If $n - m \geq SR(\Lambda)$, then $K = \operatorname{Ker} f$ is a free module.*

PROOF. Let $a_1, a_2, \dots, a_m$ be a basis in the module $m\Lambda$. It is easy to find a unimodular element $b_1 \in n\Lambda$ such that $f(b_1) = a_1$. Since $n > SR(\Lambda)$, b_1 can be included into a basis $b_1, \overline{b}_2, \dots, \overline{b}_n$ in $n\Lambda$ ($E(k, \Lambda)$ acts transitively on the set of unimodular elements in $k\Lambda$ for $k > SR(\Lambda)$). If $f(\overline{b}_i) = \lambda_1^i a_1 + \sum_{j=2}^n \lambda_j^i a_j$, we set $b_i = \overline{b}_i - \lambda_1^i b_1$ and replace $\overline{b}_i$ in our basis by b_i $(2 \leq i \leq n)$. Denote the resulting basis by $b_1, b_2, \dots, b_n$. Let $(m-1)\Lambda$ be the submodule generated by the elements $a_2, \dots, a_m$, and $(n-1)\Lambda$ be the submodule generated by the elements $b_2, \dots, b_n$. By construction, $(n-1)\Lambda$ is mapped epimorphically onto $(m-1)\Lambda$ and we can apply the preceding argument. After m steps we see that K is a free module. □

Note that if, following Bass, we require that the group $GL(n, \Lambda)$ act transitively on the set of unimodular elements of the module $n\Lambda$ (condition SR'_n), then the conclusion of the lemma is valid if the condition SR'_k is satisfied for $n - m \leq k \leq n$. This condition is a weaker one than that imposed in the hypothesis of the lemma.

If Λ is not an s-ring, we can, using the stable rank of the ring Λ, give an estimate for the minimal number of generators of a Λ-module M for which $\mu(M)$ is additive.

LEMMA 3.5. *Let M be a Λ-module. If $\mu(M) > SR(\Lambda)$, then $\mu(M)$ is additive.*

PROOF. The proof will be achieved by induction on n appearing in the equality

$$\mu(M \oplus n\Lambda) = \mu(M) + n.$$

Let us verify this equality for $n = 1$. Consider the epimorphism $g: k\Lambda \to M \oplus \Lambda$, where $k = \mu(M \oplus \Lambda)$. On the one hand, $k \leq \mu(M) + 1$ and, on the other hand, $k \geq \mu(M) \geq SR(\Lambda)$. Let $p: M \oplus \Lambda \to \Lambda$ be the natural projection onto the second summand. Consider the epimorphism $p \circ g: k\Lambda \to \Lambda$. Since $k - 1 \geq SR(\Lambda)$, we can, using Lemma 3.4, present the mapping g in the form $g = \overline{g} \oplus i: (k-1)\Lambda \oplus \Lambda \to M \oplus \Lambda$. By construction, $k - 1 \geq \mu(M)$, whence $k \geq \mu(M) + 1$. Assuming that $\mu(M \oplus (n-1)\Lambda) = \mu(M) + n - 1$, let us prove the equality $\mu(M \oplus n\Lambda) = \mu(M) + n$. Set $M \oplus (n-1)\Lambda = \overline{M}$. Evidently, $\mu(\overline{M}) = \mu(M) + n - 1 \geq SR(\Lambda) + n - 1 \geq SR(\Lambda)$ $(n \geq 2)$. Consider the epimorphism $\overline{k}\Lambda \xrightarrow{\overline{g}} \overline{M} \oplus \Lambda$, where $\overline{k} = \mu(\overline{M} \oplus \Lambda)$. It is clear that $\overline{k} \leq \mu(\overline{M}) + 1 = \mu(M) + n$ and $\overline{k} \geq \mu(\overline{M}) = \mu(M) + n - 1 \geq SR(\Lambda)$. According to the lemma, we can represent $\overline{g}$ in the form $\overline{g} = \tilde{g} \oplus i: (\overline{k} - 1)\Lambda \oplus \Lambda \to \overline{M} \oplus \Lambda$. Since $\overline{k} - 1 \geq \mu(\overline{M}) = \mu(M) + n - 1$, we have $\overline{k} \geq \mu(M) + n$. Therefore $\mu(M \oplus n\Lambda) = \mu(M) + n$. □

The next corollary follows directly from this lemma and Lemma 3.2.

COROLLARY 3.4. *Let M be a stably free module over a ring Λ and $\mu(M) > SR(\Lambda)$. Then M is a free module.* □

The above statements imply that in the definition of $d(M)$ it is sufficient to choose $k > SR(\Lambda)$ (it is assumed that $SR(\Lambda) < \infty$).

§4. Thickening of epimorphisms

DEFINITION 3.4. Two epimorphisms of Λ-modules $f: F \to M$ and $g: F \to M$ are said to be equivalent if there exists an isomorphism $\varphi: F \to F$ such that $f = g \circ \varphi$.

One can easily give examples of nonequivalent epimorphisms. Suppose that F is a free module. As Warfield states in [157], if $\mu(F) \geq \mu(M) + SR(\Lambda)$, then the epimorphisms f and g are always equivalent. In this section we discuss a procedure taking nonequivalent homomorphisms into equivalent ones.

DEFINITION 3.5. By a thickening of a homomorphism $f: F \to M$ via a module G we mean a homomorphism

$$\hat{f}: F \oplus G \to M, \qquad \hat{f}|_{F \oplus 0} = f, \quad \hat{f}|_{0 \oplus G} = 0.$$

The following lemma will be used repeatedly in what follows. In [23] Cockroft and Swan ascribe it to Schanuel (see also [88]).

LEMMA 3.6. *Let $f: F \to M$ and $g: G \to M$ be arbitrary epimorphisms, where F and G are free Λ-modules. Then the thickening of f via G is equivalent to the thickening of g via F.*

PROOF. Let $f_1, \dots, f_k$ be a basis in the module F, and $g_1, \dots, g_l$ a basis in the module G. Then* $\varphi(f_1), \dots, \varphi(f_k)$ and $\psi(g_1), \dots, \psi(g_l)$ are two systems of generators for the module M. Consider the thickenings of epimorphisms φ and ψ:

$$\hat{\varphi}: F \oplus G \to M, \qquad \hat{\varphi}(f_i) = \varphi(f_i), \quad \hat{\varphi}(g_i) = 0,$$
$$\hat{\psi}: G \oplus F \to M, \qquad \hat{\psi}(g_i) = \psi(g_i), \quad \hat{\psi}(f_i) = 0.$$

Clearly, the sets $f_1, \dots, f_k, g_1, \dots, g_l$ and $g_1, \dots, g_l, f_1, \dots, f_k$ constitute bases in the module $F \oplus G$. Consider the first basis and transform it to the required form. Choose the elements $\tilde{g}_j = g_j + h_j$ $(j = 1, \dots, l)$, where $h_i \in \varphi^{-1}(\psi(g_j))$. The elements $f_1, \dots, f_k, \tilde{g}_1, \dots, \tilde{g}_l$ evidently constitute a basis in the module $G \oplus F$. Denote by $\overline{G}$ the submodule of $G \oplus F$ generated by the elements $\tilde{g}_1, \dots, \tilde{g}_l$. By construction, the restriction of the epimorphism $\hat{\varphi}$ to the submodule $\overline{G} \oplus 0$ is an epimorphism onto the module M. Consequently, there exist elements $s_1, \dots, s_k$ in $\overline{G}$ such that $\hat{\varphi}(s_j) = \varphi(f_k)$ $(i = 1, \dots, k)$. Consider in $G \oplus F$ the basis $f_1 - s_1, \dots, f_k - s_k, \tilde{g}_1, \dots, \tilde{g}_l$. Clearly, $\hat{\varphi}(f_i - s_i) = 0$ for $1 \le i \le k$ and $\hat{\varphi}(\tilde{g}_j) = \hat{\psi}(g_j)$ for $1 \le j \le l$. Making use of this basis, we can easily construct an isomorphism $\gamma: F \oplus G \to G \oplus F$ such that the diagram

$$\begin{array}{ccc} F \oplus G & \xrightarrow{\hat{\varphi}} & M \\ {\scriptstyle\gamma}\downarrow & & \downarrow{\scriptstyle\mathrm{id}} \\ G \oplus F & \xrightarrow{\hat{\psi}} & M \end{array}$$

commutes.

We note that the transition from the first basis in the module $G \oplus F$ to the second involved only elementary transformations (i.e., transformations realized by elementary matrices). □

COROLLARY 3.5. *Let $g: C \to M$ and $f: F \to M$ be epimorphisms, where C, M, and F are Λ-modules, and F is free. Let $\hat{f}$ be the thickening of the epimorphism f via a free module of rank $k = \mu(C)$. Then there exists an epimorphism $\varphi: F \oplus k\Lambda \to C$ such that the following diagram commutes:*

$$\begin{array}{ccc} F \oplus k\Lambda & & \\ & \searrow{\scriptstyle\hat{f}} & \\ {\scriptstyle\varphi}\downarrow & & M \\ & \nearrow{\scriptstyle g} & \\ C & & \end{array}$$

*Editor's note. The author is renaming his epimorphisms.

PROOF. Let $p: k\Lambda \to C$ be an arbitrary epimorphism. Consider the epimorphisms $f: F \to M$ and $\overline{g} = g \circ p: k\Lambda \to M$. By Lemma 3.6 we have a commutative diagram

$$\begin{array}{ccc} F \oplus k\Lambda & & \\ & \searrow^{\hat{f}} & \\ \gamma \downarrow & & M, \\ & \nearrow_{\hat{g}} & \\ k\Lambda \oplus F & & \end{array}$$

where γ is an isomorphism and $\hat{g}$ is the thickening of the epimorphism g via the module F. The epimorphism $\hat{g}$ is evidently representable as a composition $\hat{g} = \overline{g} \circ \pi$, where $\pi: k\Lambda \oplus F \to k\Lambda$ is the projection on the first summand. We set $\varphi = p \circ \pi \circ \gamma$. □

We conclude this section with three definitions.

DEFINITION 3.6. Let $f: F \to M$ be an epimorphism, where F is a free module. We say that an epimorphism $\overline{f}: F \to M$ can be extracted from f if there is a decomposition

$$\begin{array}{ccc} F & & \\ & \searrow^{f} & \\ \| & & M, \\ & \nearrow_{\overline{f}} & \\ \overline{F} \oplus G & & \end{array}$$

where $f|_{\overline{F} \oplus 0} = \overline{f}|_{\overline{F} \oplus 0}$ and $f|_{0 \oplus G} = \overline{f}|_{0 \oplus G} = 0$.

DEFINITION 3.7. An epimorphism $f: k\Lambda \to M$ is said to be minimal if $k = \mu(M)$.

In general, a minimal epimorphism cannot be extracted from an arbitrary epimorphism $f: k\Lambda \to M$, but, as Lemma 3.6 shows, this can be done by thickening f via a free module of rank $\mu(M)$.

DEFINITION 3.8. Let $f: F \to G$ be a homomorphism of Λ-modules. The stabilization of the epimorphism f via a Λ-module L is the homomorphism

$$f^s: F \oplus L \xrightarrow{f \oplus \mathrm{id}} F \oplus L.$$

The operation inverse to stabilization is called contraction.

§5. Minimal epimorphisms and f-rank(N, M)

Let N be a submodule (not necessarily finitely generated) of the module M. Following Bass, we define f-rank(N, M) as the largest nonnegative integer k such that N contains a direct summand of M isomorphic to $k\Lambda$. If f-rank$(N, M) = k$, then the module M can be represented in the form $M = M' \oplus k\Lambda$; therefore N can also be written as a direct sum $N = N' \oplus k\Lambda$, where $N' \subset M'$. It is not hard to show that f-rank$(N, M) =$ f-rank$(N, M \oplus n\Lambda)$.

DEFINITION 3.9. We say that f-rank(N, M) is additive if

$$f\text{-rank}(N \oplus n\Lambda, M \oplus n\Lambda) = f\text{-rank}(N, M) + n$$

LEMMA 3.7. *Let N be a submodule (not necessarily finitely generated) of a Λ-module M. Then there exists a positive integer n_0 such that f-rank$(N \oplus n\Lambda, M \oplus n\Lambda)$ is additive for all $n > n_0$.*

PROOF. Let f-rank$(N \oplus n\Lambda, M \oplus n\Lambda) = k_n$. Obviously $\mu(M \oplus n\Lambda) \leq \mu(M) + n$ and $k_n \geq n$ for all n. The module $N \oplus n\Lambda$ can be represented in the form $N \oplus n\Lambda = N_{k_n} \oplus k_n\Lambda$, where $k_n\Lambda$ is a direct summand in the module $M \oplus n\Lambda$. If f-rank$(N \oplus n\Lambda, M \oplus n\Lambda)$ does not become additive with the growth of n, then the difference $k_n - n$ must increase unboundedly. Therefore there exists a positive integer $\overline{n}$ such that $k_{\overline{n}} - \overline{n} > \mu(M)$, which is impossible. This gives a contradiction. □

LEMMA 3.8. *Let N be a submodule (not necessarily finitely generated) of a module M. Suppose that f-rank$(N, M) = k$ and is additive. Represent the module N in the form $N = N' \oplus k\Lambda$, where $N' \subset M'$ and $M = M' \oplus k\Lambda$. Then f-rank$(N', M') = 0$ and is additive.*

PROOF. The equality f-rank$(N', M') = 0$ is obvious (otherwise f-rank$(N, M) > k$). We will show that f-rank(N', M') is also additive. Assume the contrary; then there exists a positive integer $n > k$ such that f-rank$(N' \oplus n\Lambda, M' \oplus n\Lambda) = \overline{n} > k$. Consider the following decompositions into direct sums:

$$N' \oplus n\Lambda = N' \oplus k\Lambda \oplus (n-k)\Lambda = N \oplus (n-k)\Lambda,$$
$$M' \oplus n\Lambda = M' \oplus k\Lambda \oplus (n-k)\Lambda = M \oplus (n-k)\Lambda.$$

Clearly, there exists

$$\overline{n} = f\text{-rank}(N' \oplus n\Lambda, M' \oplus n\Lambda) = f\text{-rank}(N \oplus (n-k)\Lambda, M \oplus (n-k)\Lambda).$$

Since f-rank$(N, M) = k$ and is additive, we have

$$f\text{-rank}(N \oplus (n-k)\Lambda, M \oplus (n-k)\Lambda) = k + n - k = n.$$

Therefore $\overline{n} = n$. The resulting contradiction proves the lemma. □

There is a close relation between the behavior of f-rank(N, M) and minimal epimorphisms.

LEMMA 3.9. *Let Λ be an s-ring, N a submodule (not necessarily finitely generated) of a free module F over the ring Λ, and $\mu(F) = k$. If $p: F \to F/N = M$ is a minimal epimorphism, then f-rank$(N, F) = 0$ and is additive.*

PROOF. We prove first that f-rank$(N, F) = 0$. Assuming the contrary, suppose that f-rank$(N, F) > 0$. Then there exists a representation of the module N in the form $N = N' \oplus k\Lambda$, where $N' \subset F'$ and $F = F' \oplus k\Lambda$.

Since Λ is an s-ring, F' is a free module. From the construction it follows that $F/N = F'/N'$. But $\mu(F) > \mu(F')$, so the epimorphism p is not minimal, and we have a contradiction. Now we show that f-rank(N, F) is additive. Assume the contrary. If f-rank(N, F) is not additive, then there exists a positive integer n such that $f\text{-rank}(N\oplus n\Lambda, F\oplus n\Lambda) = \overline{n} > n$. Clearly, the quotient modules F/N and $(F\oplus n\Lambda)/(N\oplus n\Lambda)$ are isomorphic. Take a decomposition of the module $N \oplus n\Lambda$ such that

$$N \oplus n\Lambda = \overline{N} \oplus \overline{n}\Lambda, \qquad \overline{N} \subset \overline{F}, \qquad \overline{F} \oplus \overline{n}\Lambda = F \oplus n\Lambda.$$

By Lemma 3.3 we have $\mu(F\oplus n\Lambda) = \mu(F)+n = \mu(\overline{F})+\overline{n}$, so $\mu(F) > \mu(\overline{F})$. By construction, $F\oplus n\Lambda/N\oplus n\Lambda = M = \overline{F}\oplus\overline{n}\Lambda/\overline{N}\oplus\overline{n}\Lambda = \overline{F}/\overline{N}$. But $\mu(F) = \mu(M)$, since p is a minimal epimorphism. Therefore $\mu(\overline{F}) \geq \mu(M) = \mu(F)$, and we have a contradiction. □

LEMMA 3.10. *Let $F \supset N$ be a submodule (not necessarily finitely generated) of a free module of rank k over an s-ring Λ. If f-rank$(N, F) = 0$ and is additive, then $p: F \to F/N = M$ is a minimal epimorphism.*

PROOF. Assume the contrary: p is not minimal. Consider the exact sequence

$$0 \to N \xrightarrow{\ i\ } F \xrightarrow{\ p\ } F/N \to 0$$

and stabilize the homomorphism i via a free module G of rank $\mu(F/N)$. By the assumption of the lemma, $f\text{-rank}(N \oplus G, F \oplus G) = \mu(G)$. Consider the exact sequence

$$0 \to N \oplus G \xrightarrow{i\oplus \mathrm{id}} G \oplus F \xrightarrow{\ \hat{p}\ } F/N \to 0.$$

Let $f: G \to M = F/N$ be a minimal epimorphism. As mentioned above, the epimorphism $\hat{p}$ and the thickening of the epimorphism f via the free module F are equivalent (Lemma 3.6). Therefore there is the exact sequence

$$0 \to \operatorname{Ker} f \oplus \overline{F} \xrightarrow{i\oplus \mathrm{id}} \overline{F} \oplus G \xrightarrow{\ f\ } F/N \to 0.$$

By construction,

$$\begin{aligned} f\text{-rank}(\operatorname{Ker} f \oplus \overline{F}, \overline{F} \oplus G) &= f\text{-rank}(\operatorname{Ker} f \oplus F, F \oplus G) \\ &\geq \mu(\overline{F}) = \mu(F) > \mu(G). \end{aligned}$$

But $\operatorname{Ker} f \oplus \overline{F} = N \oplus F$; therefore f-rank(N, F) is not additive. This gives a contradiction, which proves the lemma. □

The following theorem is an immediate consequence of these two lemmas.

THEOREM 3.1. *Let Λ be an s-ring and $f: F \to M$ an epimorphism, where F is a free module. In order for f to be a minimal epimorphism, it is necessary and sufficient that f-rank$(\operatorname{Ker} f, F) = 0$ and be additive (the module* Ker f *is not assumed to be finitely generated).* □

§6. Minimal resolutions

Suppose we are given a resolution of a Λ-module M, i.e., an exact sequence of Λ-modules of the form

$$0 \longleftarrow M \xleftarrow{f_0} F_0 \xleftarrow{f_1} F_1 \longleftarrow \cdots.$$

If all the modules F_i are free, then the resolution is said to be a free one. Any module admits a free resolution constructed by the obvious inductive procedure.

DEFINITION 3.10. Let

$$0 \longleftarrow M \xleftarrow{f_0} F_0 \xleftarrow{f_1} F_1 \longleftarrow \cdots$$

be a free resolution of the Λ-module M. The resolution is said to be minimal if $f_i : F_i \to \operatorname{Ker} f_{i-1}$ is a minimal epimorphism.

LEMMA 3.11. *Let Λ be an s-ring,*

$$0 \longleftarrow M \xleftarrow{g_0} G_0 \xleftarrow{g_1} G_1 \longleftarrow \cdots$$

an arbitrary free resolution of the Λ-module M, and

$$0 \longleftarrow M \xleftarrow{f_0} F_0 \xleftarrow{f_1} F_1 \longleftarrow \cdots$$

a minimal resolution of M. Then $\mu(F_i) \leq \mu(G_i)$.

PROOF. Consider the segments

$$0 \longleftarrow M \xleftarrow{f_0} F_0 \longleftarrow \operatorname{Ker} f_0 \longleftarrow 0$$

and

$$0 \longleftarrow M \xleftarrow{g_0} G_0 \longleftarrow \operatorname{Ker} g_0 \longleftarrow 0$$

of the resolutions. By Schanuel's lemma and Lemma 3.3, we have

$$\mu(G_0 \oplus \operatorname{Ker} f_0) = \mu(F_0 \oplus \operatorname{Ker} g_0) = \mu(G_0) + \mu(\operatorname{Ker} f_0) = \mu(F_0) + \mu(\operatorname{Ker} g_0).$$

By hypothesis, $\mu(G_0) \geq \mu(F_0)$, whence $\mu(\operatorname{Ker} f_0) \leq \mu(\operatorname{Ker} g_0)$, and so $\mu(F_1) \leq \mu(G_1)$. The relation $\mu(\operatorname{Ker} f_0) = \mu(F_1)$ implies the inequality

$$\mu(G_0) + \mu(F_1) \leq \mu(F_0) + \mu(G_1).$$

Now consider the segments

$$0 \longleftarrow M \xleftarrow{f_0} F_0 \xleftarrow{f_1} F_1 \longleftarrow \operatorname{Ker} f_1 \longleftarrow 0$$

and

$$0 \longleftarrow M \xleftarrow{g_0} G_0 \xleftarrow{g_1} G_1 \longleftarrow \operatorname{Ker} g_1 \longleftarrow 0$$

and repeat the above argument. We have

$$\mu(G_0) + \mu(F_1) + \mu(\operatorname{Ker} g_1) = \mu(F_0) + \mu(G_1) + \mu(\operatorname{Ker} f_1).$$

Using the last inequality, it is easily shown that $\mu(F_2) \le \mu(G_2)$. Since $\mu(\operatorname{Ker} f_1) = \mu(F_2)$, we have the inequality

$$\mu(G_0) + \mu(F_1) + \mu(G_2) \le \mu(F_0) + \mu(G_1) + \mu(F_2),$$

which we will now use for the next step of the proof. Repeating the argument indefinitely, we obtain the proof of the theorem. □

Observe that if there exist stably free modules over the ring Λ that are not free, the theorem is false. The following example was given by Swan in [150]. Let $G = \{a, b: aba^{-1}b^{-1} = 1, b^8a^{-2} = 1\}$ be the generalized quaternion group of order 32, and $\mathbb{Z}[G]$ its integral group ring. Consider the ideal generated by the element $a+4$ and denote it by I. Let M be the $\mathbb{Z}[G]$-module equal to $\mathbb{Z}[G]/I$. Clearly, M has a free resolution of the form

$$0 \longleftarrow M \xleftarrow{\;p\;} \mathbb{Z}[G] \xleftarrow{\;i\;} \mathbb{Z}[G] \longleftarrow 0,$$

which is a minimal one. At the same time, there exists an epimorphism $f: \mathbb{Z}[G] \to M$ such that the ideal $K = \operatorname{Ker} f$ is not a free $\mathbb{Z}[\pi]$-module, and the minimal number of generators of K is equal to 2. Now $K \oplus \mathbb{Z}[G] \approx \mathbb{Z}[G] \oplus \mathbb{Z}[G]$, which implies that the resolution

$$0 \longleftarrow M \xleftarrow{p} \mathbb{Z}[G] \longleftarrow \mathbb{Z}[G] \oplus \mathbb{Z}[G] \longleftarrow \mathbb{Z}[G] \longleftarrow 0$$

of length 3 is minimal, but does not satisfy the conclusion of the theorem.

LEMMA 3.12. *Let*

$$0 \longleftarrow M \xleftarrow{\;g_0\;} G_0 \xleftarrow{\;g_1\;} G_1 \longleftarrow \cdots$$

be an arbitrary free resolution of a Λ-module M. Then by stabilization of the epimorphisms g_i $(i \ge 1)$ one can extract from it as a direct summand an arbitrary minimal resolution.

PROOF. Consider an arbitrary minimal resolution of the module M:

$$0 \longleftarrow M \xleftarrow{\;f_0\;} F_0 \xleftarrow{\;f_1\;} F_1 \longleftarrow \cdots.$$

Stabilizing the epimorphism g_1 via the free module F_0, we obtain

$$\begin{array}{ccccccccc} & & 0 & \longleftarrow & F_0 & \xleftarrow{\text{id}} & F_0 & \longleftarrow & 0 \\ & & & & \oplus & & \oplus & & \\ 0 & \longleftarrow & M & \longleftarrow & G_0 & \longleftarrow & G_1 & \longleftarrow & \cdots. \end{array}$$

By Lemma 3.6, from the epimorphism $\hat{g}_0 = g_0 \oplus 0$ we can extract the epimorphism f_0. Then the decomposition can be written as

$$\begin{array}{ccccccccc} 0 & \longleftarrow & \hat{G}_0 & \longleftarrow & \hat{G}_0 & \longleftarrow & 0 & & \\ & & & & \oplus & & & & \\ 0 & \longleftarrow & M & \xleftarrow{f_0} & \hat{F}_0 & \xleftarrow{\tilde{g}} & \hat{G}_1 & \longleftarrow & G_2. \end{array}$$

Stabilizing each g_i $(i \geq 2)$ step-by-step, we can successively extract from the initial resolution the given minimal resolution. □

In conclusion we note that minimal resolutions have been studied by a number of authors [81, 43].

§7. n-fold extensions of modules

We first prove several facts related to extensions of modules.

Let A and C be modules over a ring Λ. An n-fold extension of A by C is an exact sequence of Λ-modules

$$E: 0 \longleftarrow C \longleftarrow E_1 \longleftarrow \cdots \longleftarrow E_n \longleftarrow A \longleftarrow 0.$$

Two n-fold extensions E and $\overline{E}$ are said to satisfy the relation $E \rightsquigarrow \overline{E}$ if there exists the following commutative diagram:

$$\begin{array}{ccccccccccccc} E: 0 & \longleftarrow & C & \longleftarrow & E_1 & \longleftarrow & \cdots & \longleftarrow & E_n & \longleftarrow & A & \longleftarrow & 0 \\ & & \downarrow{\scriptstyle \mathrm{id}} & & \downarrow & & & & \downarrow & & \downarrow{\scriptstyle \mathrm{id}} & & \\ \overline{E}: 0 & \longleftarrow & C & \longleftarrow & \overline{E}_1 & \longleftarrow & \cdots & \longleftarrow & \overline{E}_n & \longleftarrow & A & \longleftarrow & 0 \end{array}$$

Clearly, the relation $\rightsquigarrow$ is not symmetric for $n \geq 2$. The congruence relation is generated by the relation $\rightsquigarrow$. In other words, two n-fold extensions E and $\overline{E}$ are congruent if and only if there exists a chain of n-fold extensions $E = E_0, E_1, \ldots, E_k = \overline{E}$ such that $E_0 \rightsquigarrow E_1 \rightsquigarrow E_2 \rightsquigarrow \cdots \rightsquigarrow E_k$. In what follows we denote by $[E]$ the congruence class of the n-fold extension

$$E: 0 \longleftarrow C \longleftarrow E_1 \longleftarrow \cdots \longleftarrow E_n \longleftarrow A \longleftarrow 0,$$

and by $\mathrm{Ext}^n_\Lambda(C, A)$ the set of congruence classes of n-fold extensions of the module A by the module C. It is known that $\mathrm{Ext}^n_\Lambda(C, A)$ is an abelian group. Suppose that the module A is fixed and a homomorphism $f: C' \to C$ is given. Choose a representative

$$E: 0 \longleftarrow C \longleftarrow E_1 \longleftarrow \cdots \longleftarrow E_n \longleftarrow A \longleftarrow 0$$

in the group $\mathrm{Ext}^n_\Lambda(C, A)$. Then in the group $\mathrm{Ext}^n_\Lambda(C, A)$ there is a uniquely (up to congruence) defined element

$$E^f: 0 \longleftarrow C' \longleftarrow E'_1 \longleftarrow \cdots \longleftarrow E'_n \longleftarrow A \longleftarrow 0.$$

The mapping $f^*([E]) = [E^f]$ defines a group homomorphism

$$f^*: \mathrm{Ext}^n_\Lambda(C, A) \to \mathrm{Ext}^n_\Lambda(C', A).$$

Similarly, a homomorphism $g: A \to A'$ induces a group homomorphism

$$g_*: \mathrm{Ext}^n_\Lambda(C, A) \to \mathrm{Ext}^n_\Lambda(C, A').$$

To the element

$$E: 0 \longleftarrow C \longleftarrow E_1 \longleftarrow \cdots \longleftarrow E_n \longleftarrow A \longleftarrow 0,$$

there corresponds up to congruence the element

$$E_g : 0 \longleftarrow C \longleftarrow \overline{E}_1 \longleftarrow \cdots \longleftarrow \overline{E}_n \longleftarrow A' \longleftarrow 0,$$

and $g_*([E]) = E_g$.

A morphism of n-fold extensions E and $\overline{E}$ is a family of homomorphisms forming the following commutative diagram:

$$\begin{array}{ccccccccccccc} E' : 0 & \longleftarrow & C' & \longleftarrow & E'_1 & \longleftarrow & \cdots & \longleftarrow & E'_n & \longleftarrow & A' & \longleftarrow & 0 \\ & & \downarrow f & & \downarrow & & & & \downarrow & & \downarrow g & & \\ \overline{E} : 0 & \longleftarrow & C & \longleftarrow & E_1 & \longleftarrow & \cdots & \longleftarrow & E_n & \longleftarrow & A & \longleftarrow & 0 \end{array}$$

Any morphism of n-fold extensions implies that the extensions E^f and E'_g are congruent. The following statement is known.

Let $0 \to A \to B \to C \to 0$ be an exact sequence of Λ-modules and G an arbitrary Λ-module. There are the following exact sequences of abelian groups

$$\to \operatorname{Ext}^n_\Lambda(C, G) \to \operatorname{Ext}^n_\Lambda(B, G) \to \operatorname{Ext}^n_\Lambda(A, G) \xrightarrow{\delta^n} \operatorname{Ext}^{n+1}_\Lambda(C, G) \to,$$

$$\to \operatorname{Ext}^n_\Lambda(G, A) \to \operatorname{Ext}^n_\Lambda(G, B) \to \operatorname{Ext}^n_\Lambda(G, C) \xrightarrow{\overline{\delta}^n} \operatorname{Ext}^{n+1}_\Lambda(G, A) \to,$$

where the connecting homomorphisms

$$\delta^n : \operatorname{Ext}^n_\Lambda(A, G) \to \operatorname{Ext}^{n+1}_\Lambda(C, G),$$
$$\overline{\delta}^n : \operatorname{Ext}^n_\Lambda(G, C) \to \operatorname{Ext}^{n+1}_\Lambda(G, A)$$

are constructed by means of composition of the long exact sequences with the short exact sequence $0 \to A \to B \to C \to 0$.

As is known, the group $\operatorname{Ext}^n_\Lambda(C, A)$ can be computed using projective resolutions of the module C as follows. Let $0 \leftarrow C \leftarrow P_0 \leftarrow P_1 \leftarrow \cdots (P)$ be a projective resolution of the module C. Consider a sequence $E \in \operatorname{Ext}^n_\Lambda(C, A)$ as a resolution of the module C and cover 1_C by a chain transformation $g : P \to E$. Then the homomorphism $g_n : P_n \to A$ is a cocycle of the complex P. Set $\xi([E]) = g_n \in H^n(P, A)$. The mapping ξ defines the isomorphisms $\xi_* : \operatorname{Ext}^n_\Lambda(C, A) \to H^n(P, A)$ for all $n = 0, 1, \ldots$. Hence it follows that if $E \in \operatorname{Ext}^n_\Lambda(C, A)$, then E can be represented by an n-fold sequence

$$0 \longleftarrow C \longleftarrow F_0 \longleftarrow \cdots \longleftarrow F_{n-2} \longleftarrow B \longleftarrow A \longleftarrow 0$$

in which the F_i are free modules $(i = 0, 1, \ldots, n-2)$.

Definition 3.11. Let

$$E : 0 \longleftarrow C \longleftarrow F_0 \longleftarrow \cdots \longleftarrow F_{n-2} \longleftarrow B \longleftarrow A \longleftarrow 0$$

and

$$\overline{E} : 0 \longleftarrow C \longleftarrow \overline{F}_0 \longleftarrow \cdots \longleftarrow \overline{F}_{n-2} \longleftarrow \overline{B} \longleftarrow A \longleftarrow 0$$

be two congruent n-fold sequences. We say that E and $\overline{E}$ are isomorphic if there exists a commutative diagram

$$\begin{array}{ccccccccccccc} E:0 \longleftarrow & C & \longleftarrow & F_0 & \longleftarrow & \cdots & \longleftarrow & F_{n-2} & \longleftarrow & B & \longleftarrow & A & \longleftarrow 0 \\ & \downarrow{\scriptstyle \mathrm{id}} & & \downarrow{\scriptstyle f_0} & & & & \downarrow{\scriptstyle f_{n-2}} & & \downarrow{\scriptstyle f_{n-1}} & & \downarrow{\scriptstyle \mathrm{id}} & \\ \overline{E}:0 \longleftarrow & C & \longleftarrow & \overline{F}_0 & \longleftarrow & \cdots & \longleftarrow & \overline{F}_{n-2} & \longleftarrow & \overline{B} & \longleftarrow & A & \longleftarrow 0 \end{array}$$

where the f_i are isomorphisms.

Consider an n-fold sequence

$$E:0 \longleftarrow C \xleftarrow{f_0} F_0 \longleftarrow \cdots \xleftarrow{f_{n-2}} F_{n-2} \xleftarrow{f_{n-1}} B \xleftarrow{i} A \longleftarrow 0.$$

In what follows we denote by $E(S)$ the n-fold sequence obtained by a stabilization of the homomorphisms f_i $(i = 1, \dots, n-1)$.

Lemma 3.13. *Suppose that the n-fold sequences*

$$E:0 \longleftarrow C \xleftarrow{f_0} F_0 \longleftarrow \cdots \xleftarrow{f_{n-2}} F_{n-2} \xleftarrow{f_{n-1}} B \xleftarrow{i} A \longleftarrow 0$$

and

$$\overline{E}:0 \longleftarrow \overline{C} \xleftarrow{\overline{f}_0} \overline{F}_0 \longleftarrow \cdots \xleftarrow{\overline{f}_{n-2}} \overline{F}_{n-2} \xleftarrow{\overline{f}_{n-1}} \overline{B} \xleftarrow{\overline{i}} \overline{A} \longleftarrow 0$$

are congruent, where F_i $(\overline{F}_i)$ $(i = 0, 1, \dots, n-2)$ *are free modules. Then there exist stabilizations of the homomorphisms* f_i $(\overline{f}_i)$ *such that the n-fold sequences* $E(S)$ *and* $\overline{E}(S)$ *are isomorphic.*

Proof. Stabilize the homomorphism f_0 via the module $\overline{F}_0$, and $\overline{f}_0$ via F_0. Using Lemma 3.6, we can get the homomorphisms f_0^s and $\overline{f}_0^s$ to be the same. Denote by K the module $K = \operatorname{Ker} f_0^s = \operatorname{Ker} \overline{f}_0^s$. Using the exact sequence $0 \leftarrow C \leftarrow F_0 \oplus \overline{F}_0 \leftarrow K \leftarrow 0$, we show that if the n-fold sequences E and $\overline{E}$ are congruent, so are the sequences

$$0 \longleftarrow K \longleftarrow F_1 \oplus \overline{F}_0 \xleftarrow{f_2} F_2 \longleftarrow \cdots \xleftarrow{i} A \longleftarrow 0$$

and

$$0 \longleftarrow K \longleftarrow \overline{F}_1 \oplus F_0 \xleftarrow{\overline{f}_2} \overline{F}_2 \longleftarrow \cdots \xleftarrow{\overline{i}} A \longleftarrow 0.$$

Making successive repetitions of the argument, we obtain after a finite number of steps sequences that are isomorphic and congruent. □

Corollary 3.6. *Suppose that the sequences*

$$E:0 \longleftarrow C \xleftarrow{f_0} F_0 \longleftarrow \cdots \xleftarrow{f_{n-2}} F_{n-2} \xleftarrow{f_{n-1}} B \xleftarrow{i} A \longleftarrow 0$$

and

$$\overline{E}:0 \longleftarrow C \xleftarrow{\overline{f}_0} \overline{F}_0 \longleftarrow \cdots \xleftarrow{\overline{f}_{n-2}} \overline{F}_{n-2} \xleftarrow{\overline{f}_{n-1}} \overline{B} \xleftarrow{\overline{i}} A \longleftarrow 0$$

are congruent, where F_i $(\overline{F}_i)$ *are free modules. Then there exist morphisms from* E *to* $\overline{E}$ *and from* $\overline{E}$ *to* E *that are the identity on the modules* C *and* A. *There exist positive integers* k *and* s *such that the modules* $B \oplus k\Lambda$ *and* $\overline{B} \oplus s\Lambda$ *are isomorphic.* □

Suppose the Λ-modules A and C are isomorphic, respectively, to Λ-modules $\overline{A}$ and $\overline{C}$. Consider a pair of elements $a \in \operatorname{Ext}^n_\Lambda(C, A)$ and $\overline{a} \in \operatorname{Ext}^n_\Lambda(\overline{C}, \overline{A})$.

DEFINITION 3.12. The elements a and $\overline{a}$ are said to be isomorphic if there exist isomorphisms $f: A \to \overline{A}$ and $g: C \to \overline{C}$ such that $g^*(\overline{a}) = f_*(a)$ in the chain of homomorphisms

$$\operatorname{Ext}^n_\Lambda(\overline{C}, \overline{A}) \xrightarrow{g^*} \operatorname{Ext}^n_\Lambda(C, \overline{A}) \xleftarrow{f^*} \operatorname{Ext}^n_\Lambda(C, A).$$

It is easy to see that this definition is equivalent to the existence of a commutative diagram

$$\begin{array}{ccccccccccccc} E: 0 \leftarrow & C & \longleftarrow & F_0 & \longleftarrow & \cdots & \longleftarrow & F_{n-2} & \longleftarrow & B & \longleftarrow & A & \leftarrow 0 \\ & \downarrow & & \downarrow & & & & \downarrow & & \downarrow & & \downarrow & \\ \overline{E}: 0 \leftarrow & \overline{C} & \longleftarrow & \overline{F}_0 & \longleftarrow & \cdots & \longleftarrow & \overline{F}_{n-2} & \longleftarrow & \overline{B} & \longleftarrow & \overline{A} & \leftarrow 0 \end{array}$$

where E and $\overline{E}$ are arbitrary n-fold sequences representing the elements a and $\overline{a}$, respectively, for which F_i $(\overline{F}_i)$ are free modules.

Denote by $S(\Lambda)$ the abelian monoid of classes of stably equivalent modules over a ring Λ. Two modules A and B represent one and the same element in $S(\Lambda)$ if the modules $A \oplus m\Lambda$ and $B \oplus n\Lambda$ are isomorphic for some positive integers m and n. Denote by $[A]$ the class of the module A in the monoid $S(\Lambda)$. Then $[A] + [B] = [A \oplus B]$.

LEMMA 3.13. * *Suppose that the sequences*

$$E: 0 \leftarrow C \longleftarrow F_0 \longleftarrow \cdots \longleftarrow F_{n-2} \longleftarrow B \longleftarrow A \leftarrow 0$$

and

$$\overline{E}: 0 \leftarrow C \longleftarrow \overline{F}_0 \longleftarrow \cdots \longleftarrow \overline{F}_{n-2} \longleftarrow \overline{B} \longleftarrow \overline{A} \leftarrow 0$$

represent isomorphic elements $a \in \operatorname{Ext}^n_\Lambda(C, A)$ *and* $\overline{a} \in \operatorname{Ext}^n_\Lambda(\overline{C}, \overline{A})$. *Then* $[B] = [\overline{B}]$ *in* $S(\Lambda)$.

PROOF. By definition, there exist isomorphisms $f: A \to \overline{A}$ and $g: C \to \overline{C}$ such that $g^*(\overline{a}) = f_*(a)$. The sequence $g^*(E)$ is of the form

$$0 \leftarrow C \longleftarrow \overline{F}_0 \longleftarrow \cdots \longleftarrow \tilde{B} \longleftarrow \overline{A} \leftarrow 0,$$

and the sequence $f_*(\overline{E})$ is of the form

$$0 \leftarrow C \longleftarrow \overline{F}_0 \longleftarrow \cdots \longleftarrow \hat{B} \longleftarrow \overline{A} \leftarrow 0.$$

**Editor's note.* Misnumbering reflects original Russian.

By construction, $\tilde{B}$ is isomorphic to $\overline{B}$ and $\hat{B}$ is isomorphic to B. Since the sequence $g^*(E)$ is congruent to the sequence $f_*(\overline{E})$, we have by Corollary 3.6 that the modules $\tilde{B}$ and $\hat{B}$ represent the same element in $S(\Lambda)$. □

DEFINITION 3.13. Two elements $a \in \mathrm{Ext}^n_\Lambda(C, A)$ and $\overline{a} \in \mathrm{Ext}^n_\Lambda(\overline{C}, \overline{A})$ are said to be stably isomorphic if there exist positive integers k and l and isomorphisms $f: A \to \overline{A}$ and $g: C \oplus k\Lambda \to \overline{C} \oplus l\Lambda$ such that $g^*(p_1^*(\overline{a})) = f_*(p_2^*(a))$ in the chain of homomorphisms

$$\begin{array}{ccc} \mathrm{Ext}^n_\Lambda(\overline{C}, \overline{A}) \xrightarrow{p_1^*} \mathrm{Ext}^n_\Lambda(\overline{C} \oplus l\Lambda, \overline{A}) & & \\ & \searrow^{g^*} & \\ & & \mathrm{Ext}^n_\Lambda(C \oplus k\Lambda, \overline{A}), \\ & \nearrow_{f^*} & \\ \mathrm{Ext}^n_\Lambda(C, A) \xrightarrow{p_2^*} \mathrm{Ext}^n_\Lambda(C \oplus l\Lambda, A) & & \end{array}$$

where $p_1: \overline{C} \oplus l\Lambda \to \overline{C}$ and $p_2: C \oplus k\Lambda \to C$ are the canonical projections.

It is easy to see that this definition is equivalent to the existence of a morphism

$$\begin{array}{ccccccccccccc} E: 0 & \longleftarrow & C \oplus k\Lambda & \longleftarrow & F_0 & \longleftarrow \cdots \longleftarrow & F_{n-2} & \longleftarrow & B & \longleftarrow & A & \longleftarrow & 0 \\ & & \downarrow g & & \downarrow \varphi_0 & & \downarrow \varphi_{n-2} & & \downarrow \varphi_{n-1} & & \downarrow f & & \\ \overline{E}: 0 & \longleftarrow & \overline{C} \oplus l\Lambda & \longleftarrow & \overline{F}_0 & \longleftarrow \cdots \longleftarrow & \overline{F}_{n-2} & \longleftarrow & \overline{B} & \longleftarrow & \overline{A} & \longleftarrow & 0 \end{array}$$

between the n-fold sequences, representing the elements $a = [E]$ and $\overline{a} = [\overline{E}]$ in which F_i and $\overline{F}_i$ are free modules.

We also note that there is an analog of Lemma 3.13; its proof is essentially the repetition of the proof for the latter.

Consider the action of a ring isomorphism $\theta: \Lambda \to \Lambda$. Using θ, we can define a new Λ-module structure on a module C by the formula $\lambda c = \theta(\lambda)c$. We denote this structure by C_θ. There is an obvious isomorphism

$$\theta: \mathrm{Ext}^n_\Lambda(C, A) \longrightarrow \mathrm{Ext}^n_\Lambda(C_\theta, A_\theta).$$

We say that two elements $a \in \mathrm{Ext}^n_\Lambda(C, A)$ and $\overline{a} \in \mathrm{Ext}^n_\Lambda(C_\theta, A_\theta)$ are θ-isomorphic (stably θ-isomorphic) if the elements a and $\theta(\overline{a})$ are isomorphic (stably isomorphic). For θ-isomorphic (stably θ-isomorphic) elements a and $\theta(\overline{a})$ there are statements analogous to those proved in this section.

CHAPTER IV

Homotopy of Chain Complexes

In this chapter we prove a number of fundamental results relating to the structure of free chain complexes, which we will make essential use of in the next chapter in the study of Morse functions of non-simply-connected cobordisms. After recalling briefly some terms and results related to chain complexes, we introduce invariants that fully determine the homotopy type of a chain complex.

We discuss the simple homotopy type of chain complexes and the Whitehead group. Another important topic considered in this chapter is the existence of a minimal chain complex in a given homotopy type. We prove that in order for a minimal chain complex to exists in every homotopy type of a free chain complex, it is necessary and sufficient that the ring over which the chain complexes are defined be an s-ring.

We conclude this chapter with some numerical invariants of chain complexes used for computing the Morse numbers of chain complexes.

§1. Brief review

For convenience of further reference we give some definitions and results related to chain complexes. The proofs are, as a rule, omitted and can be found in the textbooks [8, 14, 30, 81].

By a graded Λ-module we mean a sequence $C = \{C_n\}_{n\in\mathbb{Z}}$ of Λ-modules. If $c \in C_n$, we say that c is of degree n. A mapping of degree k of a graded Λ-module C into a graded Λ-module $\overline{C}$ is a collection of homomorphisms $f_n: C_n \to \overline{C}_{n+k}$.

A chain complex over a ring Λ is, by definition, a pair (C, d), where C is a graded Λ-module and $d: C \to C$ is a homomorphism of degree -1 such that $d^2 = 0$.

Cycles, boundaries, and homologies are defined by the formulas $Z(C) = \operatorname{Ker} d$, $B(C) = \operatorname{Im} d$, and $H(C) = Z(C)/B(C)$, respectively. All of them are graded modules.

One often considers graded Λ-modules for which the homomorphism d is of degree $+1$. In this case we use superscripts instead of subscripts: $C = \{C^n\}_n \in \mathbb{Z}$ and $d = d^n: C^n \to C^{n+1}$. Such a pair is called a cochain complex over the ring Λ. If one considers cochain complexes, all terms acquire the 'co-' prefix. We shall consider chain (cochain) complexes for which C_n (C^n) are free modules and $n > 0$. In this case we say that $\{C, d\}$ is a free chain complex. If (C, d) and $(\overline{C}, \overline{d})$ are chain complexes, then a chain mapping from (C, d) into $(\overline{C}, \overline{d})$ is a homomorphism $\{f_n\} = f: C \to \overline{C}$ such that $\overline{d} \circ f = f \circ d$. If $\{f_n\}$ are isomorphisms, the chain complexes (C, d) and $(\overline{C}, \overline{d})$ are called chain isomorphic. Two chain mappings f and g are homotopy equivalent (we write $f \sim g$) if there exists a homomorphism $h: C \to \overline{C}$ of degree 1 such that $\overline{d}h + hd = f - g$. A chain mapping is said to induce the homomorphism of homologies $f_*: H(C) \to H(\overline{C})$, and $f_* = g_*$ if the chain mappings f and g are homotopy equivalent. A chain mapping $f: C \to \overline{C}$ is called a homotopy equivalence if there exists a chain mapping $\overline{f}: \overline{C} \to C$ such that $\overline{f} \circ f = \mathrm{id}_C$ and $f \circ \overline{f} = \mathrm{id}\,\overline{C}$. A chain complex is said to be contractible if it is homotopy equivalent to the zero complex. The cylinder of a chain mapping $f: (C, d) \to (\overline{C}, \overline{d})$ is defined as the complex (C', d'), where $C' = \overline{C} + \sum C$ (in the sense of grading) and $d'(\overline{c}, c) = (\overline{d}(\overline{c}) + f(c) - d(c))$. In matrix notation

$$d' = \begin{pmatrix} \overline{d} & f \\ 0 & -d \end{pmatrix}.$$

If $f: (C, d) \to (\overline{C}, \overline{d})$ is a chain mapping and (C', d') is its cylinder, then there exists a long exact homology sequence

$$\to H_n(C) \to H_n(\overline{C}) \to H_n(C') \to H_{n-1}(C) \to \cdots.$$

A chain mapping $f: (C, d) \to (\overline{C}, \overline{d})$ is a homotopy equivalence if and only if its cylinder is contractible.

The following proposition is due to Cockroft and Swan [23].

Proposition 4.1. *Let $f = \{f_n\}: (C, d) \to \{\overline{C}, \overline{d}\}$, $n \geq 0$, be a homotopy equivalence between free chain complexes (C, d) and $(\overline{C}, \overline{d})$. Then there exists a stabilization of boundary homomorphisms d_n and $\overline{d}_n$ such that the resulting chain complexes are chain isomorphic.*

Since this theorem will be repeatedly used in the sequel, we now present the most important parts of its proof, which is achieved by induction. Let $B_0 = d_1(C_1)$, $\overline{B}_0 = \overline{d}_1(\overline{C}_1)$, $H_0 = H_0(C)$, and $\overline{H}_0 = H_0(\overline{C})$. The homomorphism $f_0: C_0 \to \overline{C}_0$ has a homotopy inverse mapping $\overline{f}_0: \overline{C}_0 \to C_0$ such that $\mathrm{id} - f_0 \circ \overline{f}_0 = d_1 \circ \eta_0$, where $\eta_0: C_0 \to \overline{C}_1$ is the operator of homotopy deformation. Write the commutative diagram

$$\begin{array}{ccccccccc}
0 & \longleftarrow & H_0 & \xleftarrow{d_0} & C_0 & \xleftarrow{i} & B_0 & \longleftarrow & 0 \\
 & & \uparrow \overline{f}_* & & \uparrow \overline{f}_0 & & \downarrow f_0 & & \\
0 & \longleftarrow & \overline{H}_0 & \xleftarrow{\overline{d}_0} & \overline{C}_0 & \xleftarrow{\overline{i}} & \overline{B}_0 & \longleftarrow & 0.
\end{array}$$

Using stabilization, we obtain

$$\begin{array}{ccccccccc} 0 & \longleftarrow & H_0 & \xleftarrow{d_0 \oplus \mathrm{id}} & C_0 \oplus \overline{C}_0 & \longleftarrow & B_0 \oplus \overline{C}_0 & \longleftarrow & 0 \\ & & & & \downarrow f_0^s & & & & \\ 0 & \longleftarrow & \overline{H}_0 & \xleftarrow{\overline{d}_0 \oplus \mathrm{id}} & \overline{C}_0 \oplus C_0 & \longleftarrow & \overline{B}_0 \oplus C_0 & \longleftarrow & 0, \end{array}$$

where the homomorphism $f_0^s \colon C_0 \oplus \overline{C}_0 \to \overline{C}_0 \oplus C_0$ is defined by the formula

$$f_0^s(c_0 + \overline{c}_0) = c_0 - \overline{f}_0(c_0) + (f_0(c_0) + (\mathrm{id} - f_0 \overline{f}_0(d_0))),$$

$c_0 \in C_0$, $\overline{c}_0 \in \overline{C}_0$, with the inverse

$$(f_0^s)^{-1}(c_0 + \overline{c}_0) = (\mathrm{id} - \overline{f}_0 \circ f_0)c_0 + \overline{f}_0(d_0) + (-f(c_0) + \overline{c}_0)$$

(the so-called Schanuel homomorphism). Since $\mathrm{id} - \overline{f}_0 f_0 = d_1 \circ \eta_0$, we obtain a commutative diagram

$$\begin{array}{ccccccc} 0 & \longleftarrow & B_0 \oplus \overline{C}_0 & \longleftarrow & C_1 \oplus \overline{C}_0 & \longleftarrow & \cdots \\ & & & & f_1^s \downarrow & & \\ 0 & \longleftarrow & \overline{B}_0 \oplus C_0 & \longleftarrow & C_1 \oplus C_0 & \longleftarrow & \cdots \end{array}$$

and also

$$\begin{array}{ccccccccc} H_0 & \longleftarrow & C_0 \oplus \overline{C}_0 & \xleftarrow{d_1 \oplus \mathrm{id}} & C_1 \oplus \overline{C}_0 & \longleftarrow & C_2 & \longleftarrow & \cdots \\ & & f_0^s \downarrow & & f_1^s \downarrow & & & & \\ H_0 & \longleftarrow & \overline{C}_0 \oplus C_0 & \xleftarrow{\overline{d}_1 \oplus \mathrm{id}} & \overline{C}_1 \oplus C_0 & \longleftarrow & C_2 & \longleftarrow & \cdots. \end{array}$$

The homomorphisms d_1 and $\overline{d}_1$ are thus stabilized. The following argument is made by induction. Note that the matrix of the homomorphism $f_0^s \colon C_0 \oplus \overline{C}_0 \to \overline{C}_0 \oplus C_0$ is of the form

$$\begin{pmatrix} 1 & f_0 \\ -\overline{f}_0 & 1 - \overline{f}_0 \circ f_0 \end{pmatrix},$$

and one can show that it can be expressed as a product of elementary matrices. We shall make use of this fact in what follows.

Let $\theta \colon \Lambda \to \Lambda$ be an automorphism of the ring Λ such that $\theta(1) = 1$. Obviously, θ induces a mapping of a free Λ-module F, $S_\theta \colon F \to F$, given by $S_\theta(\lambda_1, \dots, \lambda_n) = (\theta(\lambda_1), \dots, \theta(\lambda_n))$. From the definition of S_θ it follows immediately that $S_{\theta^{-1}} = (S_\theta)^{-1}$ and $S_\theta \circ S_\varphi = S_{\theta \circ \varphi}$, where φ is an automorphism of the ring Λ. In addition, $S_\theta(\lambda f) = \theta(\lambda) S_\theta(f)$, where $\lambda \in \Lambda$, $f \in F$, so that S_θ is not a homomorphism of Λ-modules, although it is a homomorphism of the abelian groups.

We say that a mapping $g: F \to F$ is a semilinear homomorphism, associated with an automorphism $\theta: \Lambda \to \Lambda$, if g is a homomorphism of the underlying group and $g(\lambda, f) = \theta(\lambda)g(f)$. Obviously S_θ is a semilinear isomorphism. If $g: F_1 \to F_2$ is a semilinear homomorphism associated with $\theta: \Lambda \to \Lambda$, it can be turned into an ordinary homomorphism of Λ-modules by defining a new Λ-module structure on F_2 by the formula $\lambda f = \theta(\lambda) f$.

Let $\{C, \partial\}: C_0 \xleftarrow{\partial_1} C_1 \longleftarrow \cdots \xleftarrow{\partial_n} C_n$ be a chain complex of free modules over the ring Λ. Using S_θ, construct the semilinear isomorphism $S_\theta: C_i \to C_i$. Set $\partial_i^\theta = S_\theta \circ \partial_i \circ S_{\theta^{-1}}: C_i \to C_{i-1}$; then $\partial_i^\theta \circ \partial_{i+1}^\theta = 0$ and $\partial_i^\theta(\lambda c) = \lambda \partial_i^\theta(c)$ for $c \in C_i$, so that ∂_i^θ is a boundary operator. Let $\{C_i, \partial_i^\theta\}$ be the chain complex consisting of the modules C_i and the boundary homomorphisms ∂_i^θ. Following Whitehead, we call it the conjugate chain complex. If $g = \{g_i\}: \{C, \partial\} \to \{D, d\}$ is a chain mapping, then $g^\theta = \{g_i^\theta\}: \{C, \partial^\theta\} \to \{D, d^\theta\}$ is also a chain mapping. Furthermore, if $\{C_i, \partial_i\}$ is acyclic, so is $\{C, \partial^\theta\}$.

In what follows we shall encounter semilinear chain transformations. For these, too, one can introduce an equivalence relation of homotopy. Let us briefly review this point. Consider the case when $\Lambda = \mathbb{Z}[G]$. Any automorphism $\theta: G \to G$ of the group G induces a ring automorphism $\overline{\theta}: \mathbb{Z}[G] \to \mathbb{Z}[G]$. In what follows we consider only such automorphisms of the ring $\mathbb{Z}[G]$. Let $\{C, \partial\}$ and $\{D, d\}$ be free chain complexes over the ring $\mathbb{Z}[G]$ and suppose that $f = \{f_i\}: \{C, \partial\} \to \{D, d\}$ is a chain mapping associated with the automorphism $\theta: \mathbb{Z}[G] \to \mathbb{Z}[G]$. Following Whitehead [159, 160], we say that a chain mapping f is homotopy equivalent to g if there exist a semilinear homomorphism $\eta = \{\eta_i\}: \{C, \partial\} \to \{D, d\}$ of degree $+1$ associated with the automorphism θ and an element $\gamma \in G$ such that $\gamma g - f = d \circ \eta + \eta \circ \partial$. It can be shown that the chain transformation g is associated with the automorphism $\overline{\theta}_{\gamma^{-1}} \circ \theta$, where $\overline{\theta}_{\gamma^{-1}}$ is the inner automorphism of the group G generated by the element γ^{-1}. In this situation the definition of homotopy equivalence is evident. It is easy to see that if $f = \{f_i\}: \{C, \partial\} \to \{D, d\}$ is a homotopy equivalence associated with the automorphism $\theta: \Lambda \to \Lambda$ of the ring Λ, then $f \circ S_{\theta^{-1}} = \{f_i \circ S_{\theta^{-1}}\}: \{C, \partial^\theta\} \to \{D, \partial\}$ is a homotopy equivalence in the ordinary sense.

We observe that if $f = \{f_i\}: \{C, \partial\} \to \{D, \partial\}$ is a chain mapping associated with $\theta: \Lambda \to \Lambda$ and inducing a semilinear isomorphism of homology modules, then an analog of the Cockroft-Swan theorem holds. Indeed, since $f \circ S_{\theta^{-1}}: \{C, \partial^\theta\} \to \{D, d\}$ is a chain mapping, it is subject to the Cockroft-Swan theorem. We have the chain of mappings

$$\{C, \partial\} \xrightarrow{S_\theta} \{C, \partial^\theta\} \xrightarrow{i} \{C^s, \partial^\theta\} \xrightarrow{h} \{D^s, d^s\} \xrightarrow{p} \{D, d\},$$

where $\{C^s, \partial^\theta\}$ and $\{D^s, d^s\}$ are stabilizations of the chain complexes

$\{C, \partial^\theta\}$ and $\{D, d\}$ respectively, i is the embedding, and p is the projection.

§2. Stable invariants of chain complexes

It is known that homology determines the homotopy type of a free chain complex over the ring of integers [30, 81]. For free chain complexes over arbitrary rings this is not the case. This gives rise to the problem of describing a complete set of invariants defining the homotopy type of a free chain complex. An approach to solving this problem was proposed by Heller and Dold. Essentially, they constructed [29, 60] the algebraic analog of the natural Postnikov systems. Here we will modify this approach.

Recall that $S(\Lambda)$ denotes the abelian monoid of classes of stably equivalent modules over the ring Λ. Let

$$\{C, \partial\}: C_0 \xleftarrow{\partial_1} C_1 \longleftarrow \cdots \xleftarrow{\partial_i} C_i \xleftarrow{\partial_{i+1}} C_{i+1} \longleftarrow \cdots \longleftarrow C_n$$

be a free chain complex over the ring Λ. Following Dyer [35], we shall mean by the Swan-Wall class in dimension i the class of the module $\Gamma_i = C_i/\partial_{i+1}(C_{i+1})$ in $S(\Lambda)$. In the sequel we denote it by $\Gamma_i(C)$. If the chain complexes $\{C, \partial\}$ and $\{D, d\}$ are homotopy equivalent, then the modules $\Gamma_i(C)$ and $\Gamma_i(D)$ are stably equivalent for all i. This is a direct consequence of the Cockroft-Swan theorem. Thus, for homotopy equivalent chain complexes over the ring Λ the Swan-Wall classes coincide in all dimensions. It is not difficult to construct chain complexes with isomorphic homology modules for which the Swan-Wall classes do not coincide. It suffices to take the ring $\mathbb{Z}[\mathbb{Z}]$ and, since its homology dimension is equal to 2, the modules of cycles for each chain complex over $\mathbb{Z}[\mathbb{Z}]$ are free. Using this fact, take the free resolution of an arbitrary module whose homology dimension is equal to 2:

$$0 \longleftarrow M \xleftarrow{\partial_0} F_0 \xleftarrow{\partial_1} F_1 \xleftarrow{\partial_2} F_2 \longleftarrow 0$$

such that the module $\operatorname{Ker}\partial_1$ cannot be extracted as a direct summand in F_1 and contains a unimodular element which we will denote by a. Next, instead of the module F_2, we consider the ring $\mathbb{Z}[\mathbb{Z}]$ as a $\mathbb{Z}[\mathbb{Z}]$-module and define two monomorphisms $\tilde{\partial}_2, \overline{\partial}_2: \mathbb{Z}[\mathbb{Z}] \to \operatorname{Ker}\partial_1$ such that $\tilde{\partial}_2(1) = b$, $\overline{\partial}_2(1) = a$, where b is a basis element in $\operatorname{Ker}\partial_1$ that is not unimodular in F_1. As a result, we have two chain complexes:

$$F_0 \xleftarrow{\partial_1} F_1 \xleftarrow{\overline{\partial}_2} \mathbb{Z}[\mathbb{Z}]$$

and

$$F_0 \xleftarrow{\partial_1} F_1 \xleftarrow{\tilde{\partial}_2} \mathbb{Z}[\mathbb{Z}]$$

with isomorphic homology modules but different Swan-Wall classes in dimension 1. At the same time there exist chain complexes for which the homology modules are isomorphic, their Swan-Wall classes coincide, but the homotopy types are different.

For the chain complex $\{C, \partial\}$ we have in every dimension the exact sequence

$$0 \longleftarrow C_i/\partial_{i+1}(C_{i+1}) \longleftarrow C_i \xleftarrow{\partial_{i+1}} C_{i+1} \hookleftarrow Z_{i+1} \longleftarrow 0,$$

where $Z_i = \operatorname{Ker} \partial_{i+1}$. Factoring out the submodule of boundaries $\partial_{i+2}(C_{i+2})$, we obtain the extension

$$\alpha_i(C): 0 \longleftarrow C_i/\partial_{i+1}(C_{i+1}) \longleftarrow C_i \longleftarrow C_{i+1}/\partial_{i+1}(C_{i+2}) \longleftarrow H_{i+1}(C) \longleftarrow 0.$$

This extension defines an element of $\operatorname{Ext}^2_\Lambda(\Gamma_i(C), H_{i+1}(C))$ (for $i = 0$ an element of $\operatorname{Ext}^2_\Lambda(H_0(C), H_1(C))$). An obvious consequence of the Cockroft-Swan theorem is the fact that for homotopy equivalent chain complexes the extensions $[\alpha_i(C)]$ are stably isomorphic for $i \geq 1$ (and isomorphic for $i = 0$). Thus, we have two stable invariants of chain complexes. Observe that specifying $\alpha_i(C)$ determines $\Gamma_{i+1}(C)$.

THEOREM 4.1. *Two free chain complexes*

$$\{C, \partial\}: C_0 \xleftarrow{\partial_1} C_1 \longleftarrow \cdots \xleftarrow{\partial_n} C_n$$

and

$$\{D, d\}: D_0 \xleftarrow{d_1} D_1 \longleftarrow \cdots \xleftarrow{d_n} D_n$$

are homotopy equivalent if and only if $[\alpha_i(C)]$ *and* $[\alpha_i(D)]$ *are stably isomorphic.*

PROOF. The proof is achieved by induction. Let

$$\begin{array}{l} H_0(C) \xleftarrow{\partial_0} C_0 \overset{i_0}{\hookleftarrow} B_0 \xleftarrow{\partial_1} C_1 \\ H_0(D) \xleftarrow{d_0} D_0 \overset{i_0}{\hookleftarrow} \overline{B}_0 \xleftarrow{d_1} D_1 \end{array}$$

be segments of chain complexes. Since $H_0(C)$ and $H_0(D)$ are isomorphic, i.e., there exists an isomorphism $f_{0*}: H_0(C) \to H_0(D)$, we can use Lemma 3.6 and, by stabilizing the boundary homomorphisms ∂_1 and d_1 via the free modules D_0 and C_0, find an isomorphism $f_0: C_0 \oplus D_0 \to D_0 \oplus C_0$ for which the diagram

$$\begin{array}{ccccccc} H_0(C) & \xleftarrow{\partial\oplus 0} & C_0 \oplus D_0 & \xleftarrow{i_0\oplus \mathrm{id}} & B_0 \oplus D_0 & \longleftarrow & 0 \\ {\scriptstyle f_{0*}}\downarrow & & {\scriptstyle f_0}\downarrow & & & & \\ H_0(D) & \xleftarrow{d_0\oplus 0} & D_0 \oplus C_0 & \xleftarrow{\bar{i}_0\oplus \mathrm{id}} & \overline{B}_0 \oplus C_0 & \longleftarrow & 0 \end{array}$$

commutes. (The conclusion of Lemma 3.6 is applied to the epimorphisms $f_{0*}(\partial_0 \oplus 0)$ and $d_0 \oplus 0$.) Note that for the chain complex $\{C, \partial\}$ the sequence

$$0 \leftarrow C_1/\partial_2(C_2) \leftarrow C_1 \leftarrow C_2/\partial_3(C_3) \leftarrow H_2(C) \leftarrow,$$

that represents the element $[\alpha_1(C)]$, is transformed into the sequence

$$\tilde{\alpha}_1(C): 0 \leftarrow C_1 \oplus D_0/\partial_2(C_2) \leftarrow C_1 \oplus D_0 \leftarrow C_2/\partial_3(C_3) \leftarrow H_2(C) \leftarrow 0.$$

Similarly, for the chain complex $\{D, d\}$, the sequence

$$0 \leftarrow D_1/d_2(D_2) \leftarrow D_1 \leftarrow D_2/d_3(D_3) \leftarrow H_2(D) \leftarrow 0$$

representing $[\alpha_1(D)]$ is transformed into the sequence

$$\tilde{\alpha}_1(D): 0 \leftarrow D_1 \oplus C_0/d_2(D_2) \leftarrow D_1 \oplus C_0 \leftarrow D_2/d_3(D_3) \leftarrow H_2(D) \leftarrow 0.$$

Since, by the hypothesis, $[\alpha_1(C)]$ and $[\alpha_1(D)]$ are stably isomorphic, we can consider, by stabilizing (if necessary) the homomorphisms $\partial_1 \oplus \mathrm{id}$ and $d_1 \oplus \mathrm{id}$ via a free module of an appropriate rank, that $[\tilde{\alpha}_1(C)]$ and $[\tilde{\alpha}_1(D)]$ are also isomorphic. Suppose that we have constructed a mapping of segments of the chain complexes $\{C, \partial\}$ and $\{D, d\}$ (more precisely, of their stabilizations)

$$\begin{array}{ccccccc} \{C^s, \partial^s\}: C_0 \oplus D_0 & \xleftarrow{\partial_0 \oplus \mathrm{id}} & C_1 \oplus n_1\Lambda & \longleftarrow & \cdots & \xleftarrow{\partial_{i-1} \oplus \mathrm{id}} & C_{i-1} \oplus n_{i-1}\Lambda \\ \downarrow f_0 & & \downarrow f_1 & & & & \downarrow f_{i-1} \\ \{D^s, d^s\}: D_0 \oplus C_0 & \xleftarrow{d_0 \oplus \mathrm{id}} & D_1 \oplus m_1\Lambda & \longleftarrow & \cdots & \xleftarrow{\partial_{i-1} \oplus \mathrm{id}} & D_{i-1} \oplus m_{i-1}\Lambda \end{array}$$

satisfying the conditions:

(1) $f_{i-1}(\partial_j \oplus \mathrm{id}) = (d_j \oplus \mathrm{id}) f_j$;

(2) $f_j(Z_j) = \overline{Z}_j$, where $Z_j = \mathrm{Ker}(\partial_j \oplus \mathrm{id})$, $\overline{Z}_j = \mathrm{Ker}(d_j \oplus \mathrm{id})$;

(3) $f_j(B_j) = \overline{B}_j$, where $B_j = \mathrm{Im}(\partial_{i+1} \oplus \mathrm{id})$, $\overline{B}_j = \mathrm{Im}(d_{j+1} \oplus \mathrm{id})$, $0 \le j \le i-1$;

(4) the sequences

$$\tilde{\alpha}_{i-1}(C): 0 \leftarrow \frac{C_{i-1} \oplus n_{i-1}\Lambda}{\partial_i \oplus \mathrm{id}(C_i \oplus n_i\Lambda)} \xleftarrow{p_{i-1}} C_{i-1} \oplus n_{i-1}\Lambda \leftarrow \frac{C_i \oplus n_i\Lambda}{\partial_{i+1}(C_{i+1})} \leftarrow H_i(C) \leftarrow 0,$$

$$\tilde{\alpha}_{i-1}(D): 0 \leftarrow \frac{D_{i-1} \oplus m_{i-1}\Lambda}{d_i \oplus \mathrm{id}(D_i \oplus m_i\Lambda)} \xleftarrow{\overline{p}_{i-1}} D_{i-1} \oplus m_{i-1}\Lambda \leftarrow \frac{D_i \oplus m_i\Lambda}{d_{i+1}(D_{i+1})} \leftarrow H_i(D) \leftarrow 0$$

are such that the elements $[\tilde{\alpha}_{i-1}(C)]$ and $[\tilde{\alpha}_{i-1}(D)]$ are isomorphic.

We now show how to construct an isomorphism $f_i: C_i \oplus n_i\Lambda \to D_i \oplus m_i\Lambda$ satisfying conditions (1)–(4). Observe that although the elements $[\tilde{\alpha}_{i-1}(C)]$ and $[\tilde{\alpha}_{i-1}(D)]$ are isomorphic, there exists a commutative diagram

$$\begin{array}{ccccccc} 0 \leftarrow \frac{C_{i-1} \oplus n_{i-1}\Lambda}{\partial_i \oplus (C_i \oplus n_i\Lambda)} & \longleftarrow & C_{i-1} \oplus n_{i-1}\Lambda & \longleftarrow & \frac{C_i \oplus n_i\Lambda}{\partial_{i+1}(C_{i+1})} & \longleftarrow & H_i(C) \leftarrow 0 \\ \downarrow \tilde{g}_{i-1} & & \downarrow g_{i-1} & & \downarrow \tilde{d}_i & & \downarrow h_{i*} \\ 0 \leftarrow \frac{D_{i-1} \oplus m_{i-1}\Lambda}{d_i \oplus (D_i \oplus m_i\Lambda)} & \longleftarrow & D_{i-1} \oplus m_{i-1}\Lambda & \longleftarrow & \frac{D_i \oplus m_i\Lambda}{d_{i+1}(D_{i+1})} & \longleftarrow & H_i(D) \leftarrow 0 \end{array}$$

in which the homomorphism g_{i-1} does not coincide with the isomorphism f_{i-1}. Using the isomorphisms f_{i-1}, h_{i*}, and the element $[\tilde{\alpha}_{i-1}(D)]$, we can construct a new element in the group

$$\mathrm{Ext}^2_\Lambda(C_{i-1} \oplus n_{i-1}\Lambda)/\partial_i \oplus \mathrm{id}(C_i \oplus n_i\Lambda, H_i(C))$$

isomorphic to the element $[\tilde{\alpha}_{i-1}(C)]$.

We now recall two constructions related to the extensions of modules. Given an extension

$$E: 0 \leftarrow A \leftarrow C \leftarrow B \leftarrow 0$$

of a module B via a module A, and an isomorphism $\alpha : A' \to A$, one can canonically construct the extension of the module B via the module A' (the pullback of E):

$$E^\alpha: 0 \leftarrow A' \leftarrow C' \leftarrow B \leftarrow 0,$$

for which there exists a commutative diagram

$$\begin{array}{ccccccccc} 0 & \longleftarrow & A' & \longleftarrow & C' & \longleftarrow & B & \longleftarrow & 0 \\ & & \Big\downarrow{\scriptstyle \alpha} & & \Big\downarrow{\scriptstyle f_1} & & \Big\downarrow{\scriptstyle \mathrm{id}} & & \\ 0 & \longleftarrow & A & \longleftarrow & C & \longleftarrow & B & \longleftarrow & 0, \end{array}$$

where f_1 is an isomorphism [81].

Given an isomorphism $\beta: B \to B'$, one can canonically construct an extension of the B' via a module A (the pushout of E):

$$E^\beta: 0 \leftarrow A \leftarrow C'' \leftarrow B' \leftarrow 0,$$

and there exists a commutative diagram

$$\begin{array}{ccccccccc} 0 & \longleftarrow & A & \longleftarrow & C & \longleftarrow & B & \longleftarrow & 0 \\ & & \Big\downarrow{\scriptstyle \mathrm{id}} & & \Big\downarrow{\scriptstyle f_2} & & \Big\downarrow{\scriptstyle \beta} & & \\ 0 & \longleftarrow & A & \longleftarrow & C'' & \longleftarrow & B' & \longleftarrow & 0, \end{array}$$

where f_2 is an isomorphism [81].

In the sequence $\tilde{\alpha}_{i-1}(D)$ choose an exact subsequence

$$\beta_{i-1}: 0 \leftarrow \mathrm{Ker}\,\overline{p}_{i-1} \leftarrow D_i \oplus m_i\Lambda/d_{i+1}(D_{i+1}) \leftarrow H_i(D) \leftarrow 0.$$

Since $\mathrm{Ker}\,\overline{p}_{i-1} = \overline{B}_{i-1}$, consider the restriction of the isomorphism f_{i-1} to the submodule B_{i-1},

$$f_{i-1}|B_{i-1} = \overline{f}_{i-1}: B_{i-1} \to \overline{B}_{i-1}$$

and construct the pullback for β_{i-1}:

$$\beta_{i-1}^{\overline{f}_{i-1}}: 0 \leftarrow B_{i-1} \xleftarrow{q} N \leftarrow H_i(D) \leftarrow 0.$$

Now consider the isomorphism $h_{i*}^{-1}: H_i(D) \to H_i(C)$ and construct the push-out for $\beta_{i-1}^{\overline{f}_{i-1}}$:

$$h_{i*}^{-1}\beta_{i-1}^{\overline{f}_{i-1}}: 0 \leftarrow B_{i-1} \xleftarrow{q} N \leftarrow H_i(C) \leftarrow 0.$$

There is a commutative diagram

$$\begin{array}{ccccccccc}
0 & \longleftarrow & B_{i-1} & \xleftarrow{q} & N & \longleftarrow & H_i(C) & \longleftarrow & 0 \\
 & & \downarrow{\scriptstyle \overline{f}_{i-1}} & & \downarrow{\scriptstyle f} & & \downarrow{\scriptstyle h_{i*}} & & \\
0 & \longleftarrow & \operatorname{Ker}\overline{p}_{i-1} & \longleftarrow & \frac{D_i \oplus m_i\Lambda}{d_{i+1}(D_{i+1})} & \longleftarrow & H_i(D) & \longleftarrow & 0,
\end{array}$$

where f is an isomorphism. Replacing the last two terms in the sequence $\tilde{\alpha}_{i-1}(C)$, we obtain:

$$\overline{\alpha}_{i-1}(C): 0 \leftarrow \frac{C_{i-1} \oplus n_{i-1}\Lambda}{\partial_i \oplus \mathrm{id}(C_i \oplus n_i\Lambda)} \leftarrow C_{i-1} \oplus n_{i-1}\Lambda \xleftarrow{q} N \leftarrow H_i(C) \leftarrow 0.$$

It is clear that the sequences $[\overline{\alpha}_{i-1}(C)]$ and $[\alpha_{i-1}(D)]$ are isomorphic and that there exists the following commutative diagram:

$$\begin{array}{ccccccc}
0 \leftarrow \frac{C_{i-1} \oplus n_{i-1}\Lambda}{\partial_i \oplus \mathrm{id}(C_i \oplus n_i\Lambda)} & \longleftarrow & C_{i-1} \oplus n_{i-1}\Lambda & \xleftarrow{q} & N & \longleftarrow & H_i(C) \leftarrow 0 \\
\downarrow{\scriptstyle \hat{f}_{i-1}} & & \downarrow{\scriptstyle f_{i-1}} & & \downarrow{\scriptstyle s} & & \downarrow{\scriptstyle h_{i*}} \\
0 \leftarrow \frac{D_{i-1} \oplus m_{i-1}\Lambda}{d_i \oplus \mathrm{id}(D_i \oplus m_i\Lambda)} & \longleftarrow & D_{i-1} \oplus m_{i-1}\Lambda & \xleftarrow{\overline{q}} & \frac{D_i \oplus m_i\Lambda}{d_{i+1}(D_{i+1})} & \longleftarrow & H_i(D) \leftarrow 0
\end{array}$$

We observe that our construction involves no distinguished homomorphism from the module $C_i \oplus n_i\Lambda$ into the module N. But since there is an epimorphism

$$\partial_i \oplus \mathrm{id}: C_i \oplus n_i\Lambda \to B_{i-1},$$

we can, using Lemma 3.5 and stabilizing the homomorphism ∂_{i+1} via a free module of an appropriate rank $\overline{n}_i$, construct an epimorphism

$$t_i: C_i \oplus n_i\Lambda \oplus \overline{n}_i\Lambda \to N$$

such that $q \circ t_i = \partial_i \oplus \mathrm{id} \oplus 0$. This results in the following commutative diagram:

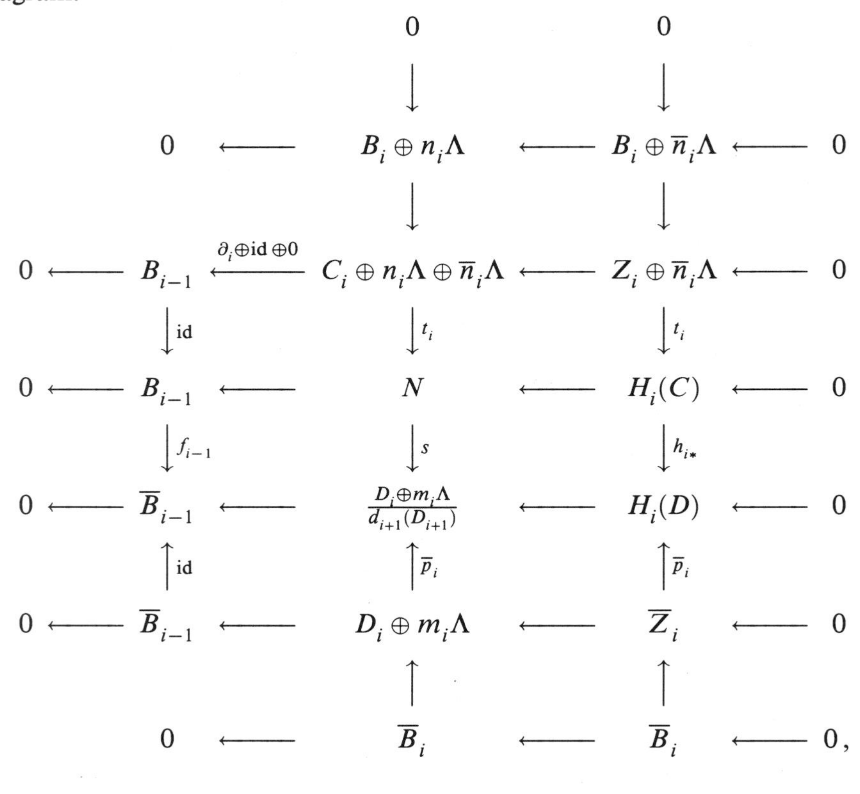

where $q \circ \overline{p}_i = d_i \oplus \mathrm{id}$. Without loss of generality we can assume that the modules $C_i \oplus n_i\Lambda \oplus \overline{n}_i\Lambda$ and $D_i \oplus m_i\Lambda$ are isomorphic. If this condition does not hold, we can stabilize the required boundary homomorphism in dimension $i+1$. Applying Lemma 3.6 to the epimorphisms $\overline{p}_i$ and $s \circ t_i$ and stabilizing the homomorphisms $\partial_{i+1} \oplus \mathrm{id}$ and d_{i+1} via the free modules $D_i \oplus m_i\Lambda$ and $C_i \oplus n_i\Lambda \oplus \overline{n}_i\Lambda$, respectively, we can find an isomorphism

$$f_i \colon C_i \oplus n_i\Lambda \oplus \overline{n}_i\Lambda \oplus D_i \oplus m_i\Lambda \to D_i \oplus m_i\Lambda \oplus C_i \oplus n_i\Lambda \oplus \overline{n}_i\Lambda$$

that satisfies conditions (1)–(4) because the diagram commutes. A consecutive application of the argument yields, after finitely many steps, a homotopy equivalence between our chain complexes. □

Thus, in the category of free chain complexes, the homotopy type of a chain complex (C, ∂) is completely defined by the homology module $H_i(C)$, the Swan-Wall class $\Gamma_i(C)$, and the extension $\alpha_i \in \mathrm{Ext}^2_\Lambda(\Gamma_i, H_{i+1}(C))$. The case of a semilinear homotopy equivalence is very similar to the one above.

§3. The Whitehead group

Let $GL(n, \Lambda)$ be the full linear group over a ring Λ. Embed $GL(n, \Lambda)$ in $GL(n+1, \Lambda)$ by identifying each matrix $A \in GL(n, \Lambda)$ with the matrix

$\left(\begin{smallmatrix} A & 0 \\ 0 & 1 \end{smallmatrix}\right)$ in $GL(n+1, \Lambda)$ and put $GL(\Lambda) = \varinjlim(GL(n, \Lambda))$. Similarly, $GL(\Lambda)$ can be considered as the group of all infinite invertible matrices of the form

$$\left|\begin{matrix} A & & & 0 \\ & 1 & & \\ & & \ddots & \\ 0 & & & 1 \\ & & & & \ddots \end{matrix}\right| .$$

Let $E(n, \Lambda)$ be the subgroup of $GL(n, \Lambda)$ generated by all elementary matrices, $E(\Lambda) = \varinjlim(E(n, \Lambda))$. A lemma due to Whitehead [8, 95, 97, 25] asserts that if $A, B \in GL(n, \Lambda)$, then the matrix is of the form

$$\left|\begin{matrix} A & B \\ 0 & (BA)^{-1} \end{matrix}\right| \in E(2n, \Lambda).$$

This implies that $E(\Lambda)$ coincides with the commutator group of $GL(\Lambda)$, and therefore the quotient group $K_1(\Lambda) = GL(\Lambda)/E(\Lambda)$ is abelian. The group $K_1(\Lambda)$ is called the Whitehead group of the ring Λ [8, 95, 97, 25].

Let g be the subgroup of units (invertible elements) of the ring Λ. Denote by E_G the group generated by $E(\Lambda)$ and all matrices of the form

$$\left|\begin{matrix} 1 & & & & & & 0 \\ & \ddots & & & & & \\ & & 1 & & & & \\ & & & g & & & \\ & & & & 1 & & \\ & & & & & \ddots & \\ 0 & & & & & & 1 \\ & & & & & & & \ddots \end{matrix}\right| ,$$

where $g \in G$. Set $K_G(\Lambda) = GL(\Lambda)/E_G$. Clearly, $K_G(\Lambda)$ is an abelian group and if $G = 1$, then $K_G(\Lambda) = K_1(\Lambda)$. Let $G = \{+1, -1\}$. In this case $\overline{K}_1(\Lambda) = K_G(\Lambda)$ is called the reduced Whitehead group.

If $\Lambda = \mathbb{Z}[G]$ for some multiplicative group G, then each element in G is a unit in $\mathbb{Z}[G]$. Let $T = G \cup (-G)$; then $\mathrm{Wh}(G) = K_T(\mathbb{Z}[G])$ is called the Whitehead group of the group G.

There is a natural group homomorphism $\tau\colon GL(n, \Lambda) \to \overline{K}_G(\Lambda)$. For each $A \in GL(n, \Lambda)$, $\tau(A)$ is called the torsion of the matrix A. Since $K_G(\Lambda)$ is an abelian group, $\tau(AB) = \tau(A) + \tau(B)$.

If G and G' are the subgroups of units in the rings Λ and Λ' and $f\colon \Lambda \to \Lambda'$ is a ring homomorphism such that $f(G) = G'$, then there is an induced group homomorphism $f_*\colon K_G(\Lambda) \to K_{G'}(\Lambda)$ $(f_*(\tau(A)) = \tau(f(A)))$. Similarly, a group homomorphism $G \to H$ induces a homomorphism of the Whitehead groups $\mathrm{Wh}(G) \to \mathrm{Wh}(H)$. It is known that if $f\colon G \to G$ is an inner automorphism, then $f_*\colon \mathrm{Wh}(G) \to \mathrm{Wh}(G)$ is the identity mapping

[95, 25]. Let Λ be a commutative ring, G a subgroup of the group U of all units in Λ. Denote by $SL(n, \Lambda)$ the group of invertible matrices with determinant equal to 1. Set $SL(\Lambda) = \varinjlim SL(n, \Lambda)$. Clearly, $SL(\Lambda)$ is contained in $GL(\Lambda)$ and the determinant defines an epimorphism $K_G(\Lambda) \to U/G$ with the kernel $SK_1(\Lambda) = SL(\Lambda)/E(\Lambda)$.

The short exact sequence

$$0 \longrightarrow SK_1(\Lambda) \longrightarrow K_G(\Lambda) \longrightarrow U/G \longrightarrow 0$$

splits via the homomorphism

$$U/G \longrightarrow GL(1, \Lambda) \hookrightarrow GL(n, \Lambda) \longrightarrow K_G(\Lambda),$$

whence $K_G(\Lambda) = SK_1(\Lambda) \oplus U/G$.

It is known that if Λ is a field or a commutative local ring, then $SK_1(\Lambda) = 0$. We now state some results on the structure of Whitehead groups:

(1) $\mathrm{Wh}(\mathbb{Z}) = 0$ [25],
(2) $\mathrm{Wh}(\mathbb{Z} \oplus \cdots \oplus \mathbb{Z}) = 0$ [9],
(3) $\mathrm{Wh}(\mathbb{Z}_k) = 0$, $k = 2, 3, 4, 6$ [8, 95],
(4) $\mathrm{Wh}(\mathbb{Z}_5) = \mathbb{Z}$ [95],
(5) if G is an abelian group, U the group of units in the ring $\mathbb{Z}[G]$, and T the subgroup of trivial units, then $\mathrm{Wh}(G) = SK_1(\mathbb{Z}[G]) \oplus U/T$.

Proposition 4.2 [25]. *Let G be a finite group. Then $SK_1(\mathbb{Z}[G]) = 1$ if and only if either*

(1) $G \approx \mathbb{Z}_2 \oplus \cdots \oplus \mathbb{Z}_2$, *or*
(2) *every Sylow p-subgroup of the group G is of the form $\mathbb{Z}_{p^n}$ or $\mathbb{Z}_p \oplus \mathbb{Z}_{p^n}$ for some $n \geq 1$.* □

Thus, in this case we have $\mathrm{Wh}(G) = U/T$. If $j > 1$, $k > 1$, and $j + k < q$, $jk = aq + 1$, then the nontrivial units in the ring $\mathbb{Z}[G]$ are of the form

$$u = (1 + x + \cdots + x^{j-1})(1 + x + \cdots + x^{k-1}) - a(1 + x + \cdots + x^{k-1}),$$

where $x \in G$ and the order of x is not equal to $1, 2, 3, 4, 6$.

Proposition 4.3 [149]. *Let $G_1 * G_2$ be the free product of the groups G_1 and G_2. Then* $\mathrm{Wh}(G_1 * G_2) = \mathrm{Wh}(G_1) \oplus \mathrm{Wh}(G_2)$. □

The following example belonging to Whitehead demonstrates the importance of stabilization in the definition of a Whitehead group.

Let G be a (noncommutative) group generated by the elements a and b and the relation $b^2 = 1$. Consider the elements $x = 1 - b$, $y = a(1 + b)$ in $\mathbb{Z}(G)$. Clearly, $yx = 0$ but $xy \neq 0$. The (1×1)-matrix $1 - xy$ is not an elementary matrix but it represents the zero element in the Whitehead group, since

$$\begin{vmatrix} 1 - ab & 0 \\ 0 & 1 \end{vmatrix} = \begin{vmatrix} 1 & 0 \\ b & 1 \end{vmatrix} \begin{vmatrix} 1 & a \\ 0 & 1 \end{vmatrix} \begin{vmatrix} 1 & 0 \\ b & 1 \end{vmatrix}^{-1} \begin{vmatrix} 1 & a \\ 0 & 1 \end{vmatrix}.$$

DEFINITION 4.1 [38]. We say that the representable dimension of the group G is less than or equal to i (r-dim $G \leq i$) if every element of $\mathrm{Wh}(G)$ can be realized as a matrix in $GL(\mathbb{Z}[G], i)$.

We list here some known results on r-dim(G).

If G is a finite group, then r-dim $G \leq 2$.

If G is an abelian group, then r-dim $G \leq 2$. This result belongs to Suslin.

If G is a finite abelian group, then r-dim $G \leq 1$ if and only if either $G \approx \mathbb{Z}_2 \oplus \cdots \oplus \mathbb{Z}_2$, or every Sylow p-subgroup of G is of the form $\mathbb{Z}_{p^n}$ or $\mathbb{Z}_p \oplus \mathbb{Z}_p \oplus \mathbb{Z}_{p^n}$ [2].

If G is a finite (nonabelian) group, then r-dim $G \leq 1$ if and only if G admits no epimorphic mapping onto the generalized quaternion group, the binary tetrahedral, octahedral, or icosahedral groups, and the following groups:

$$\mathbb{Z}_{p^2} \oplus \mathbb{Z}_{p^2}, \quad \mathbb{Z}_p \oplus \mathbb{Z}_p \oplus \mathbb{Z}_p,$$
$$\mathbb{Z}_p \oplus \mathbb{Z}_2 \oplus \mathbb{Z}_2 \oplus \mathbb{Z}_2,$$
$$\mathbb{Z}_4 \oplus \mathbb{Z}_2 \oplus \mathbb{Z}_2, \quad \mathbb{Z}_4 \oplus \mathbb{Z}_4,$$

where p is a prime.

Thus r-dim $G \leq 1$ for all finite simple groups. As of this writing, the author knows of no examples of finitely representable groups for which r-dim $G \geq 2$. On the other hand, it is unknown whether for a group G there exists an integer n such that r-dim $G < n$.

We conclude this section with the proof of a simple lemma which we shall need in what follows.

LEMMA 4.1. *Let $A \in GL(n, \Lambda)$, and let B be a matrix of dimension $n \times m$. Then the matrix*

$$\begin{bmatrix} A & 0 \\ 0 & B \end{bmatrix} = A \oplus B$$

can be reduced by elementary transformations to the form

$$\begin{bmatrix} E & 0 \\ 0 & \overline{B} \end{bmatrix},$$

where E is the $n \times n$ identity matrix.

PROOF. The matrix

$$\begin{vmatrix} E & A^{-1} \\ 0 & E \end{vmatrix}$$

is obtained from the identity matrix by elementary transformations. The following equality is evident:

$$\begin{vmatrix} A & 0 \\ 0 & B \end{vmatrix} \begin{vmatrix} E & A^{-1} \\ 0 & E \end{vmatrix} = \begin{vmatrix} A & E \\ 0 & B \end{vmatrix}.$$

The matrix $\left|\begin{smallmatrix} A & E \\ 0 & B \end{smallmatrix}\right|$ is reduced by elementary transformations to the form $\left|\begin{smallmatrix} 0 & E \\ B & 0 \end{smallmatrix}\right|$. Finally, permuting the rows, we obtain the form desired. □

§4. Torsion of homotopy equivalence

In this section we give a brief overview of the algebraic theory of simple homotopy type developed by Whitehead in [159, 160]. Throughout this section we assume that G is a subgroup of units in the ring Λ containing -1.

Suppose that F is a free module over the ring Λ. Let $\overline{b} = (b_1, \dots, b_k)$ and $\overline{c} = (c_1, \dots, c_k)$ be two different bases of the module F. By setting $c_i = \sum_{j=1}^{k} \alpha_{ij} b_j$ $(1 \le i \le k)$ we obtain a nondegenerate matrix (α_{ij}) with elements in Λ. Denote the corresponding element of the Whitehead group $K_G(\Lambda)$ by $[\overline{c}, \overline{b}]$. If $[\overline{c}, \overline{b}] = 0$, we shall say that $\overline{b}$ is equivalent to $\overline{c}$ $(\overline{b} \sim \overline{c})$. The identities $[\overline{e}, \overline{c}] + [\overline{c}, \overline{b}] = [\overline{e}, \overline{b}]$ and $[\overline{b}, \overline{b}] = 0$ demonstrate that this is indeed an equivalence relation.

LEMMA 4.2. *Let F be a free module over a ring Λ, and $f_1, \dots, f_k$ and $\overline{f}_1, \dots, \overline{f}_k$ two bases for it. Then there exists a basis $\overline{f}_1, \dots, \overline{f}_k, g_1, \dots, g_k$ in the module $F \oplus k\Lambda$ such that the bases $f_1, \dots, f_k, e_1, \dots, e_k$ and $\overline{f}_1, \dots, \overline{f}_k, g_1, \dots, g_k$ are equivalent, where*

$$e_1 = (1, 0, \dots, 0), \; e_2 = (0, 1, 0, \dots, 0), \; \dots, \; e_k = (0, \dots, 0, 1),$$

$1 \in \Lambda$, and $g_1, \dots, g_k$ is a basis in the module $k\Lambda$.

PROOF. Let A be the transition matrix from the basis $f_1, \dots, f_k$ to the basis $\overline{f}_1, \dots, \overline{f}_k$. Consider a basis $g_1, \dots, g_k$ in the module $k\Lambda$ for which the transition matrix from the basis $e_1, \dots, e_k$ to the basis $g_1, \dots, g_k$ is equal to A^{-1}. Then the transition matrix from $f_1, \dots, f_k, e_1, \dots, e_k$ to $\overline{f}_1, \dots, \overline{f}_k, g_1, \dots, g_k$ is of the form

$$\begin{vmatrix} A & 0 \\ 0 & A^{-1} \end{vmatrix}.$$

In the group $K_G(\Lambda)$ this element is equal to zero. Therefore these bases are equivalent. □

Let F_1 and F_2 be two free Λ-modules with chosen bases, and $f: F_1 \to F_2$ an isomorphism. By definition, the torsion of the isomorphism f (denoted by $\tau(f)$) is the element $\tau(A) \in K_G(\Lambda)$, where A is the matrix of the isomorphism f with respect to the chosen bases in the modules F_1 and F_2. We say that f is a simple homomorphism if $\tau(f) = 0$. In this case we write $f: F_1 \to F_2(\Sigma)$. Clearly, $\tau(f)$ assumes the same value for the bases of the same class of equivalence.

If a free chain complex is considered with a fixed basis for each module, we say that we have a based chain complex.

A based chain complex $\{C, \partial\}: 0 \to C_{i+1} \xrightarrow{\partial_{i+1}} C_i \to 0$ is called an elementary chain complex in dimension i. By definition, the torsion of an elementary chain complex $\tau(C)$ is the element $(-1)^i \tau(\partial_{i+1}) \in K_G(\Lambda)$.

An elementary chain complex $\{C, \partial\}$ is called an elementary trivial chain complex if $\tau(C) = 0$. If a chain complex is a direct sum of elementary trivial chain complexes in the category of based chain complexes, it is said to be a trivial chain complex.

We now define the torsion of an acyclic based chain complex

$$\{C, \partial\}: C_0 \xleftarrow{\partial_1} C_1 \longleftarrow \cdots \xleftarrow{\partial_n} C_n.$$

Since $\{C, \partial\}$ is acyclic, it is contractible, i.e., there exists a chain homotopy $\eta = \{\eta_i\}: C_i \to C_{i+1}$ such that $\eta_{i-1} \circ \partial_i + \partial_{i+1} \circ \eta_i = \mathrm{id}_{C_i}$. Note that $\eta_i \circ \eta_{i-1} = 0$. Put $\delta = \{\delta_i = \eta_i \circ \partial_{i+1} \circ \eta_i\}$. Clearly, δ is also a chain homotopy, since

$$\delta_{i-1} \circ \partial_i + \partial_{i+1} \circ \delta_i = \eta_{i-1} \circ \partial_i \circ \partial_i + \partial_{i+1} \circ \eta_i \circ \partial_{i+1} \circ \eta_i = \eta_{i-1} \circ \partial_i + \partial_{i+1} \circ \eta_i.$$

Consider the homomorphism $d = \{d_i = \partial_i \oplus \delta_i\}: \{C, \partial\} \to \{C, \partial\}$.

Clearly, d is an automorphism, since $d^2 = 1_C$. Consider the modules

$$C_{\text{odd}} = C_1 \oplus C_3 \oplus \cdots, \qquad C_{\text{even}} = C_0 \oplus C_2 \oplus \cdots,$$

whose bases are defined as unions of bases in C_i. Consider the restriction of the automorphism d to C_{odd}:

$$\sum_i \partial_i \oplus \delta_i: C_{\text{odd}} \to C_{\text{even}} \qquad (i = 1, 3, \ldots).$$

It is easy to see that $\sum_i \partial_i \oplus \delta_i$ $(i = 1, 2, \ldots)$ is an isomorphism. By definition, the torsion of a chain complex $\{C, \partial\}$ is the element $\tau(C) = \tau(\sum_i \partial_i \oplus \delta_i)$. One can show that the torsion $\tau(C)$ does not depend on the choice of the deformation operator $\delta = \{\delta_i\}$.

Let $f = \{f_i\}: \{C, \partial\} \to \{D, d\}$ be a homotopy equivalence between two based chain complexes. The torsion of the homotopy equivalence $\tau(f)$ is defined as the torsion of the cylinder of the mapping $\{M(f)\}$. An acyclic chain complex $\{M(f)\}$ inherits the bases of the modules in the chain originating from the bases in $\{C, \partial\}$ and $\{D, d\}$.

If $f = \{f_i\}: (C, \partial) \to (D, d)$ and $g = \{g_i\}: (D, d) \to (E, \delta)$ are homotopy equivalences, then $\tau(g \circ f) = \tau(g) + \tau(f)$. If $f = \{f_i\}$, $g = \{g_i\}: (C, \partial) \to (D, d)$, and f is homotopy equivalent to g, then $\tau(f) = \tau(g)$. A homotopy equivalence f is said to be simple if $\tau(f) = 0$.

Let $f = \{f_i\}: (C, \partial) \to (D, d)$ be a chain isomorphism of based chain complexes such that $f_i: C_i \to D_i(\Sigma)$. In this case we say that $f = \{f_i\}$ is a simple isomorphism, and that the complexes (C, ∂) and (D, d) are simply isomorphic. We denote $(C, \partial) \approx (D, d)(\Sigma)$.

Two based chain complexes (C, ∂) and (D, d) are simple homotopy equivalent, $(C, \partial) \overset{s}{\sim} (D, d)$, if there exist trivial chain complexes (T, t)

and $(\overline{T}, \bar{t})$ such that the chain complexes $(C \oplus T, \partial \oplus t)$ and $(D \oplus \overline{T}, d \oplus \bar{t})$ are simply isomorphic. Clearly, for the homotopy equivalences $i: (C, \partial) \to (C \oplus T, \partial \oplus t)$ and $p: (C \oplus T, \partial \oplus t) \to (C, \partial)$, we have $\tau(f) = \tau(p) = 0$.

It can be shown that if (C, ∂) is an acyclic based chain complex, then (C, ∂) is simple homotopy equivalent to an elementary chain complex.

We now list some basic properties of the torsion in the category of acyclic based complexes.

If $(C, \partial) \approx (D, d)(\Sigma)$, then $\tau(C) = \tau(D)$. Hence $\tau(C \oplus D, \partial \oplus d) = \tau(C) + \tau(D)$.

Suppose that

$$0 \longrightarrow C' \xrightarrow{\quad i \quad} C \xrightarrow{\quad j \quad} C'' \longrightarrow 0$$

is an exact sequence of acyclic chain complexes and let $\sigma: C'' \to C$ be a section of degree 0 (not necessarily a chain mapping). For fixed bases in C, C', and C'' we have

$$\tau(C) = \tau(C') + \tau(C'') + \sum_i (-1)^i \tau[C_i' C_i'', C_i],$$

where $c_i' c_i'' \equiv i(c_i') \cup \sigma(c_i'')$. For instance, if $i(c_i') \cup \sigma(c_i'')$ is equivalent to the basis c_i for all i, then $\tau(C) = \tau(C') + \tau(C'')$.

The following proposition holds.

PROPOSITION. *Let $f = \{f_i\}: (C, \partial) \to (D, d)$ be a homotopy equivalence between based chain complexes (C, ∂) and (D, d). Then $f = \{f_i\}$ is homotopy equivalent to a simple homotopy equivalence if and only if $\tau(f) = 0$.* □

There is an equivalent way of defining the torsion of a homotopy equivalence, which we will now briefly describe.

Let $f = \{f_i\}: (C, \partial) \to (D, d)$ be a homotopy equivalence between based chain complexes such that $f_i: C_i \to D_i$ is an isomorphism for all i. Define the torsion of a homotopy equivalence in this situation by the equality

$$\tau(f) = \sum_{i=0}^{n} (-1)^{i+1} \tau(f_i).$$

If the homotopy equivalence $f = \{f_i\}: (C, \partial) \to (D, d)$ is not an isomorphism on chain modules, then, making use of the Cockroft-Swan theorem, we can find trivial chain complexes (T, t) and $(\overline{T}, \bar{t})$ and chain isomorphisms

$$h = \{h_i\}: (C \oplus T, \partial \oplus t) \to (D \oplus \overline{T}, d \oplus \bar{t})$$

such that $f = p \circ h \circ i$, where $i: (C, \partial) \to (C \oplus T, \partial \oplus t)$ is an embedding and $p: (D \oplus \overline{T}, d \oplus \bar{t}) \to (D, d)$ is a projection.

We define the torsion of the homotopy equivalence f by the formula

$$\tau(f) = \tau(p \circ h \circ i) = \tau(p) + \tau(h) + \tau(i) = \tau(h),$$

because $\tau(p) = \tau(i) = 0$. Now

$$\tau(f) = \tau(h) = \sum_{i=0}^{n}(-1)^{i+1}\tau(h_i) = (-1)^{n+1}\tau(h_n),$$

since in the Cockroft-Swan theorem all but the last isomorphisms on chain modules are simple.

Note that Milnor defined the torsion of a based chain complex in the situation where the homology modules are free [95]. We do not give this definition here. In the case when the chain complex is acyclic, Milnor's definition is equivalent to the one above.

Let $\theta\colon \Lambda \to \Lambda$ be an automorphism of the ring Λ such that $\theta(1) = 1$. As before, denote by s_θ the induced θ-semilinear isomorphism of the free module $k\Lambda$. Let Θ_G be the group of automorphisms of the ring Λ consisting of the automorphisms f of this ring such that $f(G) = G$, where G is the subgroup of units in the ring Λ. Then the group Θ_G acts on the group $K_G(\Lambda)$ (as an operator group) by the rule $\theta_\tau(g) = \tau(g^\theta)$, where $g\colon k\Lambda \to k\Lambda$ is an isomorphism, $k\Lambda$ is a free module, and $g^\theta = s_\theta \circ g \circ s_{\theta^{-1}}$.

Let (C, ∂) be a based chain complex over Λ and (C, ∂^θ) the conjugate chain complex. If (C, ∂) is acyclic, so is (C, ∂^θ). If $\tau(C, \partial) = 0$, then $\tau(C, \partial^\theta) = 0$.

As mentioned above, if $f = \{f_i\}\colon (C, \partial) \to (D, d)$ is a homotopy equivalence between based chain complexes associated with an automorphism $\theta\colon \Lambda \to \Lambda$, $\theta \in \Theta_G$, then $f \circ s_{\theta^{-1}} = \{f_i \circ s_{\theta^{-1}}\}\colon (C, \partial^\theta) \to (D, d)$ is a homotopy equivalence in the ordinary sense. In accordance with the definition, we have $\tau(f) = \tau(f \circ s_{\theta^{-1}})$.

§5. Minimal complexes

This is one of the key sections. In it we prove that if Λ is an s-ring, then in the homotopy type of a free chain complex over Λ there always exists a minimal chain complex.

DEFINITION 4.2. A chain complex $(C, \partial)\colon C_0 \xleftarrow{\partial_1} C_1 \longleftarrow \cdots \xleftarrow{\partial_n} C_n$ is called minimal in dimension i if for any chain complex $(D, d)\colon D_0 \xleftarrow{d_1} D_1 \longleftarrow \cdots \xleftarrow{d_n} D_n$ homotopy equivalent to (C, ∂) we have $\mu(C_i) \le \mu(D_i)$. A chain complex (C, ∂) is said to be minimal if it is minimal in all dimensions.

In what follows we assume that each C_i is a free module with a fixed basis.

LEMMA 4.3. *Let Λ be an s-ring. Then for any chain complex*

$$(C, \partial)\colon C_0 \xleftarrow{\partial_1} C_1 \longleftarrow \cdots \xleftarrow{\partial_n} C_n$$

over Λ there exists a homotopy equivalent chain complex $(\overline{C}, \overline{\partial})$ such that f-rank$(\overline{\partial}_{i+1}(\overline{C}_{i+1}), \overline{C}_i) = 0$ and is additive.

Proof. Consider a chain complex $C_0 \xleftarrow{\partial_1} C_1 \longleftarrow \cdots \xleftarrow{\partial_n} C_n$. Let $\partial_0 \colon C_0 \to H_0(C)$. Stabilize the boundary homomorphism ∂_1 via the free module $n_0\Lambda$, $n_0 = \mu(H_0(C))$, and then extract from the thickening of ∂_0 a minimal epimorphism $\overline{\partial}_0 \colon \overline{C}_0 \to H_0(C)$, obtaining a decomposition

$$\begin{array}{ccccccccc} 0 & \longleftarrow & H_0 & \xleftarrow{\overline{\partial}_0} & \overline{C}_0 & \xleftarrow{\hat{\partial}_1} & \hat{C}_1 & \xleftarrow{\partial_2} & C_2 \\ & & & & \oplus & & \oplus & & \\ & & 0 & \longleftarrow & \tilde{C}_0 & \xleftarrow{\text{id}} & \tilde{C}_0 & \longleftarrow & 0, \end{array}$$

where $\hat{C}_1$ is a free module. The fact that $\overline{\partial}_0 \colon \overline{C}_0 \to H_0(C)$ is a minimal epimorphism implies that $f\text{-rank}(\hat{\partial}_1(\hat{C}_1), \overline{C}_0) = 0$ and is additive (Theorem 3.1). Consider the submodule $\partial_2(C_2) \subset \hat{C}_1$. Lemma 3.7 ensures the existence of an integer m_1, starting from which

$$f\text{-rank}(\partial_2(C_2) \oplus m_1\Lambda, \hat{C}_1 \oplus m_1\Lambda) = n_1$$

and is additive. The submodule $\tilde{C}_0$ has no influence on the value of $f\text{-rank}(\cdot, \cdot)$. Stabilizing the boundary homomorphism ∂_2 via a free module of rank m_1, we obtain a chain complex

$$\begin{array}{ccccccccccccc} & & & & & & 0 & \longleftarrow & m_1\Lambda & \xleftarrow{\text{id}} & m_1\Lambda & \longleftarrow & 0 \\ & & & & & & & & \oplus & & \oplus & & \\ 0 & \longleftarrow & H_0(C) & \longleftarrow & \overline{C}_0 & \xleftarrow{\hat{\partial}_1} & \hat{C}_1 & \xleftarrow{\partial_2} & C_2 & \xleftarrow{\partial_3} & C_3 & \xleftarrow{\partial_4} & \\ & & & & \oplus & & \oplus & & & & & & \\ & & 0 & \longleftarrow & \tilde{C}_0 & \xleftarrow{\text{id}} & \tilde{C}_0 & \longleftarrow & 0. & & & & \end{array}$$

The change in notation is evident.

Now find a decomposition of the module $\partial_2(C_2) \oplus m_1\Lambda$ such that

$$\partial_2(C_2) \oplus m_1\Lambda = B_1 \oplus \tilde{C}_1, \qquad \mu(\tilde{C}_1) = n_1, \qquad B_1 \subset \overline{C}_1,$$

where $\overline{C}_1 \oplus \tilde{C}_1$ is a new decomposition of the module $\hat{C}_1 \oplus m_1\Lambda$. Write the resulting representation of the chain complex:

$$\begin{array}{ccccccccccccc} & & & & & & 0 & \longleftarrow & \tilde{C}_1 & \xleftarrow{\text{id}} & \tilde{C}_1 & \longleftarrow & 0 \\ & & & & & & & & \oplus & & & & \\ 0 & \longleftarrow & H_0(C) & \longleftarrow & \overline{C}_0 & \xleftarrow{\overline{\partial}_1} & \overline{C}_1 & \xleftarrow{\hat{\partial}_2} & \hat{C}_2 & \xleftarrow{\partial_3} & C_3 & \xleftarrow{\partial_4} & \\ & & & & \oplus & & \oplus & & & & & & \\ & & 0 & \longleftarrow & \tilde{C}_0 & \xleftarrow{\text{id}} & \tilde{C}_0 & \longleftarrow & 0. & & & & \end{array}$$

All direct summands in this decomposition are free. Consider the submodule $B_1 = \hat{\partial}_2(\hat{C}_2)$. By the construction and by Lemma 3.8, f-rank$(B_1, \overline{C}_1) = 0$ and is additive. We note that the choice of B_1 is not unique; this fact will be examined below. Thus, stabilizing the boundary homomorphisms ∂_i $(i \geq 3)$ in succession, we arrive at a diagram in which f-rank$(\overline{\partial}_i(C_i), \overline{C}_{i-1}) = 0$ and is additive. As a result, we obtain a representation of the chain complex in the form:

$$\begin{array}{ccccccccccccc}
 & & & & 0 & \longleftarrow & \tilde{C}_1 & \xleftarrow{\text{id}} & \tilde{C}_1 \longleftarrow 0 \longleftarrow \tilde{C}_3 & & & \longleftarrow & \cdots \\
 & & & & & & \oplus & & \oplus & & \oplus & & \\
0 & \longleftarrow & H_0(C) & \longleftarrow & \overline{C}_0 & \xleftarrow{\overline{\partial}_1} & \overline{C}_1 & \xleftarrow{\overline{\partial}_2} & \overline{C}_2 & \xleftarrow{\overline{\partial}_3} & \overline{C}_3 & \longleftarrow & \cdots \\
 & & & & \oplus & & \oplus & & \oplus & & \oplus & & \\
 & & 0 & \longleftarrow & \tilde{C}_0 & \xleftarrow{\text{id}} & \tilde{C}_0 \longleftarrow 0 \longleftarrow \tilde{C}_22 & & & \xleftarrow{\text{id}} & \tilde{C}_2 & &
\end{array}$$

Clearly, the subcomplex

$$\begin{array}{ccccccccc}
0 & \longleftarrow & \tilde{C}_0 & \xleftarrow{\text{id}} & \tilde{C}_0 \longleftarrow 0 \longleftarrow \tilde{C}_2 & \xleftarrow{\text{id}} & \tilde{C}_2 & \longleftarrow & 0 \\
 & & \oplus & & & & & & \oplus \\
0 & \longleftarrow & \tilde{C}_1 & \xleftarrow{\text{id}} & \tilde{C}_1 \longleftarrow 0 \longleftarrow \tilde{C}_3 & \xleftarrow{\text{id}} & \tilde{C}_3 & \longleftarrow & 0
\end{array}$$

is acyclic. Discarding it, we obtain the chain complex

$$\overline{C}_0 \xleftarrow{\overline{\partial}_1} \overline{C}_1 \longleftarrow \cdots \longleftarrow \overline{C}_{n-1} \xleftarrow{\overline{\partial}_n} \overline{C}_n$$

in which f-rank$(\overline{\partial}_i(\overline{C}_i), \overline{C}_{i-1}) = 0$ and is additive. $\square$

It will be shown below that any chain complex for which

$$f\text{-rank}(\partial_i(C_i), C_{i-1})) = 0$$

and is additive is a minimal chain complex. Let us examine its construction in more detail. Our proof of Lemma 4.3 failed to take into consideration the bases in the chain modules. We shall now fill in this omission. When the boundary homomorphisms are stabilized via certain free modules, we can always suppose the latter to be already furnished with canonical bases $e_1, \dots, e_n$, where $e_i = (0, \dots, 0, 1, 0, \dots, 0)$ with 1 in ith position. The situation is more complicated when the chain modules are decomposed into direct sums of free submodules. For example, if a module C with fixed basis $(c_1, \dots, c_k)$ is represented as a direct sum of free modules A and B, it does not necessarily follow that A has a basis $(a_1, \dots, a_m)$ and B has a basis $(b_1, \dots, b_n)$ such that the bases $(a_1, \dots, a_m, b_1, \dots, b_n)$ and $(c_1, \dots, c_k)$ are equivalent. We will show that by an appropriate stabilization we can get around this difficulty.

Choose arbitrary bases in the modules $\overline{C}_i \oplus \tilde{C}_i \oplus \tilde{C}_{i-1}$; for example, such that the homomorphisms id are given by the identity matrices. Stabilizing the boundary homomorphisms $\overline{\partial}_i \oplus \mathrm{id}$ via free modules R_i of an appropriate rank, and selecting bases in them in accordance with Lemma 4.2, we can obtain that the new bases are equivalent to the original ones. Observe that since the homomorphisms id are given by identity matrices, the modules $\tilde{C}_i$ can be discarded without changing the simple homotopy type of our chain complex. In addition, it follows from the theory of simple homotopy type that the subcomplex formed by the modules R_i can be contracted to the complex

$$0 \longleftarrow \overline{R}_j \xleftarrow{\varphi} \overline{R}_{j+1} \longleftarrow 0.$$

This and Lemma 4.2 imply the following corollary.

COROLLARY 4.1. *Under the conditions of Lemma* 4.3 *the chain complex is simple homotopy equivalent to the chain complex*

$$\begin{array}{ccccccccc} \overline{C}_0 & \xleftarrow{\partial_1} & \overline{C}_1 \longleftarrow \cdots \longleftarrow & \overline{C}_j & \xrightarrow{\overline{\partial}_{j+1}} & \overline{C}_{j+1} & \longleftarrow \cdots \xleftarrow{\overline{\partial}_n} & \overline{C}_n \\ & & & \oplus & & \oplus & & \\ & 0 & \longleftarrow & \overline{R}_j & \xleftarrow{\varphi} & \overline{R}_{j+1} & \longleftarrow & 0, \end{array}$$

where $f\text{-rank}(\overline{\partial}_i(\overline{C}_i), \overline{C}_{i-1}) = 0$ *and is additive.* □

LEMMA 4.4. *Let* Λ *be an* s*-ring and*

$$\begin{array}{l} (C_i, \partial_i): C_0 \xleftarrow{\partial_1} C_1 \xleftarrow{\partial_2} \cdots \xleftarrow{\partial_n} C_n, \\ (D, d): \overline{D}_0 \xleftarrow{d_1} D_1 \xleftarrow{d_2} \cdots \xleftarrow{d_n} D_n \end{array}$$

two homotopy equivalent chain complexes over Λ *for which*

$$f\text{-rank}(\partial_i(C_i), C_{i-1}) = f\text{-rank}(d_i(D_i), D_{i-1}) = 0$$

and is additive for all i. *Then* $\mu(C_i) = \mu(D_i)$.

PROOF. Assume the contrary. Then there exists an integer k such that $\mu(C_k) > \mu(D_k)$, $\mu(C_i) = \mu(D_i)$ $(0 \le i < k)$. Obviously $k > 0$, since $\partial_0: C_0 \to H_0$ and $d_0: D_0 \to H_0$ are minimal epimorphisms (a consequence of Lemma 3.10). Using the Cockroft-Swan statement, we find acyclic complexes

$$\begin{array}{l} (T, t): T_0 \xleftarrow{t_1} T_1 \longleftarrow \cdots \xleftarrow{t_k} T_k, \\ (S, s): S_0 \xleftarrow{s_1} S_1 \longleftarrow \cdots \xleftarrow{s_k} S_k, \end{array}$$

such that up to and including dimension $k-1$ the complexes

$$\begin{array}{ccccccc} C_0 \oplus T_0 \longleftarrow \cdots \xleftarrow{\partial_{k-1}\oplus t_{k-1}} & C_{k-1} \oplus T_{k-1} & \xleftarrow{\partial_k \oplus t_k} & C_k \oplus T_k & \xleftarrow{\partial_{k+1}} & C_{k+1} \\ \downarrow \psi_0 \qquad & \downarrow \psi_{k-1} & & & & \\ D_0 \oplus S_0 \longleftarrow \cdots \xleftarrow{d_{k-1}\oplus s_{k-1}} & D_{k-1} \oplus S_{k-1} & \xleftarrow{d_k \oplus s_k} & D_k \oplus S_k & \xleftarrow{d_{k+1}} & D_{k+1} \end{array}$$

are isomorphic. Note that $\mu(T_i) = \mu(S_i)$, since $\mu(C_j) = \mu(D_j)$ for all $0 \le j \le k-1$. Stabilize ∂_{k+1} via the free module $D_k \oplus S_k$, and d_{k+1} via the free module $C_k \oplus T_k$; then, as in the proof of the Cockroft-Swan theorem, construct an isomorphism

$$\psi_k : C_k \oplus T_k \oplus D_k \oplus S_k \to D_k \oplus S_k \oplus C_k \oplus T_k$$

such that $\psi_{k-1} \circ (\partial_k \oplus t_k) = (d_k \oplus s_k) \circ \psi_k$. Since $f\text{-rank}(\partial_{k+1}(C_{k+1}), C_k) = 0$ and is additive, we have

$$f\text{-rank}(\partial_{k+1}(C_{k+1}) \oplus D_k \oplus S_k, C_k \oplus T_k \oplus D_k \oplus S_k) = \mu(D_k) + \mu(S_k).$$

Similarly

$$f\text{-rank}(d_{k+1}(D_{k+1}) \oplus C_k \oplus T_k, D_k \oplus S_k \oplus C_k \oplus T_k) = \mu(C_k) + \mu(T_k).$$

Since ψ_k is an isomorphism, we have

$$\begin{aligned} &f\text{-rank}(\partial_{k+1}(C_{k+1}) \oplus D_k \oplus S_k, C_k \oplus T_k \oplus D_k \oplus S_k) \\ &\quad = f\text{-rank}(d_{k+1}(D_{k+1}) \oplus C_k \oplus T_k, D_k \oplus S_k \oplus C_k \oplus T_k). \end{aligned}$$

But, by assumption, $\mu(C_k) > \mu(D_k)$, and we arrive at a contradiction. □

LEMMA 4.5. *Let Λ be an s-ring and*

$$(C, \partial) : C_0 \xleftarrow{\partial_1} C_1 \xleftarrow{\partial_2} \cdots \xleftarrow{\partial_n} C_n$$

a based chain complex in which $f\text{-rank}(\partial_i(C_i), C_{i-1}) = 0$ and is additive for all i. Then (C, ∂) is a minimal chain complex.

PROOF. Assume the contrary. Then in the class of chain complexes homotopy equivalent to (C, ∂) there exists a complex

$$(D, d) : D_0 \xleftarrow{d_1} D_1 \xleftarrow{d_2} \cdots \xleftarrow{d_n} D_n$$

such that $\mu(C_j) > \mu(D_j)$ for some j. Obviously $j > 0$. Let k be the first value for which this inequality holds. By stabilization and contraction of the boundary homomorphisms d_i $(0 < i < k-1)$, and an argument similar to that in the proof of Lemma 4.4, construct a chain complex

$$\overline{D}_0 \xleftarrow{\overline{d}_1} \overline{D}_1 \longleftarrow \cdots \xleftarrow{\overline{d}_{k-2}} \overline{D}_{k-2} \xleftarrow{\tilde{d}_{k-1}} \tilde{D}_{k-1} \xleftarrow{d_k} D_k \xleftarrow{d_{k+1}} D_{k+1} \longleftarrow \cdots,$$

for which $f\text{-rank}(\overline{d}_i(\overline{D}_i), \overline{D}_{i-1}) = 0$ and is additive $(0 < i < k-1)$. If

$$f\text{-rank}(d_k(D_k), \tilde{D}_{k-1}) > 0,$$

find a submodule $E \subset d_k(D_k)$ such that the complex takes the form

$$\begin{array}{ccccccccccc} \overline{D}_0 \overset{\overline{d}_1}{\longleftarrow} \overline{D}_1 \longleftarrow \cdots \overset{\overline{d}_{k-2}}{\longleftarrow} \overline{D}_{k-2} & \overset{\overline{d}_{k-1}}{\longleftarrow} & \overline{D}_{k-1} & \overset{d_k}{\longleftarrow} & \tilde{D}_k & \overset{d_{k+1}}{\longleftarrow} & D_{k+1} \longleftarrow \cdots \\ & & \oplus & & \oplus & & \\ 0 & \longleftarrow & E & \overset{\varphi}{\longleftarrow} & E & \longleftarrow & 0, \end{array}$$

where

$$\overline{D}_{k-1} \oplus E = \tilde{D}_{k-1} \quad \text{and} \quad \tilde{D}_k \oplus E = D_k .$$

Cancelling the module E, we obtain a chain complex in which $\mu(\tilde{D}_k) < \mu(D_k)$. If

$$f\text{-rank}(d_{k+1}(D_{k+1}), \tilde{D}_k) > 0,$$

again pull out a "superfluous" module in $\tilde{D}_k$ and D_{k+1} and cancel it. Note that if $f\text{-rank}(d_k(\overline{D}_k), \tilde{D}_{k-1})$ is not additive, then under stabilization of the homomorphism d_k it can only increase, and we can therefore only decrease $\mu(D_k)$. The situation is similar in dimension $k+1$. Thus we can assume, without loss of generality, that we have constructed a chain complex in which $f\text{-rank}(\overline{d}_i(\overline{D}_i), \overline{D}_{i-1}) = 0$ and is additive for all i, and $\mu(\overline{D}_k) < \mu(D_k)$. But the preceding lemma implies that $\mu(\overline{D}_k) = \mu(C_k)$. The resulting contradiction completes the proof. □

THEOREM 4.2. *Let $(C, \partial)_\Lambda$ be the class of chain complexes over a ring Λ. Then every homotopy type of a free chain complex in $(C, \partial)_\Lambda$ contains a minimal chain complex if and only if Λ is an s-ring.*

PROOF. *Necessity.* Assume the contrary. Then there exist stably free modules over the ring Λ that are not free. Let M be such a module, with $M \oplus \Lambda = n\Lambda$, and construct the chain complex of the form

$$\begin{array}{ccccccc} (C, \partial): 0 & \longleftarrow & C_i & \overset{\partial_{i+1}}{\longleftarrow} & C_{i+1} & \longleftarrow & 0 \\ & & \| & & \| & & \\ 0 \longleftarrow \quad M & \longleftarrow & n\Lambda & \overset{i}{\longleftarrow} & \Lambda & \longleftarrow & 0. \end{array}$$

Obviously $H_i(C) = M$ and by construction

$$f\text{-rank}(\partial_{i+1}(C_{i+1}), C_i) = 1 .$$

But we cannot perform a contraction, since the module complementary to $\partial_{i+1}(C_{i+1})$ in C_i is not free. Stabilizing the chain complex (C, ∂) in dimension i, we obtain

$$\begin{array}{ccccccc}
 & & & & C_i & \xleftarrow{\partial_{i+1}} & C_{i+1} \\
 & & & & \oplus & & \\
0 & \longleftarrow & \Lambda & \xleftarrow{\text{id}} & \Lambda & \longleftarrow & 0 \\
 & & \| & & \| & & \\
 & & \overline{C}_{i-1} & \xleftarrow{\overline{\partial}_i} & \overline{C}_i & \xleftarrow{\overline{\partial}_{i+1}} & C_{i+1},\quad \overline{\partial}_{i+1} = \partial_{i+1}.
\end{array}$$

Here a contraction of the homomorphism $\overline{\partial}_{i+1}$ is possible. Thus we obtain the chain complex

$$\begin{array}{ccccccccc}
(\tilde{C}, \tilde{\partial}): & & \tilde{C}_{i-1} & \xleftarrow{\tilde{\partial}_i} & \tilde{C}_i & & & & \\
 & & \| & & \| & & & & \\
0 & \longleftarrow & \Lambda & \longleftarrow & n\Lambda & \longleftarrow & M & \longleftarrow & 0
\end{array}$$

which is homotopy equivalent to the chain complex (C, ∂). Obviously, the chain complex (C, ∂) is minimal in dimension i but not in dimension $i+1$. The chain complex $(\tilde{C}, \tilde{\partial})$ is minimal in dimensions i and $i+1$ but not in dimension $i-1$. Therefore the homotopy type of the chain complex (C, ∂) has no minimal chain complex. The necessity is proved.

Sufficiency follows from Lemmas 4.3 and 4.5. □

We see that there is an obstruction to the existence of a minimal chain complex in a simple homotopy type of a chain complex over an s-ring (see Corollary 4.1).

Consider this question in more detail. Fix the subgroup G of units in the ring Λ. Let $\tau: GL(n, \Lambda) \to \overline{K}_G(\Lambda)$ be the natural homomorphism. Denote by $\overline{K}_G(\Lambda, n)$ the quotient group $\overline{K}_G(\Lambda)/\tau(GL(n, \Lambda))$. Let $(\overline{C}, \overline{\partial})$ be a minimal chain complex and

$$(R, \delta): 0 \longleftarrow R_i \xleftarrow{\delta} R_{i+1} \longleftarrow 0$$

an elementary chain complex with nontrivial torsion τ_0 lying in the group $\overline{K}_G(\Lambda)$. Then the simple homotopy type of the chain complex $(\overline{C} \oplus R, \partial \oplus \delta)$ does not necessarily contain a minimal chain complex. At the same time the following lemma holds.

LEMMA 4.6. *Suppose that the homotopy type of a chain complex (C, ∂) over $\Lambda = \mathbb{Z}[G]$ contains the minimal chain complex*

$$(\overline{C}, \overline{\partial}): \overline{C}_0 \xleftarrow{\overline{\partial}_1} C_1 \longleftarrow \cdots \xleftarrow{\overline{\partial}_n} \overline{C}_n.$$

If $\mu(\overline{C}_{i_0}) \geq k$ *for some* i_0, *where* r-dim $G \leq k$, *then the chain complex* (C, ∂) *is simple homotopy equivalent to the chain complex* $(\overline{C}, \overline{\partial})$.

PROOF. By the Cockroft-Swan theorem and simple homotopy type theory, the chain complex (C, ∂) is simple homotopy equivalent to a chain complex of the form

$$\begin{array}{ccccccccc} \overline{C}_0 & \xleftarrow{\overline{\partial}_1} & \overline{C}_1 \leftarrow \cdots \leftarrow & \overline{C}_{i_0} & \xleftarrow{\overline{\partial}_{i_0+1}} & C_{i_0+1} & \leftarrow \cdots \leftarrow C_n \\ & & & \oplus & & \oplus & \\ & & 0 \leftarrow & R_{i_0} & \xleftarrow{\delta} & R_{i_0+1} & \leftarrow 0. \end{array}$$

Since r-dim $G \leq k$, the matrix of the isomorphism δ can be reduced by elementary transformations to the form

$$\begin{vmatrix} A & 0 \\ 0 & E \end{vmatrix}, \qquad A \in GL(\mathbb{Z}[G], k).$$

We can now apply Lemma 4.1 to the matrix $A \oplus \overline{\partial}_{i_0+1}$ and cancel the "redundant" submodule. □

Lemma 4.6 and Corollary 4.1 make it possible to associate with each chain complex over an s-ring an element of the group $\overline{K}_G(\Lambda, n)$ which is an obstruction to the existence of a minimal chain complex in simple homotopy type. Here n is the maximum value of rank taken over all chain modules in a minimal chain complex. Let

$$(C, \partial): C_0 \xleftarrow{\partial_1} C_1 \longleftarrow \cdots \xleftarrow{\partial_n} C_n$$

be an arbitrary chain complex over an s-ring. Write it in the form

$$\begin{array}{ccccccccc} \overline{C}_0 & \xleftarrow{\overline{\partial}_1} & \overline{C}_1 \leftarrow \cdots \leftarrow & \overline{C}_{i_0} & \xleftarrow{\overline{\partial}_{i_0+1}} & \overline{C}_{i_0+1} & \leftarrow \cdots \leftarrow \overline{C}_n \\ & & & \oplus & & \oplus & \\ & & 0 \leftarrow & R_{i_0} & \xleftarrow{\delta} & R_{i_0+1} & \leftarrow 0, \end{array}$$

where i_0 is the maximal value of $\mu(\overline{C}_i)$. Consider the value of the isomorphism δ in the group $K_G(\Lambda, \mu(\overline{C}_{i_0}))$ and associate the resulting element in $\tau(\delta)$ with the chain complex (C, ∂). Lemma 4.6 evidently guarantees that if the obstruction is equal to zero, then (C, ∂) is simple homotopy equivalent to the minimal chain complex.

LEMMA 4.7. *Let* $\{C, \partial\}_*$ *denote the class of homotopy equivalent chain complexes over an* s-*ring* Λ *such that for each representative*

$$C_0 \xleftarrow{\partial_1} C_1 \longleftarrow \cdots \xleftarrow{\partial_n} C_n$$

in $\{C, \partial\}_*$, f-rank$(\partial_{k+1}(C_{k+1}), C_k) = 0$ *and is additive. Then the number*

$$\mu(C_0) - \mu(C_1) + \cdots + (-1)^k \mu(C_k) = \chi_k$$

is the same for all chain complexes in $\{C, \partial\}_*$.

PROOF. It is well known that the Euler characteristic

$$\sum (-1)^i \mu(C_i) = \chi(C_i, \partial_i)$$

is an invariant of $\{C, \partial\}_*$. Stabilize the homomorphisms ∂_i $(1 \leq i \leq k, i \geq k+2)$ via free modules of suitable rank so that the resulting chain complex has as a direct summand a chain complex $\{\overline{C}_i, \overline{\partial}_i\}$ in which f-rank$(\overline{\partial}_i(C_i), C_{i-1}) = 0$ and is additive for all i $(1 \leq i \leq k, i \geq k+2)$, while the complementary chain complex is acyclic. Observe that it is not necessary to stabilize the homomorphism ∂_{k+1} because of the hypothesis of the lemma. By Lemma 4.5 $\{\overline{C}_i, \overline{\partial}_i\}$ is a minimal chain complex. By construction,

$$\sum_{l \geq k+1} (-1)^l \mu(C_l) = \kappa.$$

By Lemmas 4.4 and 4.5, the ranks of the chain modules in the minimal chain complex are invariants of the homotopy type. Therefore the number κ is constant for all chain complexes $\{C_i, \partial_i\}$ in $\{C, \partial\}$. Thus the equality

$$\chi_k + \kappa = \chi(C_i, \partial_i)$$

implies the conclusion of the lemma. □

§6. Morse numbers of complexes

Let $(C, \partial): C_0 \xleftarrow{\partial_1} C_1 \leftarrow \cdots \xleftarrow{\partial_n} C_n$ be a free chain complex over a ring Λ. In what follows, when speaking of the homotopy type (simple homotopy type) of a chain complex (C, ∂), we consider free chain complexes homotopy equivalent (simple homotopy equivalent) to (C, ∂) whose length does not exceed $n+1$.

DEFINITION 4.3. The ith Morse number of a chain complex (C, ∂) is the number $\mathscr{M}_i(C, \partial) = \mu(\overline{C}_i)$, where

$$(\overline{C}, \overline{\partial}): \overline{C}_0 \xleftarrow{\overline{\partial}_1} \overline{C}_1 \longleftarrow \cdots \longleftarrow \overline{C}_n$$

is the minimal chain complex in dimension i simple homotopy equivalent to (C, ∂).

DEFINITION 4.4. Let $(C, \partial): C_0 \xleftarrow{\partial_1} C_1 \leftarrow \cdots \leftarrow C_n$ be a chain complex of length $n+1$. Set $m(C, \partial) = \sum_{i=0}^n \mu(C_i)$. The Morse number of the chain complex (C, ∂) is the number

$$\mathscr{M}(C, \partial) = \min(m(\tilde{C}, \tilde{\partial})),$$

where the minimum is taken over all chain complexes $(\tilde{C}, \tilde{\partial})$ simple homotopy equivalent of (C, ∂).

Let us discuss the question of homotopy invariance and the value of the ith Morse numbers (Morse numbers) of chain complexes. The theory of simple homotopy type implies that the Morse numbers of homotopy equivalent chain complexes may be different. With the use of Corollary 4.1 it is not difficult to construct examples of homotopy equivalent chain complexes whose length does not exceed 3 and the ith Morse numbers are different. For that it is sufficient to consider a ring Λ with $\overline{K}_1(\Lambda) \neq 0$. It follows directly from the Cockroft-Swan proposition that if the length of the chain complex is greater than 3, then the ith Morse numbers are invariants of homotopy type.

Let

$$(C, \partial): C_0 \xleftarrow{\partial_1} C_1 \longleftarrow \cdots \xleftarrow{\partial_n} C_n$$

be a chain complex over the ring $\mathbb{Z}[G]$. Denote, as before,

$$\Gamma_i = C_i/\partial_{i+1}(C_{i+1}).$$

DEFINITION 4.5. For each chain complex (C, ∂) over the ring $\mathbb{Z}[G]$ we set

$$S_i(C, \partial) = S(\Gamma_i) = d(\Gamma_i) - \mu(\mathbb{Z} \otimes_{\mathbb{Z}[G]} \Gamma_i)$$

(the number $d(\Gamma_i)$ is defined in §1 of the present chapter).

LEMMA 4.8. *Let (C, ∂) and (D, d) be two homotopy equivalent chain complexes. Then $S_i(C, \partial) = S_i(D, d)$ for all i.*

PROOF. By the Cockroft-Swan theorem there exist acyclic chain complexes (T, t) and (R, r) such that the chain complexes $(C \oplus T, \partial \oplus t)$ and $(D \oplus R, d \oplus r)$ are isomorphic. Therefore the equality $S_i(C \oplus T, \partial \oplus t) = S_i(D \oplus R, d \oplus r)$ holds. By construction

$$C_i/\partial_{i+1}(C_{i+1}) \oplus n_i\mathbb{Z}[G] \approx C_i \oplus T_i/\partial_{i+1}(C_{i+1}) \oplus t_{i+1}(T_{i+1})$$

for some positive integer i. By virtue of Corollary 3.2 the equality

$$S_i(C, \partial) = S_i(C \oplus T, \partial \oplus t) = S_i(D, d).$$

holds. □

Let us estimate the ith Morse numbers of chain complexes over the ring $\mathbb{Z}[G]$. Consider the chain complex

$$(C, \partial): C_0 \xleftarrow{\partial_1} C_1 \longleftarrow \cdots \xleftarrow{\partial_n} C_n.$$

Denote by (C^G, ∂^G) the chain complex of free abelian groups:

$$(C^G, \partial^G): \mathbb{Z} \otimes_{\mathbb{Z}[G]} C_0 \xleftarrow{\mathrm{id} \otimes \partial_1} \mathbb{Z} \otimes_{\mathbb{Z}[G]} C_1 \longleftarrow \cdots \xleftarrow{\mathrm{id} \otimes \partial_n} \mathbb{Z} \otimes_{\mathbb{Z}[G]} C_n.$$

Its homology groups will be denoted by $H_i(C^G)$.

Theorem 4.3. *Let (C, ∂) be a free chain complex over the ring $\mathbb{Z}[G]$. Its ith Morse numbers satisfy the following inequalities:*

$$\mathscr{M}_0(C, \partial) \geq \mu(H_0(C))$$

(if $n > 1$, then $\mathscr{M}_0(C, \partial) = \mu(H_0(C))$);

$$\mathscr{M}_i(C, \partial) \geq S_i(C, \partial) + S_{i-1}(C, \partial) + \mu(H_i(C^G)) + \mu(\operatorname{Tors} H_{i-1}(C^G)).$$

Proof. The first inequality is evident. If the length of the chain complex allows, then the preceding argument makes it possible to achieve, via stabilization and contraction of the boundary homomorphism ∂_1, the equality $\mu_0(C, \partial) = \mu(H_0(C))$.

Let

$$(\tilde{C}, \tilde{\partial}): \tilde{C}_0 \xleftarrow{\tilde{\partial}_1} \tilde{C}_1 \longleftarrow \cdots \xleftarrow{\tilde{\partial}_n} \tilde{C}_n$$

be an arbitrary free chain complex that is simple homotopy equivalent to the chain complex (C, ∂). Consider a section of this chain complex

$$\tilde{C}_i \xleftarrow{\tilde{\partial}_{i+1}} \tilde{C}_{i+1} \longleftarrow \cdots \xleftarrow{\tilde{\partial}_n} \tilde{C}_n.$$

Clearly, the module $\Gamma_i = \tilde{C}_i/\tilde{\partial}_{i+1}(\tilde{C}_{i+1})$ may be interpreted as the first homology module of this chain complex, and the group $\mathbb{Z} \otimes_{\mathbb{Z}[G]} \Gamma_i$ as the first homology group of the chain complex

$$\mathbb{Z} \otimes_{\mathbb{Z}[G]} \tilde{C}_i \xleftarrow{\mathrm{id} \otimes \tilde{\partial}_{i+1}} \mathbb{Z} \otimes_{\mathbb{Z}[G]} \tilde{C}_{i+1} \longleftarrow \cdots \xleftarrow{\mathrm{id} \otimes \tilde{\partial}_n} \mathbb{Z} \otimes_{\mathbb{Z}[G]} \tilde{C}_n.$$

The equality

$$S_i(\tilde{C}, \tilde{\partial}) = d(\Gamma_i) - \mu(\mathbb{Z} \otimes_{\mathbb{Z}[G]} \Gamma_i)$$

implies that

$$\mu(\tilde{C}_i) \geq S_i(C, \partial) + \mu(\mathbb{Z} \otimes_{\mathbb{Z}[G]} \Gamma_i),$$

because $\mu(\tilde{C}_i) \geq d(\Gamma_i)$. Therefore, in the chains of dimension i of our chain complex, there is a subgroup A_i that can be extracted as a directed summand and whose rank is not less than $S_i(\tilde{C}, \partial)$. But since this group does not contribute to the homology groups of the chain complex

$$(\tilde{C}^G, \tilde{\partial}^G): \mathbb{Z} \otimes_{\mathbb{Z}[G]} \tilde{C}_0 \xleftarrow{\mathrm{id} \otimes \tilde{\partial}_1} \mathbb{Z}_0 \otimes_{\mathbb{Z}[G]} \tilde{C}_1 \longleftarrow \cdots \xleftarrow{\mathrm{id} \otimes \tilde{\partial}_n} \mathbb{Z}_0 \otimes_{\mathbb{Z}[G]} \tilde{C}_n,$$

there exists a subgroup A_{i+1} in the group $\mathbb{Z} \otimes_{\mathbb{Z}[G]} \tilde{C}_{i+1}$ that can be mapped on the group A_i isomorphically. Clearly, the ith Morse number of the chain complex $(\tilde{C}^G, \tilde{\partial}^G)$ is equal to

$$\mathscr{M}_i(\tilde{G}^G, \tilde{\partial}^G) = \mu(H_i(\tilde{C}^G)) + \mu(\operatorname{Tors} H_{i-1}(\tilde{C}^G)).$$

Thus the ith Morse number of (C, ∂) is not less than

$$S_i(C, \partial) + S_{i-1}(C, \partial) + \mu(H_i(C^G)) + \mu(\operatorname{Tors} H_{i-1}(C^G)). \qquad \square$$

COROLLARY 4.2. *If G is an s-group and $n > 3$ (where n is the length of the chain complex), then the ith Morse number of the chain complex (C, ∂) is equal to*

$$\mathcal{M}_i(C, \partial) = S_{i-1}(C, \partial) + S_i(C, \partial) + \mu(H_i(C^G)) + \mu(\operatorname{Tors} H_{i-1}(C^G)).$$

PROOF. By virtue of Corollary 4.1 the chain complex (C, ∂) is simple homotopy equivalent to a chain complex of the form

$$\begin{array}{ccccccccccccc} \overline{C}_0 & \xleftarrow{\overline{\partial}_1} & \overline{C}_1 & \longleftarrow & \cdots & \xleftarrow{\overline{\partial}_i} & \overline{C}_i & \xleftarrow{\overline{\partial}_{i+1}} & \overline{C}_{i+1} & \longleftarrow & \cdots & \xleftarrow{\overline{\partial}_n} & \overline{C}_n \\ & & & & & & \oplus & & & & & & \\ & & & & 0 & \longleftarrow & R_i & \xleftarrow{\delta} & R_{i+1} & \longleftarrow & 0, & & \end{array}$$

where

$$\overline{C}_0 \xleftarrow{\overline{\partial}_1} \overline{C}_1 \longleftarrow \cdots \xleftarrow{\overline{\partial}_n} \overline{C}_n$$

is the minimal chain complex. Because of the restrictions on the length of the chain complex (C, ∂) we can shift the subcomplex

$$0 \longleftarrow R_i \xrightarrow{\delta} R_{i+1} \longleftarrow 0$$

in such a way that it no longer appears in dimensions $i-1$ and i. As noted above, for modules over s-rings, the number $S(M)$ can be defined by the formula $S(M) = \mu(M) - \mu(\mathbb{Z} \otimes_{\mathbb{Z}[G]} M)$. Making use of the same argument as in the proof of Theorem 4.3, we can replace inequality by equality, which completes the proof of the corollary. □

Observe that if $\mathbb{Z}[G]$ is not an s-ring, then there exists a minimal chain complex (C, ∂) whose second Morse number is greater than

$$S_2(C, \partial) + S_1(C, \partial) + \mu(H_2(C^G)) + \mu(\operatorname{Tors}(H_1(C^G))).$$

Indeed, let M be a stably free module constructed by Swan over the integral group ring of the generalized quaternion group $\mathbb{Z}[G]$, which satisfies the conditions

$$\mu(M) = 2, \qquad \mathbb{Z} \otimes_{\mathbb{Z}[G]} M \approx \mathbb{Z}, \qquad M \oplus \mathbb{Z}[G] \approx \mathbb{Z}[G] \oplus \mathbb{Z}[G],$$

where $M \not\approx \mathbb{Z}[G]$ [150]. Let

$$(C, \partial): C_1 \xleftarrow{\partial_2} C_2$$

be a chain complex in which $C_1 = \mathbb{Z}[G] \oplus \mathbb{Z}[G] = M_1 \oplus \mathbb{Z}[G]$ and $C_2 = \mathbb{Z}[G] \oplus \mathbb{Z}[G] = M_2 \oplus \mathbb{Z}[G]$, where $M_1 \approx M_2 \approx M$. The boundary homomorphism $\partial_2 \colon C_2 \to C_1$ is of the form $\partial_2(\mathbb{Z}[G] \oplus 0) = \mathbb{Z}[G] \oplus 0$, $\partial_2(M_2 \oplus 0) = 0$. Clearly, $S_1(C, \partial) = 0$, $S_2(C, \partial) = 0$, and $H_1(C^G) \approx H_2(C^G) \approx \mathbb{Z}$. By construction, (C, ∂) is a minimal chain complex.

Now consider the dual situation and let us discuss cochain complexes.

Let $C^i = \mathrm{Hom}_{\mathbb{Z}[G]}(C_i, \mathbb{Z}[G])$ be a right free $\mathbb{Z}[G]$-module. Making use of the involution in the group ring $\mathbb{Z}[G]$ $(\bar{-}: g \to g^{-1})$ we can turn the right $\mathbb{Z}[G]$-module C_i into a left one $(c\lambda = \bar{\lambda}c)$. Write the cochain complex

$$(C^*, \partial^*): C^0 \xrightarrow{\partial^0} C^1 \xrightarrow{\partial^1} \cdots \xrightarrow{\partial^{n-1}} C^n.$$

The following relation between the Hom functor and the tensor product are known [14]:

$$\mathrm{Hom}_{\mathbb{Z}[G]}(C_i, \mathbb{Z}) \approx \mathrm{Hom}_{\mathbb{Z}}(\mathbb{Z} \otimes_{\mathbb{Z}[G]} C_i, \mathbb{Z}),$$
$$\mathrm{Hom}_{\mathbb{Z}[G]}(C_i, \mathbb{Z} \otimes_{\mathbb{Z}[G]} \mathbb{Z}[G]) \approx \mathrm{Hom}_{\mathbb{Z}[G]}(C_i, \mathbb{Z}),$$

which must be used in order to define the numbers $S^i(C^*, \partial^*)$:

$$S^i(C^*, \partial^*) = d(\Gamma^i) - \mu(\mathbb{Z} \otimes_{\mathbb{Z}[G]} \Gamma^i),$$

where $\Gamma^i = C^i/\partial^{i-1}(C^{i-1})$.

COROLLARY 4.3. *The ith Morse number of the cochain complex (C^*, ∂^*) satisfies the inequality*

$$\mathscr{M}_i(C^*, \partial^*) \geq S^i(C^*, \partial^*) + S^{i+1}(C^*, \partial^*) + \mu(H_i(C^{*G})) + \mu(\mathrm{Tors}\, H^{i+1}(C^{*G})),$$

where $H^i(C^{*G})$ are the cohomology groups of the cochain complex

$$(C^{*G}, \partial^{*G}): \mathbb{Z} \otimes_{\mathbb{Z}[G]} C^0 \xrightarrow{\mathrm{id} \otimes \partial^0} \mathbb{Z} \otimes_{\mathbb{Z}[G]} C^1$$
$$\to \cdots \xrightarrow{\mathrm{id} \otimes \partial^{n-1}} \mathbb{Z} \otimes_{\mathbb{Z}[G]} C^n. \quad \square$$

Passing to the cochain complex (C^*, ∂^*) yields nothing essentially new, but the segments

$$\begin{array}{ccccccc} \mathbb{Z} \otimes_{\mathbb{Z}[G]} C^0 & \xrightarrow{\mathrm{id} \otimes \partial^0} & \mathbb{Z} \otimes_{\mathbb{Z}[G]} C^1 & \longrightarrow \cdots & \xrightarrow{\mathrm{id} \otimes \partial^{i-1}} & \mathbb{Z} \otimes_{\mathbb{Z}[G]} C^i, \\ C^0 & \xrightarrow{\partial^0} & C^1 & \xrightarrow{\partial^1} \cdots & \xrightarrow{\partial^{i-1}} & C^i \end{array}$$

in the cochain complexes $(C^{G_*}, \partial^{G_*})$ and (C^*, ∂^*) have a quite definite meaning in the geometrical context. They represent the cochain complexes of the ith skeleton of the universal covering space and its base for some triangulation of the given manifold.

The following proposition is evident.

PROPOSITION 4.4. *Let (C, ∂) be a free chain complex over the ring $\mathbb{Z}[G]$. Then its Morse number satisfies the inequality*

$$\mathscr{M}(C, \partial) \geq 2\sum_{i=0}^{n} S_i(C, \partial) + \sum_{i=0}^{n} \mu(H_i(C^G)) + \sum_{i=0}^{n} \mu(\mathrm{Tors}\, H_i(C^G)).$$

The equality holds if G is an s-group and $\mathrm{Wh}(G) = 0$.

PROOF. The proposition follows directly from Theorem 4.3 and Corollary 4.3. $\square$

CHAPTER V

Morse Numbers and Minimal Morse Functions

It is known that there is the following estimate for the number of critical points of index i for a Morse function f on a smooth manifold M^n:

$$N_i(f) \geq p_i + q_i + q_{i-1},$$

where p_i is the rank and q_i the torsion coefficients of the group $H_i(M^n, \mathbb{Z})$ [113]. Smale proved that if $\pi_1(M^n, x) = 0$ and $n \geq 6$, then there exists a Morse function g on the manifold M^n for which $N_i(g) = p_i + q_i + q_{i-1}$ for all i. A similar result for 5-dimensional simply connected manifolds was proved by Barden [7]. The non-simply-connected case turned out to be much more complicated [11, 127, 129, 134, 135, 137, 57]. There are examples of manifolds such that arbitrary Morse functions on them have the property $N_i(f) > p_i + q_i + q_{i-1}$ (homology spheres). In the works of S. P. Novikov [106, 107] and S. P. Novikov and M. A. Shubin [109] the number of critical points of a Morse function on a non-simply-connected manifold is estimated via homologies of the local system constructed from the representation of the fundamental group of the manifold in $GL(n, \mathbb{C})$ or the von Neumann Π_1-factor.

In this chapter we study Morse functions on non-simply-connected manifolds and introduce new numerical invariants of a manifold, with the aid of which we estimate the number of critical points of a Morse function.

Let $p: \hat{M}^n \to M^n$ be the universal covering manifold for M^n, and M_i the ith skeleton of a sufficiently fine triangulation of M^n, $\hat{M}_i = p^{-1}(M_i)$. Set

$$S_i(M^n) = \mu(H^i(\hat{M}_i, \mathbb{Z})) - \mu(H^i(M_i, \mathbb{Z})),$$

where $H^i(\hat{M}_i, \mathbb{Z})$ is considered as a $\mathbb{Z}[\pi_1(M^n)]$-module; $\mu(H)$ is the minimal number of generators of the module (group) H. Suppose that $f: M^n \to [0, 1]$ is an arbitrary Morse function. The number of critical points of index i satisfies the following inequality:

$$N_i(f) \geq S_i(M^n) + S_{i+1}(M^n) + p_i + q_i + q_{i-1}.$$

Suppose that $(W^n, V_0^{n-1}, V_1^{n-1})$ is a cobordism for which

$$\pi_1(V_0^{n-1}) \hookrightarrow \pi_1(W^n) \hookleftarrow \pi_1(V_1^{n-1})$$

are isomorphisms, $\pi_1(W^n)$ is an s-group (i.e., every stably free $\mathbb{Z}[\pi_1(W^n)]$-module is free). If $\mathrm{Wh}(\pi_1(M^n)) = 0$ $(n \geq 6)$, then there exists a minimal Morse function on W^n (i.e., a function with the minimal possible number of critical points of each index) with the number of critical points of index i equal to

$$N_i(f) = S_i(M^n) + S_{i+1}(M^n) + p_i + q_i + q_{i-1}.$$

§1. Numerical invariants

In this section we construct new numerical invariants of non-simply-connected manifolds using the numerical invariants of chain complexes introduced earlier. For that it is necessary to construct an algebraic chain complex corresponding to the geometrical situation. This can be done in a number of ways. Here we consider a construction based on the theory of the Whitehead C^1-triangulation, and a construction involving Morse functions.

Let $t: |K| \to M^n$ be a C^1-triangulation in the sense of Whitehead (see [95, 72]). Here $|K|$ denotes the topological space of the simplicial complex K. Consider the universal covering space $\hat{M}^n$ of the manifold M^n and let $\hat{t}: |\hat{K}| \to |\hat{M}^n|$ be the $\hat{t}$-triangulation induced by the triangulation t. Clearly, there is the commutative diagram

$$\begin{array}{ccc} |\hat{K}| & \xrightarrow{\hat{t}} & \hat{M}^n \\ {\scriptstyle \bar{p}}\downarrow & & \downarrow{\scriptstyle p} \\ |K| & \xrightarrow{t} & M^n. \end{array}$$

We shall identify the fundamental group $\pi_1(M^n, x)$ with the group of automorphisms of the covering. Thus, each element $\sigma \in \pi_1(M^n, x)$ defines a simplicial mapping $\sigma: \hat{K} \to \hat{K}$, and therefore a chain mapping $\sigma_*: C_i(K, \hat{\mathbb{Z}}) \to C_i(\hat{K}, \mathbb{Z})$. This action turns each chain group $C_i(\hat{K}, \mathbb{Z})$ into a module over the group ring $\mathbb{Z}[\pi_1(M^n, x)]$. It is evident that the resulting chain module $C_i(\hat{K}, \mathbb{Z})$ is free with the i-simplex in K as a generator. Since the complex K is finite, the module $C_i(K, \mathbb{Z})$ is finitely generated over $\mathbb{Z}[\pi_1(M^n, x)]$. As a result we obtain a free chain complex

$$C(\hat{K}): C_0(\hat{K}, \mathbb{Z}) \xleftarrow{\partial_1} C_1(\hat{K}, \mathbb{Z}) \longleftarrow \cdots \xleftarrow{\partial_n} C_n(\hat{K}, \mathbb{Z}).$$

There is an arbitrariness in the procedure for constructing this chain complex which we will discuss in what follows.

Now consider the construction that makes use of Morse functions.

Let $(W^n, V_0^{n-1}, V_1^{n-1})$ be a compact smooth manifold with boundary $\partial W^n = V_0^{n-1} \cup V_1^{n-1}$ (V_0^{n-1}, V_1^{n-1} may be empty). Let $\pi = \pi_1(W^n, x)$ be the fundamental group of the manifold W^n. Denote by $p: (\hat{W}^n, \hat{V}_0^{n-1}, \hat{V}_1^{n-1}) \to (W^n, V_0^{n-1}, V_1^{n-1})$ the universal covering space. Here $\hat{V}_i^{n-1} = p^{-1}(V_i^{n-1})$ $(i = 0, 1)$. On W^n, choose an ordered Morse function $f: W^n \to [0, 1]$,

$f^{-1}(0) = V_0$, $f^{-1}(1) = V_1$ and a gradient-like vector field ξ. Using the mapping p, lift f and ξ to $\hat{W}^n$, and denote the lifted function and vector field by $\hat{f}$ and $\hat{\xi}$, respectively. Using f, ξ $(\hat{f}, \hat{\xi})$, construct chain complexes of abelian groups:

$$C(W^n, f, \xi): C_0 \xleftarrow{\partial_1} C_1 \longleftarrow \cdots \xleftarrow{\partial_n} C_n,$$

$$C(\hat{W}^n, \hat{f}, \hat{\xi}): \hat{C}_0 \xleftarrow{\hat{\partial}_1} \hat{C}_1 \longleftarrow \cdots \xleftarrow{\hat{\partial}_n} \hat{C}_n,$$

where $C_i = H_i(W_i, W_{i-1}, \mathbb{Z})$, $\hat{C}_i = H_i(\hat{W}_i, \hat{W}_{i-1}, \mathbb{Z})$, and $W_i = f^{-1}[0, a_i]$, $\hat{W}_i = \hat{f}^{-1}[0, a_i]$ are submanifolds containing all the critical points of indices less than or equal to i. For the generators of the chain groups C_i $(\hat{C}_i)$ one can take middle disks of critical points of index i constructed by the field ξ $(\hat{\xi})$. The fundamental group $\pi = \pi_1(W^n, x)$ acts on the manifold $\hat{W}^n$. Making use of this action, we can turn the chain groups $\hat{C}_i$ into finitely generated modules over the ring $\mathbb{Z}[\pi]$. This construction gives a distinguished basis in the module $\hat{C}_i$. For each critical point $w \in W^n$ of the function f fix a middle disk constructed by the field ξ and choose a middle disk of the critical point $\hat{w} \in p^{-1}(w)$ covering it. The boundary homomorphism $\hat{\partial}_i: \hat{C}_i \to \hat{C}_{i-1}$ can be described by the matrix of incidence coefficients $a_{ij} \in \mathbb{Z}[\pi]$. Since we shall be carrying out elementary operations over the chain complex $C(\hat{W}^n, \hat{f}, \hat{\xi})$, we must single out critical points of the function, i.e., take fixed paths joining the critical points to the base point x. Without loss of generality we can assume that the paths to critical points of index i lie in the submanifold W_i. Choose distinguished points on the middle and comiddle spheres, and join them by paths on middle and comiddle disks to the corresponding critical points. We can now alter the Morse function f and the field ξ so as to impose on the matrix of the boundary homomorphism $\hat{\partial}_i$ elementary operations over the ring $\mathbb{Z}[\pi]$. See [123] for details.

The construction of the chain complexes $C(\hat{K})$ and $C(\hat{W}^n, \hat{f}, \hat{\xi})$ involves a certain arbitrariness, depending on the choice of the base point $x \in W^n$, the identification isomorphism $\varphi: \pi \to \pi_1(W^n, x)$, and the bases in the chains C_i. Let us examine this in detail. Suppose $y \in W^n$ is another base point, $\hat{W}_1^n$ the corresponding universal covering space (which is defined as the set of equivalence classes of paths joining a fixed point to the points of W^n), $\hat{p}_1: \hat{W}_1^n \to W^n$ the projection, $\hat{f}_1$ and $\hat{\xi}_1$ the liftings of f and ξ to $\hat{W}_1^n$, and $C(\hat{W}_1^n, \hat{f}_1, \hat{\xi}_1)$ the chain complex of $\mathbb{Z}[\pi_1(W^n, y)]$-modules. Similarly, let $C(\hat{K}_1)$ be the chain complex of $\mathbb{Z}[\pi_1(W^n, y)]$-modules constructed with the use of the triangulation $t: |K| \to W^n$. Choose a path $\gamma: [0, 1] \to W^n$

joining the points. It induces a diffeomorphism $\hat{\gamma}$ in the commutative diagram

$$\begin{array}{ccc} \hat{W}_1^n & \xleftarrow{\hat{\gamma}} & \hat{W}^n \\ {\scriptstyle p_1}\downarrow & & \downarrow{\scriptstyle p} \\ W^n & \xleftarrow{\text{id}} & W^n \end{array}$$

and an isomorphism $\gamma_*: \pi_1(W^n, y) \to \pi_1(W^n, x)$. Clearly, $\hat{\gamma}$ induces a chain isomorphism $\hat{\gamma}_*: C(\hat{W}_1^n, \hat{f}_1, \hat{\xi}_1) \to C(\hat{W}^n, \hat{f}, \hat{\xi})$ associated with the isomorphism

$$\overline{\gamma}: \mathbb{Z}[\pi_1(W^n, y)] \to \mathbb{Z}[\pi_1(W^n, x)]$$

and induced by γ_*, i.e., $\hat{\gamma}(\alpha \cdot c) = \overline{\gamma}(\alpha)\hat{\gamma}_*(c)$, $\alpha \in \pi_1(W^n, y)$, and $c \in C(\hat{W}_1^n, \hat{f}_1, \hat{\xi}_1)$. Choosing an isomorphism $\varphi_1: \pi \to \pi_1(W^n, y)$, we can regard $C(\hat{W}_1^n, \hat{f}_1, \hat{\xi}_1)$ $(C(\hat{K}_1))$ as a chain complex of $\mathbb{Z}[\pi]$-modules. There exists an isomorphism $\theta: \pi \to \pi$ that makes the diagram

$$\begin{array}{ccc} \pi_1(W^n, y) & \xrightarrow{\gamma_*} & \pi_1(W^n, x) \\ \uparrow{\scriptstyle \varphi_1} & & \uparrow{\scriptstyle \varphi} \\ \pi & \xrightarrow{\theta} & \pi \end{array}$$

commutative. Thus, the chain complexes $C(\hat{W}^n, \hat{f}, \hat{\xi})$ and $C(\hat{K})$ are θ-chain isomorphic to $C(\hat{W}_1^n, \hat{f}_1, \hat{\xi}_1)$ and $C(\hat{K}_1)$, respectively. If we choose another path γ_1 joining the points x and y, then the isomorphism γ_* is replaced by $\omega \cdot \gamma_*$, where ω is the isomorphism of conjugation in the group $\pi_1(W^n, x)$ via the element $\gamma \cdot \gamma_1^{-1}$. The semilinear isomorphism is also modified. Thus, the arbitrariness in the choice of the base point determines the chain complexes $C(\hat{W}^n, \hat{f}, \hat{\xi})$ and $C(\hat{K})$ up to a semilinear isomorphism. It is easily shown that the choice of the identification isomorphism $\varphi: \pi \to \pi_2(W^n, x)$ likewise determines our chain complexes up to a semilinear isomorphism. Finally, the bases in the modules $\hat{C}_i$ are determined up to multiplication by elements of the fundamental group, which allows us to choose them so that the torsion of the corresponding semilinear isomorphisms vanishes. Consequently, the simple homotopy type of the chain complex constructed from a Morse function f and a gradient-like vector field ξ or triangulation is determined uniquely. It is proved in [72] that the chain complexes constructed from Morse functions on the manifold $\hat{W}^n$ via a triangulation (or cellular decomposition) of $\hat{W}^n$ have the same simple homotopy type. We can therefore speak of the simple homotopy type of a chain complex of the manifold $\hat{W}^n$ which will be denoted by $C(\hat{W}^n)$.

Let

$$C(\hat{W}^n): \hat{C}_0 \xleftarrow{\hat{\partial}_1} \hat{C}_1 \longleftarrow \cdots \xleftarrow{\hat{\partial}_n} \hat{C}_n$$

be the chain complex of the manifold $\hat{W}^n$. Denote by $w: \pi_1(W^n, x) \to \{-1, 1\}$ the homomorphism of orientation (the first Stiefel-Whitney class). Define an involution on the group ring $\mathbb{Z}[\pi_1(W^n, x)]$ by the formula $g \to w(g)g^{-1}$. This involution makes it possible to turn any right $\mathbb{Z}[\pi_1(W^n, x)]$-module into a left $\mathbb{Z}[\pi_1(W^n, x)]$-module. Making use of this involution, we turn the right $\mathbb{Z}[\pi]$-module $\hat{C}^i = \operatorname{Hom}_{\mathbb{Z}[\pi]}(\hat{C}_i, \mathbb{Z}[\pi])$ into a left one and consider the cochain complex

$$C^*(\hat{W}^n): \hat{C}^0 \xrightarrow{\hat{\partial}^0} C^1 \xrightarrow{\hat{\partial}^1} \cdots \xrightarrow{\hat{\partial}^{n-1}} C^n$$

of free modules. Taking the tensor product of $C^*(\hat{W}^n)$ and the trivial $\mathbb{Z}[\pi]$-module $\mathbb{Z}$, we obtain the cochain complex of groups

$$C^*(W^n): \mathbb{Z} \otimes_{\mathbb{Z}[\pi]} \hat{C}^0 \xrightarrow{\mathrm{id} \otimes \partial^0} \mathbb{Z} \otimes_{\mathbb{Z}[\pi]} \hat{C}^1 \longrightarrow \cdots \xrightarrow{\mathrm{id} \otimes \partial^{n-1}} \mathbb{Z} \otimes_{\mathbb{Z}[\pi]} \hat{C}^n.$$

If the cochain complex $C^*(\hat{W}^n)$ is constructed from a triangulation of the manifold $\hat{W}^n$, then the segments of the cochain complexes

$$\hat{C}^0 \xrightarrow{\partial^0} \hat{C}^1 \longrightarrow \cdots \xrightarrow{\partial^{i-1}} \hat{C}^i,$$

$$\mathbb{Z} \otimes_{\mathbb{Z}[\pi]} \hat{C}^0 \xrightarrow{\mathrm{id} \otimes \hat{\partial}^0} \mathbb{Z} \otimes_{\mathbb{Z}[\pi]} \hat{C}^1 \longrightarrow \cdots \xrightarrow{\mathrm{id} \otimes \hat{\partial}^{i-1}} \mathbb{Z} \otimes_{\mathbb{Z}[\pi]} \hat{C}^i$$

are evidently the cochain complexes of the ith skeleton of the triangulations of $\hat{W}^n$ and W^n, respectively. Therefore the $\mathbb{Z}[\pi]$-module

$$\hat{\Gamma}^i(\hat{W}^n) = \hat{C}^i / \hat{\partial}^{i-1}(\hat{C}^{i-1})$$

(the abelian group $\Gamma^i(W^n) = \mathbb{Z} \otimes_{\mathbb{Z}[\pi]} \hat{C}^i / \mathrm{id} \otimes \hat{\partial}^{i-1}(\mathbb{Z} \otimes_{\mathbb{Z}[\pi]} \hat{C}^{i-1})$) can be interpreted as the ith cohomology module (ith cohomology group) of the ith skeleton of $\hat{W}^n$ (ith skeleton of W^n).

DEFINITION 5.1. For a manifold W^n, set

$$S_i(W^n) = d(\hat{\Gamma}^i(W^n)) - \mu(\Gamma^i(W^n)).$$

Recall that $\mu(\Gamma^i(W^n))$ is the minimal number of generators of the group $\Gamma^i(W^n)$, and that $d(\hat{\Gamma}^i(\hat{W}^n))$ is defined in Chapter III, §1. It follows directly from Lemma 4.7 that $S_i(W^n)$ is an invariant of the homotopy type of the manifold W^n. Observe that for a sufficiently small triangulation of the manifold W^n, we can, by virtue of Lemma 3.1 and Corollary 3.1, define $S_i(W^n)$ by the formula

$$S_i(W^n) = \mu(H^i_{\mathbb{Z}[\pi]}(\hat{W}_i, \mathbb{Z})) - \mu(H^i(W_i), \mathbb{Z}),$$

where $\hat{W}_i$ and W_i are the ith skeletons of the triangulation of $\hat{W}^n$ and W^n, respectively. This formula is apparently valid for an arbitrary triangulation of the manifold W^n.

We also note that this construction holds not only for a triangulation of a manifold, but also for its cellular decomposition.

§2. Morse numbers

DEFINITION 5.2. The ith Morse number of a manifold W^n is the minimal number of critical points of index i taken over all Morse functions on W^n.

In the works of Bogoyavlenskiĭ and Hajduk [11, 57] the question of homotopy invariance of ith Morse numbers is discussed. The following proposition holds.

PROPOSITION 5.1. *If M^n and N^n are two closed manifolds of the same homotopy type, $n > 6$, then they have the same ith Morse numbers. In dimension 6, the ith Morse numbers are invariants of simple homotopy type.*

As of this writing, it is not known whether the ith Morse numbers are topological invariants.

DEFINITION 5.3. The Morse number of a manifold M^n is the minimal number of critical points of all indices taken over all Morse functions on M^n.

As proved by Hajduk [57], the Morse numbers for closed manifolds of dimension greater than five are invariants of simple homotopy type.

THEOREM 5.1. *Let W^n be a smooth compact manifold with boundary $\partial W^n = V_0^{n-1} \cup V_1^{n-1}$. The following inequality holds for the ith Morse numbers:*

$$\mathscr{M}_i W^n \geq S_i(W^n) + S_{i+1}(W^n) + \mu(H_i(W^n, V_0^{n-1}, \mathbb{Z})) + \mu(\operatorname{Tors} H_{i-1}(W^n, V_0^{n-1}, \mathbb{Z})).$$

PROOF. Consider an arbitrary ordered Morse function f and gradient-like vector field ξ on W^n. Construct the chain complex of $\mathbb{Z}[\pi]$-modules corresponding to them. The inequality follows from Theorem 4.3. □

THEOREM 5.2. *Let W^n be a smooth manifold with boundary $\partial W^n = V_0^{n-1} \cup V_1^{n-1}$ $(n \geq 8)$, and $\pi = \pi_1(W^n, x)$ an s-group. Then for $4 \leq i \leq n-4$ the following equality holds:*

$$\mathscr{M}_i(W^n) = S_i(W^n) + S_{i+1}(W^n) + \mu(H_i(W^n, V_0^{n-1}\mathbb{Z})) + \mu(\operatorname{Tors}(H_{i-1}(W^n, V_0^{n-1}, \mathbb{Z}))).$$

PROOF. The chain complex of the manifold $C(\hat{W}^n)$ contains in its simple homotopy type a chain complex of the form

$$\begin{array}{ccccccc} & \overline{C}_0 & \xleftarrow{\bar{\partial}_1} & \overline{C}_1 & \longleftarrow \cdots \xleftarrow{\bar{\partial}_n} & \overline{C}_n \\ \tilde{C}(\hat{W}^n): & \oplus & & \oplus & & \oplus \\ & D_0 & \xleftarrow{d_1} & D_1 & \longleftarrow \cdots \xleftarrow{d_n} & D_n, \end{array}$$

where

$$(\overline{C}, \overline{\partial}): \overline{C}_0 \xleftarrow{\overline{\partial}_1} \overline{C}_1 \longleftarrow \cdots \longleftarrow \xleftarrow{\overline{\partial}_n} \overline{C}_n$$

is the minimal chain complex and

$$D_0 \xleftarrow{d_1} D_1 \longleftarrow \cdots \xleftarrow{d_n} D_n$$

is an acyclic chain complex. This follows from the results of Chapter IV, §6. Note that the chain complex $\check{C}(\hat{W}^n)$ is obtained from the chain complex $C(\hat{W}^n)$ by the operations of choosing a new basis and stabilization. No operation of contraction is used. For indices satisfying the inequality $4 \leq i \leq n-4$ all elementary operations on the chain complex $C(\hat{W}^n)$, including the contraction, can be realized geometrically by altering the Morse function and the gradient-like vector field. It is then evident that $C(\hat{W}^n)$ can be transformed to the form:

$$\begin{array}{ccccccccccc}
\overline{C}_0 \xleftarrow{\overline{\partial}_1} & \overline{C}_1 \leftarrow \cdots \xleftarrow{\overline{\partial}_{i-1}} & \overline{C}_{i-1} & \xleftarrow{\overline{\partial}_i} & \overline{C}_i & \xleftarrow{\overline{\partial}_{i+1}} & \overline{C}_{i+1} \leftarrow \cdots \xleftarrow{\overline{\partial}_n} \overline{C}_n \\
\oplus & \oplus & \oplus & & & & \oplus \\
\overline{D}_0 \xleftarrow{\overline{d}_1} & \overline{D}_1 \leftarrow \cdots \xleftarrow{\overline{d}_{i-1}} & \overline{D}_{i-1} & \longleftarrow & 0 & \longleftarrow & \overline{D}_{i+1} \leftarrow \cdots \xleftarrow{\overline{d}_n} \overline{D}_n
\end{array}$$

Making use of Corollary 4.1, we obtain the proof of the theorem. □

Let us now make several comments. If $V_0^{n-1} \neq \varnothing$, then $\mathscr{M}_0(W^n) = 0$; if $V_1^{n-1} \neq \varnothing$, then $\mathscr{M}_n(W^n) = 1$. There are sharper estimates for the Morse numbers $\mathscr{M}_i(W^n)$ ($i = 1, 2, 3, n-1, n-2, n-3$) of a closed manifold W^n. They are connected with the use of the minimal representation of the fundamental group $\pi_1(W^n)$ and the first k-invariant of the manifold W^n, $k \in H^3(\pi_1(W^n), \pi_2(W^n))$ to be considered in Chapter VII.

PROPOSITION 5.2. *The Morse number of the manifold* $(W^n, V_0^{n-1}, V_1^{n-1})$ *satisfies the inequality*

$$\mathscr{M}(W^n) \geq 2\sum_{i=0}^{n} S_i(W^n) + \sum_{i=0}^{n} \mu(H_i(W^n, V_0^{n-1}, \mathbb{Z})) + \sum_{i=0}^{n} \mu(\operatorname{Tors} H_{i-1}(W^n, V_0^{n-1}, \mathbb{Z})).$$

PROOF. The statement is a consequence of Proposition 4.1.

§3. Minimal Morse functions on cobordisms

Recall that $\mathfrak{F}_i(W^n)$ denotes the set of Morse functions on a manifold M^n with the minimal number of critical points of index i. A Morse function $f \in \bigcap_{i=0}^n \mathfrak{F}_i(W^n)$ is called a minimal Morse function on the manifold W^n.

There are manifolds on which there are no minimal Morse functions. Consider, by way of example, an h-cobordism, i.e., a smooth manifold W^n with boundary $\partial W^n = V_0^{n-1} \cup V_1^{n-1}$ such that V_i^{n-1} is a deformation retract of W^n $(i = 0, 1)$. According to the Stallings theorem, if $n \geq 6$ and $\pi = \pi_1(W^n)$ is nontrivial, then the h-cobordism is uniquely determined by the submanifold V_0^{n-1} and the torsion invariant $\tau_0 = \tau(W^n, V_0^{n-1})$ (see Chapter VII). It is known that if $\tau_0 \neq 0$, then there exists a Morse function on W^n having critical points of indices i and $i+1$, $2 \leq i \leq n-3$, i.e., the indices of critical points are not defined uniquely and can be altered.

Denote by R the class of cobordisms $(W^n, V_0^{n-1}, V_1^{n-1})$ such that $\partial W^n = V_0^{n-1} \cup V_1^{n-1}$, and $\pi_1(V_0^{n-1}) \hookrightarrow \pi_1(W^n) \hookleftarrow \pi_1(V^{n-1})$ are isomorphisms $(n \geq 6)$. In this section we shall discuss the question of when does a cobordism $(W^n, V_0^{n-1}, V_1^{n-1})$ of class R admit a minimal Morse function.

It is known that if the inclusions $V_0^{n-1} \hookrightarrow W^n \hookleftarrow V_1^{n-1}$ induce the isomorphisms $\pi_1(V_0^{n-1}) \hookrightarrow \pi_1(W^n) \hookleftarrow \pi_1(V_1^{n-1}) = \pi$, then there exists on the manifold W^n a Morse function without critical points of indices $0, 1, n-1, n$ [96]. We can therefore restrict ourselves to chain complexes of length $n-4$. All the elementary operations over the chain complexes over $\mathbb{Z}[\pi]$ can be realized geometrically by alterations on the Morse function and the gradient-like vector field. We stress one feature of this process, connected with application of the Whitney lemma in codimension 2; it occurs when we are eliminating critical points of indices $n-2$ and $n-3$.

Let $V = f^{-1}(c)$, where $f(x_{n-2}^i) > c > f(y_{n-3}^j)$, y_{n-3}^j is a critical point of index $n-3$ and x_{n-2}^i is a critical point of index $n-2$. Obviously, $\pi_1(V) \approx \pi_1(V_1^{n-2})$, since the middle spheres of critical points of index $n-3$ have codimension 3 in the submanifold V and so induce no change in the fundamental group of the level surface of the function f. But since the middle spheres of critical points of index $n-2$ have codimension 2 in V, we need an additional argument to allow us to eliminate critical points of indices $n-2$ and $n-3$ in the case when they satisfy an algebraic annihilation condition. Suppose that the intersection index for a pair of middle and comiddle spheres is $\varepsilon(S_R^2, S_L^{n-2}) = \pm g$, where $g \in \pi$. We prove that the corresponding critical points of indices $n-2$ and $n-3$ can be eliminated. Here S_R^2 is the middle sphere of a critical point of index $n-2$, and S_L^{n-2} is a comiddle sphere of a critical point of index $n-3$. For this it is necessary to verify all the conditions for applying the Whitney lemma [96]; i.e., to verify that $\pi_1(V \setminus S_L^{n-2}) \hookrightarrow \pi_1(V)$ is a monomorphism. But this is indeed the case, since the trajectories of the gradient-like vector field ξ of the function f determine a diffeomorphism $V_1 \setminus S_R^1 \approx V \setminus S_L^{n-2}$, where S_R^1 is the comiddle sphere of the critical point of index $n-2$. The fact that $\pi_1(V_1^{n-2}) \hookrightarrow \pi_1(W^n)$ is an isomorphism means that S_R^1 realizes the trivial element of $\pi_1(V_1^{n-1})$.

Therefore $\pi_1(V_1^{n-1}) \approx \pi_1(V_1^{n-1} \backslash S_R^1)$; and so $\pi_1(V_1^{n-1} \backslash S_R^1) \approx \pi_1(V \backslash S_L^{n-2}) \approx \pi_1(V)$.

PROPOSITION 5.3. *Let $C(\hat{W}, \hat{F}, \hat{\xi})$ and $C(\hat{W}^n, \hat{g}, \hat{\eta})$ be two based chain complexes constructed from ordered Morse functions f and g and gradient-like vector fields ξ and η on a cobordism $(W^n, V_0^{n-1}, V_1^{n-1})$ of class R. Then they are simple homotopy equivalent. Furthermore, any free based chain complex of $\mathbb{Z}[\pi]$-modules of the form*

$$C_2 \xleftarrow{\partial_3} C_3 \longleftarrow \cdots \xleftarrow{\partial_{n-2}} C_{n-2}$$

simple homotopy equivalent to $C(\hat{W}^n, \hat{f}, \hat{\xi})$ can be realized as a chain complex associated with some ordered Morse function and a gradient-like vector field defined on W^n.

PROOF. The first assertion is proved in [72]. Suppose that the based chain complex $C(\hat{W}^n, \hat{f}, \hat{\xi})$ associated with f and ξ is simple homotopy equivalent to (D, d). We can assume that (D, d) has length $n-4$, i.e., that f has no critical points of indices $0, 1, n-1, n$, since $\pi_1(V_0^{n-1}) \hookrightarrow \pi_1(W^n) \hookleftarrow \pi_1(V_1^{n-1})$ is an isomorphism. There exists a sequence of elementary operations taking $C(\hat{W}^n, \hat{f}, \hat{\xi})$ into the chain complex (D, d). Since each elementary operation can be realized geometrically by suitably altering the function f and the field ξ, we obtain our assertion. The existence of such a sequence of elementary operations is ensured by the Cockroft-Swan theorem applied to the chain complexes $C(\hat{W}^n, \hat{f}, \hat{\xi})$ and (D, d). □

We remark that the restriction on the length of the chain complex in the hypothesis cannot be omitted. It is known that if M^n is a closed manifold with the fundamental group $\pi_1(M^n) \approx \pi$, then any element τ_0 of the Whitehead group $\mathrm{Wh}(\pi)$ (the Whitehead group will be considered in more detail in the following chapter) can be realized by an h-cobordism $(\hat{W}^{n+1}, M_0^n, M_1^n)$ such that $\tau(W^{n+1}, M_0^n) = \tau_0$ $(n > 4)$. The h-cobordism (W^{n+1}, M_0^n, M_1^n) admits a Morse function $f: (W^{n+1}, M_0^n, M_1^n) \to [0, 1]$, $f^{-1}(0) = M_0^n$, $f^{-1}(1) = M_1^n$ with critical points of index 2 and 3 only. At the same time, as shown by Cohen, there exist examples of groups (e.g., the dihedral group D_{2k+1}) and elements $\tau_0 \in \mathrm{Wh}(D_{2k+1})$ such that h-cobordisms with torsion equal to τ_0 do not admit Morse functions with critical points of indices 1 or 2 only [26].

PROPOSITION 5.4. *Let $(W^n, V_0^{n-1}, V_1^{n-1})$ be a cobordism of class R. In order that there exist on W^n a minimal Morse function, it is necessary and sufficient that the chain complex $C(\hat{W}^n)$ of the cobordism be simple homotopy equivalent to a minimal chain complex.*

PROOF. *Necessity.* Obvious.

Sufficiency. Let f be an arbitrary ordered Morse function, ξ a gradient-like vector field, and $C(\hat{W}^n, \hat{f}, \hat{\xi})$ the based chain complex associated with them. Denote by $(\overline{C}, \overline{\partial})$ a minimal chain complex simple homotopy equivalent to $C(\hat{W}^n, \hat{f}, \hat{\xi})$. Proposition 5.3 ensures the existence of a Morse function g and a gradient-like vector field η that realize the complex $(\overline{C}, \overline{\partial})$. Therefore g is a minimal Morse function. □

COROLLARY 5.1. *Let (W^n, V_0, V_1) be a cobordism of class R, $\pi_1(W^n) = \pi$ an s-group, and* $\mathrm{Wh}(\pi) = 0$. *Then on W^n there exists a minimal Morse function $f\colon W^n \to [0, 1]$, $f^{-1}(0) = V_0^{n-1}$, $f^{-1}(1) = V_1^{n-1}$ for which the number of critical points of index i is equal to*

$$\mathscr{M}_i(W^n) = S_i(W^n) + S_{i+1}(W^n) + \mu(H_i(W^n, V_0^{n-1}, \mathbb{Z})) + \mu(H_{i-1}(W^n, V_0^{n-1}, \mathbb{Z})).$$

These conditions are satisfied if π is a free or free abelian group, or $\mathbb{Z}_k$ $(k = 2, 3, 4, 6)$.

PROOF. Lemma 4.6 ensures that any chain complex associated with an arbitrary Morse function and gradient-like vector field on W^n is simple homotopy equivalent to a minimal chain complex since $\mathrm{Wh}(\pi) = 0$. The number of critical points for the minimal Morse function can be computed with the use of Corollary 4.2. □

PROPOSITION 5.5. *Let $(W^n, V_0^{n-1}, V_1^{n-1})$ be a cobordism of class R, and let $\pi_1(W^n) = \pi$ be an s-group. Suppose that* $\text{r-dim}\,\pi \le k$ *and that the ith Morse number of the manifold W^n satisfies the inequality $\mathscr{M}_i(W^n) \ge k$ $(2 \le i \le n-2)$. Then on W^n there exists a minimal Morse function.*

PROOF. This follows from Lemma 4.6 and Proposition 5.3. □

From these results, and the results of Chapter III, §2, concerning s-groups and of Chapter IV, §3, concerning $\text{r-dim}\,\pi$, we have the following propositions.

PROPOSITION 5.6. *Let $(W^n, V_0^{n-1}, V_1^{n-1})$ be a cobordism of class R that is not an h-cobordism. Suppose that $\pi_1(W^n) = \pi$, $n \ge 6$. Then*

(1) *π is a finite abelian group isomorphic to $(\mathbb{Z}_2)^n$, or*

(2) *every Sylow p-subgroup is of the form $\mathbb{Z}_p$ or $\mathbb{Z}_p \oplus \mathbb{Z}_p$, or*

(3) *π is a finite nonabelian group that has no epimorphic mapping onto*

$$\mathbb{Z}_{p^2} \oplus \mathbb{Z}_{p^2},\ \mathbb{Z}_p \oplus \mathbb{Z}_p \oplus \mathbb{Z}_p,\ \mathbb{Z}_4 \oplus \mathbb{Z}_4,\ \mathbb{Z}_4 \oplus \mathbb{Z}_2 \oplus \mathbb{Z}_2,\ \mathbb{Z}_p \oplus \mathbb{Z}_2 \oplus \mathbb{Z}_2 \oplus \mathbb{Z}_2,$$

where p is prime.

Then on W^n there exists a minimal Morse function with the number of critical points of index i equal to

$$\mathscr{M}_i(W^n) = S_i(W^n) + S_{i+1}(W^n) + \mu(H_i(W^n, V_0^{n-1}, \mathbb{Z})) + \mu(\operatorname{Tors} H_{i-1}(W^n, V_0^{n-1}, \mathbb{Z})). \quad \square$$

PROPOSITION 5.7. *Let* $(W^n, V_0^{n-1}, V^{n-1})$ *be a cobordism of class* R, $\pi_1(W^n) = \pi$ *a finite abelian group, and let for some* i_0 $(2 \le i_0 \le n-2)$ *the Morse number satisfy the inequality*

$$\mu(H_{i_0}(W^n, V_0^{n-1}, \mathbb{Z})) + \mu(\operatorname{Tors} H_{i_0}(W^n, V_0)) \geq 2.$$

Then on W^n *there exists a minimal Morse function whose number of critical points of index* i *is*

$$\mathscr{M}_i(W^n) = S_i(W^n) + S_{i+1}(W^n) + \mu(H_i(W^n, V_0^{n-1}, \mathbb{Z})) + \mu(\operatorname{Tors} H_{i-1}(W^n, V_0^{n-1}, \mathbb{Z})). \qquad \square$$

We note that Corollary 5.7 guarantees that on a cobordism of class R for which $\pi_1(W^n) = \pi$ is an s-group, there exists a Morse function for which, with the possible exception of two adjacent indices, the number of critical points of index i $(2 \le i \le n-3)$ is equal to the ith Morse number of the cobordism $(W^n, V_0^{n-1}, V_1^{n-1})$.

PROPOSITION 5.8. *In the homotopy type of a cobordism* $(W^n, V_0^{n-1}, V_1^{n-1})$ *of class* R *for which* $\pi_1(W^n) = \pi$ *is an* s*-group, there always exists a cobordism admitting a minimal Morse function.*

PROOF. Corollary 4.1 allows us to construct on W^n an ordered Morse function f and a gradient-like vector field ξ such that the associated chain complex is of the form

$$\begin{array}{ccccccc} \overline{C}_2 & \xleftarrow{\overline{\partial}_3} \cdots \longleftarrow & \overline{C}_{n-3} & \xleftarrow{\overline{\partial}_{n-2}} & \overline{C}_{n-1} & & \\ & & \oplus & & \oplus & & \\ & 0 \longleftarrow & T_{n-3} & \xleftarrow{\delta} & T_{n-2} & \longleftarrow & 0, \end{array}$$

where

$$0 \longleftarrow T_{n-3} \xleftarrow{\delta} T_{n-2} \longleftarrow 0$$

is an elementary chain complex generated by middle disks of critical points of indices $n-3$ and $n-2$. Let Ω be a cobordism with boundary $\partial\Omega = V_0^{n-1} \cup V_1^{n-1}$ containing only those critical points of indices $n-3$ and $n-2$ whose middle disks give the generators in T_{n-3} and T_{n-2}. Obviously, Ω is an h-cobordism. Consider the manifold $\overline{W}^n = \overline{W^n \setminus \Omega}$. Clearly, $\overline{W}^n$ and W^n have the same homotopy type. By construction, there exists on $\overline{W}^n$ a minimal Morse function. $\square$

THEOREM 5.3. *For any cobordism of class* R *the Morse numbers are invariants of homotopy type.*

PROOF. Let $(W^n, V_0^{n-1}, V_0^{n-1})$ and $(\overline{W}^n, \overline{V}_0^{n-1}, \overline{V}_1^{n-1})$ be cobordisms of class R of the same simple homotopy type (i.e., the chain complexes

$C(\hat{W}^n)$ and $C(\hat{\overline{W}}^n)$ have the same simple homotopy type). Suppose that $f: (W^n, V_0^{n-1}, V_1^{n-1}) \to ([0, 1], 0, 1)$ is a Morse function for which the number of critical points equals the Morse number of the cobordism $(W^n, V_0^{n-1}, V_1^{n-1})$. We can assume that f has no critical points of indices 1 and $n-1$. Otherwise, by introducing an appropriate number of mutually cancelling critical points of indices 2 and 3, we can replace critical points of index 1 by those of index 3 without altering the Morse numbers. A similar procedure can be applied to critical points of index $n-1$ if any. Clearly, one can also eliminate critical points of indices 0 and n if there are any. From the Morse function f and the gradient-like vector field ξ construct the chain complex

$$C(\hat{W}^n, \hat{f}, \hat{\xi}): C_2 \xleftarrow{\partial_3} C_3 \longleftarrow \cdots \xleftarrow{\partial_{n-2}} \hat{C}_{n-2}.$$

Let $g: (\overline{W}^n, \overline{V}_0^{n-1}, \overline{V}_1^{n-1}) \to [0, 1]$ be an arbitrary ordered Morse function without critical points of indices $0, 1, n-1, n$, and let η be its gradient-like vector field. Denote by $C(\hat{\overline{W}}^n, \hat{g}, \hat{\eta})$ the chain complex associated with g and η. Since the chain complexes $C(\hat{W}^n, \hat{f}, \hat{\xi})$ and $C(\hat{\overline{W}}^n, \hat{g}, \hat{\eta})$ are simple homotopy equivalent, by virtue of the proposition there exist on $(\overline{W}^n, \overline{V}_0^{n-1}, \overline{V}_1^{n-1})$ a Morse function $\overline{g}$ and vector field $\overline{\eta}$ realizing the chain complex $C(\hat{W}^n, \hat{f}, \hat{\xi})$. □

As noted in Chapter IV, §6, if the length of a chain complex is greater than 3, then the ith Morse numbers are invariants of homotopy type. This is a direct consequence of the Cockroft-Swan theorem and implies the following proposition.

PROPOSITION 5.9. *For any cobordism of class R whose dimension is greater than 6, the ith Morse numbers are invariants of homotopy type.*

§4. Minimal Morse functions on cobordisms of class R

Denote by R_π the set of cobordisms of class R whose fundamental group is isomorphic to a fixed group π.

THEOREM 5.4. *In order that each representative in R_π admit a minimal Morse function, it is necessary and sufficient that π be an s-group and* $\mathrm{Wh}(\pi) = 0$.

PROOF. *Sufficiency* follows directly from Corollary 5.1.

Necessity. Examples of h-cobordisms with nonzero torsion demonstrate that the condition $\mathrm{Wh}(\pi) = 0$ is necessary, since such h-cobordisms admit no minimal Morse functions. Let us show that the second condition is also necessary. We shall make use of the argument used in the proof of Theorem 4.2. Construct a cobordism of class R_π and a Morse function on it that defines the chain complex given in Theorem 4.2. Suppose that π is not an s-group, M is a stably free but not free $\mathbb{Z}[\pi]$-module such that $M \oplus$

$\mathbb{Z}[\pi] \approx k\mathbb{Z}[\pi]$. Consider a closed manifold V^{n-1} for which $\pi_1(V^{n-1}) \approx \pi$ $(n \geq 10)$. Let $N = V^{n-1} \times [0, 1]$. Instead of Morse functions, we shall use the handlebody decomposition of the manifold. Attach k pairs of mutually cancelling handles of indices 4 and 5 to $V^{n-1} \times 1$ and one handle of index 5, the latter corresponding to the trivial embedding. Fix bases in the chain complex associated with this handlebody decomposition:

$$\begin{array}{ccc} C_4 & \xleftarrow{\partial_5} & C_5 \\ \| & & \| \\ k\mathbb{Z}[\pi] & \longleftarrow & k\mathbb{Z}[\pi] \oplus \mathbb{Z}[\pi], \end{array} \qquad \partial_5(0 \oplus \mathbb{Z}[\pi]) = 0.$$

In the module C_4, let us extract as direct summands two modules M and S such that $M \oplus S = C_4$, $S \approx \mathbb{Z}[\pi]$. Let s be a generator of the module S, and g an element of $k\mathbb{Z}[\pi]$ such that $\partial_5(g) = s$. Add to the handle of index 5 attached to the manifold via a trivial embedding a linear combination of the remaining handles of index 5 realizing the element g. Discarding the handles of index 5 corresponding to the summand $k\mathbb{Z}[\pi] \oplus 0$, we obtain a cobordism for which the chain complex is of the form

$$\begin{array}{ccc} C_4 & \xleftarrow{\overline{\partial}_5} & \overline{C}_5 \\ \| & & \| \\ k\mathbb{Z}[\pi] & \xleftarrow{\overline{\partial}_5} & \mathbb{Z}[\pi], \end{array} \qquad k\mathbb{Z}[\pi]/\overline{\partial}_5(\mathbb{Z}[\pi]) \approx M.$$

Since the dimension of the manifold allows us to perform elementary operations, the proof is concluded on the lines given in Theorem 4.2. □

THEOREM 5.5. *Let W^n be a manifold with boundary $\partial W^n = V_0^{n-1} \cup V_1^{n-1}$, $\pi_1(V_0^{n-1}) \hookrightarrow \pi_1(W^n) \hookleftarrow \pi_1(V_1^{n-1})$ epimorphisms, $n \geq 6$. Suppose that $\pi_1(W^n) = \pi$ is an s-group and the ith Morse number satisfies the inequality $\mathscr{M}_i(W^n) \geq k$ $(3 \leq i \leq n-3)$. Then there exists on W^n a minimal Morse function with the number of critical points of index i equal to*

$$\mathscr{M}_0(W^n) = \mathscr{M}_1(W^n) = \mathscr{M}_{n-1}(W^n) = \mathscr{M}_n(W^n) = 0,$$
$$\mathscr{M}_2(W^n) = \mu(\pi_2(W^n, V_0^{n-1})),$$
$$\mathscr{M}_{n-2}(W^n) = \mu(\pi_2(W^n, V_1^{n-1})),$$
$$\mathscr{M}_i(W^n) = S_i(W^n) + S_{i+1}(W^n) + \mu(H_i(W^n, V_0^{n-1}, \mathbb{Z})) + \mu(\operatorname{Tors} H_{i-1}(W^n, V_0^{n-1}, \mathbb{Z})) \qquad (3 \leq i \leq n-3),$$

where $\mu(\pi_2(W^n, V_0^{n-1}))$ is the minimal number of generators of the crossed $\pi_1(V_0)$-module $\pi_2(W^n, V_0^{n-1})$.

PROOF. Consider $\pi_2(W^n, V_0^{n-1}) = G_1$ and $\pi_2(W^n, V_1^{n-1}) = G_2$, and choose minimal systems $a_1, \dots, a_s$ of generators in G_1 and $b_1, \dots, b_t$ in G_2 as crossed modules, realizing them by disjoint embeddings

$$\alpha_i: (D_i^2, S_i^1) \hookrightarrow (W^n, V_0^{n-1}),$$
$$\beta_j: (D_j^2, S_j^2) \hookrightarrow (W^n, V_1^{n-1}).$$

Tubular neighborhoods of $\alpha(D_i^2, S_i^1) = A_i$ and $\beta_j(D_j^2, S_j^1) = B_j$ can be taken as handles of index 2 attached to collars of V_0^{n-1} and V_1^{n-1}, respectively. Denote them by Ω_1 and Ω_2. Let $f_1: \Omega_1 \to [0, 1/3]$, $f_1^{-1}(0) = V_0^{n-1}$, $f_3: \Omega_2 \to [2/3, 1]$, $f_3^{-1}(1) = V_1^{n-1}$, be Morse functions on Ω_1 and Ω_2, constructed by handlebody decomposition. It is easily shown that the cobordism

$$\overline{W}^n = \overline{W^n \setminus (\Omega_1 \cup \Omega_2)}, \qquad \partial \overline{W}^n = \overline{V}_1^{n-1} \cup \overline{V}_2^{n-1}$$

has the property that $\pi = \pi_1(\overline{V}_1^{n-1}) \hookrightarrow \pi_1(W^n) \hookleftarrow \pi_1(\overline{V}_2^{n-1})$ are isomorphisms and consequently satisfy the hypothesis of Proposition 5.7. Let f_2: $\overline{W}^n \to [1/2, 2/3]$ be a minimal Morse function on the manifold $\overline{W}^n$. Consider the function $f_1 \cup f_2 \cup f_3$, and let us prove that it is a minimal Morse function on W^n. Construct on W^n an arbitrary ordered Morse function f and a gradient-like vector field ξ. Let $C(\hat{W}^n, \hat{f}, \hat{\xi})$ be the associated chain complex

$$C_2 \xleftarrow{\partial_3} C_3 \longleftarrow \cdots \xleftarrow{\partial_{n-2}} C_{n-2}.$$

We may assume that f has no critical points of indices $0, 1, n-1$, or n since $\pi_1(W^n, V_0^{n-1}) = \pi_1(W^n, V_1^{n-1}) = 0$. Furthermore, in view of the fact that the critical points of indices i $(3 \le i \le n-3)$ have no effect on the fundamental group of the level surface of the Morse function, we can also assume that f is chosen so that

$$f\text{-rank}(\partial_4(C_4)C_3) = 0$$

and is additive. By Lemma 4.7, the quantity $\chi = \mu(C_2) - \mu(C_3)$ is constant for all Morse functions on W^n for which the associated chain complexes satisfy this condition. Therefore the number of critical points of index 3 attains its minimum whenever the number of critical points of index 2 does. Since the middle disks of critical points of index 2 give generators of the crossed module $\pi_2(W^n, V_0^{n-1})$, there cannot be less than s of them. Similarly, using the function f, we can show that f is a minimal Morse function for index $n-3$. The same argument (which follows from Theorem 5.5) gives the number of critical points of index i. $\square$

Suppose that the boundary of the cobordism $(W^n, \partial W^n)$ consists of several components $V_0^{n-1}, \dots, V_k^{n-1}, \overline{V}_0^{n-1}, \dots, \overline{V}_l^{n-1}$. Group them as: $A = \bigcup_i V_i^{n-1}$, $B = \bigcup_j \overline{V}_j^{n-1}$. Using the preceding theorem, it is easy to state conditions for W^n to have a minimal Morse function $f: W^n \to [0, 1]$, $f^{-1}(0) = A$, $f^{-1}(1) = B$. By the van Kampen-Seifert theorem, we have

$$\pi_1(V_0^{n-1} \# \cdots \# V_k^{n-1}) = \pi_1(V_0^{n-1}) * \cdots * \pi_1(V_k^{n-1}),$$
$$\pi_1(\overline{V}_0^{n-1} \# \cdots \# \overline{V}_l^{n-1}) = \pi_1(\overline{V}_0^{n-1}) * \cdots * \pi_1(\overline{V}_l^{n-1}).$$

Using handles of index 1, join the components $V_0^{n-1}, \ldots, V_k^{n-1}$ and then the components $\overline{V}_0^{n-1}, \ldots, \overline{V}_l^{n-1}$. Denote the resulting manifolds by V^{n-1} and $\overline{V}^{n-1}$, respectively.

PROPOSITION 5.10. *Suppose the inclusions* $\pi_1(V^{n-1}) \hookrightarrow \pi_1(W^n) \hookleftarrow \pi_1(\overline{V}^{n-1})$ *induce epimorphisms,* $n \geq 6$, $\pi_1(W^n) \approx \pi$ *is an s-group,* r-dim $\pi \leq k$, *and the* i*th Morse number satisfies the inequality* $\mathscr{M}_i(W^n) \geq k$ $(3 \leq k \leq n-3)$. *Then on the manifold* W^n *there exists a minimal Morse function.*

We remark that similar results hold for noncompact cobordisms admitting adjunction of a boundary.

CHAPTER VI

Elements of the Homotopy Theory of Non-Simply-Connected CW-Complexes

In the present chapter we develop a technique based on the homotopy theory of CW-complexes needed for the analysis of Morse functions on non-simply-connected closed manifolds and manifolds having just one connected component of the boundary. Some algebraic topics are also considered. Section 1 provides the information we shall need from the homotopy theory of CW-complexes. Section 2 includes a brief review of the theory of simple homotopy type of CW-complexes. In §3 we study the structure of two-dimensional CW-complexes. As is known, the classification of such complexes up to homotopy type is closely connected with a number of topics of combinatorial group theory. Accordingly, §3 studies some questions related to presentations of finitely defined groups. In §4 we consider crossed modules arising naturally in the study of non-simply-connected manifolds, as well as in the group cohomology theory. In §5 we introduce homotopy systems in the sense of Whitehead; these systems will appear in the next chapter in the study of Morse functions. In §6 an analog of the Cockroft-Swan theorem for homotopy systems is considered. In §7 we give conditions for the existence of minimal homotopy systems. Section 8 discusses the same question in the case when the homotopy type of a homotopy system is fixed.

§1. Brief review

Let us make our terminology more precise and consider some results obtained in the theory of cellular spaces.

A topological space K is said to be a CW-complex if it is represented as a union of disjoint sets e^n_α called cells (n is the dimension of the cell), i.e., if $K = \bigcup_\alpha e^n_\alpha$. Furthermore, for each cell there is a continuous mapping $\varphi_\alpha : D^n \to K$ of a closed n-dimensional ball D^n into the space K. This mapping is said to be characteristic and has the following properties:

(1) the restriction of φ_α to the open ball $\mathring{D}^n$ (whose closure is the ball D^n) is a homeomorphism of this open ball onto the cell e^n_α;

(2) the boundary of each cell, i.e., the set $\bar{e}^n_\alpha \setminus e^n_\alpha$ (where $\bar{e}^n_\alpha$ denotes the

closure of e^n_α in K) is contained in the union of finitely many cells of lesser dimension;

(3) a set $A \subset K$ is closed if and only if for all cells e^n_α the inverse image $\varphi^{-1}_\alpha(A \cap e^n_\alpha)$ is closed in the ball D^n.

A subset $L \subset K$ is said to be a subcomplex if L is closed in K, is a CW-complex, and all its cells and characteristic mappings are at the same time cells and characteristic mappings in K. Among subcomplexes in K are the i-skeletons of K, which we will denote by K^i, i.e., the subcomplexes consisting of the union of all cells whose dimension does not exceed i.

Let K and L be two CW-complexes. A continuous mapping $f: K \to L$ is said to be a cellular mapping if $f(K^i) \subset L^i$ for all i. If a continuous mapping $f: K \to L$ is homotopy equivalent to a cellular mapping $g: K \to L$, then the latter is called a cellular approximation of f. It is known that each continuous mapping between two CW-complexes can be approximated by a cellular mapping.

Let $f, g: X \to Y$ be two continuous mappings of topological spaces X and Y. These mappings are called homotopy equivalent (we write $f \sim g$) if there exists a continuous mapping $F: X \times I \to Y$ (where $I = [0, 1]$) such that $F(x, 0) = f(x)$, $F(x, 1) = g(x)$. Two spaces X and Y are said to be homotopy equivalent if there exist continuous mappings $f: X \to Y$, $g: Y \to X$ such that $g \circ f \approx \mathrm{id}_X$ and $f \circ g = \mathrm{id}_Y$, where id_X is the identity mapping of the space X onto itself. An example of homotopy equivalence is a strong deformation retraction. Let $X \subset Y$; then a continuous mapping $D: Y \to X$ is said to be a strong deformation retraction if there exists a mapping $F: Y \times I \to Y$ such that $F(y, 0) = \mathrm{id}_Y$ and $F_t(x) = x$ for all $(x, t) \in X \times I$ and $F(y, 1) = D(y)$ for all $y \in Y$. Here $F_t(y)$ is equal to $F(y, t)$. In what follows we write $Y \rightsquigarrow X$ if there exists a strong deformation retraction from Y to X.

Let $f: X \to Y$ be a continuous mapping. By definition, the cylinder of a mapping f is the topological space M_f obtained from the disjoint union of $X \times I$ and Y by identifying the points $(x, 1)$ and $f(x)$. Clearly, Y as a subspace M_f is a strong deformation retraction. If K and L are CW-complexes and $f: K \to L$ is a cellular mapping, then M_f is also a CW-complex.

Let X be a path-connected topological space. The universal covering space $\hat{X}$ of the space X is defined as the set of classes of mappings $f: (I, 0) \to (X, x)$ homotopy equivalent with respect to the point 1. The formula $p(f) = f(1)$ defines a mapping $p: \hat{X} \to X$, where $\hat{X}$ is endowed with the weakest topology in which this mapping is continuous. The fundamental group $\pi_1 = \pi_1(X, x)$ acts on the space $\hat{X}$ by the formula $g \circ f = g + f$. Suppose K is a CW-complex, and $p: \hat{K} \to K$ is its universal covering space; then $\hat{K}$ is also a CW-complex. The cellular structure of K can be lifted to the space $\hat{K}$ in such a way that the mapping $p: \hat{K} \to K$ is a cellular mapping.

Let $L \subset K$ be a pair of connected CW-complexes such that $i: \pi_1(L, x) \to \pi_1(K, x)$ is an epimorphism. Suppose that $p: \hat{K} \to K$ is the universal covering space and $\hat{L} = p^{-1}(L)$. Then the restriction of p to $\hat{L}$, i.e., the mapping $p|_{\hat{L}}: \hat{L} \to L$ is the universal covering space for L. Furthermore, if there exists a strong deformation retraction of K onto L, then $\hat{K} \rightsquigarrow \hat{L}$.

Let $f: K \to L$ be a cellular mapping between two connected CW-complexes such that the induced mapping $f_*: \pi_1(K, x) \to \pi_1(L, f(x))$ is an isomorphism. Let $\hat{K}$ and $\hat{L}$ be the universal covering spaces and $\hat{f}: \hat{K} \to \hat{L}$ the covering mapping for f. Then the cylinder $M_{\hat{f}}$ of the mapping $\hat{f}$ is the universal covering space for the cylinder M_f of the mapping f.

It is known that with each pair $X \subset Y$ of topological spaces one can associate the abelian groups $H_i(Y, X, \mathbb{Z})$, called singular homology groups. However, for CW-complexes one can also define so-called cellular homologies which turn out to coincide with the singular ones but have an important advantage: their computation is much easier. This is why cellular homologies are mainly used in practice. We now briefly review their construction.

Let K^i be the i-dimensional skeleton of a CW-complex K. Fix an orientation of all its cells. It is easy to compute singular homology for the pair of spaces (K^i, K^{i-1}):

$$H_j(K^i, K^{i-1}, \mathbb{Z}) = \begin{cases} 0 & \text{for } j \neq i, \\ \underbrace{\mathbb{Z} \oplus \cdots \oplus \mathbb{Z}}_{s} & \text{for } j = i. \end{cases}$$

Here s is the number of i-dimensional cells e^i, and the generators of $H_i(K^i, K^{i-1}, \mathbb{Z})$ are in one-to-one correspondence with the i-dimensional cells. We denote the groups $H_i(K^i, K^{i-1}, \mathbb{Z})$ by $C_i(K, \mathbb{Z})$ and call them i-dimensional cellular chains of the complex K. There is the boundary homomorphism $\partial_i: C_i(K, \mathbb{Z}) \to C_{i-1}(K, \mathbb{Z})$ constructed from the exact sequence of the triple $(K^i \supset K^{i-1} \supset K^{i-2})$ for singular homology of the spaces. Thus there arises a chain complex of free abelian groups

$$C_0(K, \mathbb{Z}) \xleftarrow{\partial_1} C_1(K, \mathbb{Z}) \longleftarrow \cdots \xleftarrow{\partial_n} C_n(K, \mathbb{Z}).$$

Its homology groups are called the cellular homologies of the complex K. Relative cellular homology for a pair of CW-complexes $(L \subset K)$ are constructed in a similar manner.

Let $p: \hat{K} \to K$ be the universal covering space for K. By choosing a base point $x \in K$ and $\hat{x} \in p^{-1}(x)$ we obtain a standard identification of the group of covering transformations G (where G is the set of all homeomorphisms $f: \hat{K} \to \hat{K}$ such that $p = p \cdot f$) with the fundamental group $\pi_1(K, x)$. This identification is of the following form. For each mapping $\alpha: (I, \mathring{I}) \to (K, x)$ denote by $\hat{\alpha}$ the lifting of α with $\hat{\alpha}(0) = \hat{x}$. Let $g_{[\alpha]}: \hat{K} \to \hat{K}$ be the only homeomorphisms such that $g_{[\alpha]}(\hat{x}) = \hat{\alpha}(1)$. It is easy to see that the mapping $\theta = \theta(x, \hat{x}): \pi_1(K, x) \to G$ does not depend on the choice of representative in the class α and is an isomorphism.

Suppose that $L \subset K$ is a subcomplex and $\hat{L} = p^{-1}(L)$. Consider the CW-complex K. Each element $g \in \pi_1(K, x)$ generates a cellular homeomorphism $g_{[\alpha]}: \hat{K} \to \hat{K}$ which induces a homomorphism of groups

$$g_*: C_i(\hat{K}, \hat{L}, \mathbb{Z}) \to C_i(\hat{K}, \hat{L}, \mathbb{Z})$$

satisfying the condition $\partial \circ g_* = g_* \circ \partial$, where ∂ is the boundary homomorphism in the CW-complex $C(\hat{K}, \hat{L})$. Define the action of the group G, and, consequently, that of the fundamental group $\pi_1(K, x)$ on $C_i(\hat{K}, \hat{L}, \mathbb{Z})$ by the formula $gc = g_*(c)$ $(g \in G,\ c \in C_i(\hat{K}, L, \mathbb{Z}))$. This turns the groups $C_i(\hat{K}, \hat{L}, \mathbb{Z})$ into $\mathbb{Z}[\pi_1(K, x)]$-modules. The corresponding action is induced in the homology groups $H_i(K, L, \mathbb{Z})$ which also become $\mathbb{Z}[\pi_1(K,x)]$-modules. More details on the topics considered in this section can be found in [95, 123, 25].

§2. Torsion and simple homotopy type

Suppose that a CW-complex K is obtained from a CW-complex L by attaching two cells e^i and e^{i+1} for which there exist characteristic mappings φ_1 and φ_2 such that $\varphi_1 = \varphi_2 \circ r$, where $r: I^{i+1} \to I^i$ is the standard projection. Then we say that K is an elementary extension of the complex L, and that K is elementarily contracted to the complex L, $K \searrow L$. Note that the retraction of the disk I^{i+1} to I^i evidently defines a strong deformation retraction of K to L. A complex K is said to be contractible to a subcomplex L, or L is said to be extendible to the complex K, if there exists a finite sequence of elementary contractions such that

$$K = K_0 \searrow K_1 \searrow \cdots \searrow K_j = L.$$

A finite sequence of operations consisting of elementary extensions or contractions is called a formal deformation. If there is a formal deformation from the complex K to the complex L, we write $K \curvearrowright L$ and assume that K and L have the same simple homotopy type. If both K and L have a common subcomplex M that is not affected by a formal deformation, we write $K \curvearrowright L \operatorname{rel} M$. Suppose that $K = K_0 \curvearrowright K_1 \curvearrowright \cdots \curvearrowright K_j = L$ is a formal deformation. Obviously, if $K_i \searrow K_{i+1}$ then there exists a strong deformation retraction $f_i: K_i \to K_{i+1}$, and if $K_{i+1} \searrow K_i$, then there is an inclusion $f_i: K_i \to K_{i+1}$ that is also a homotopy equivalence. The mapping $f = f_{i-1} \circ \cdots \circ f_1 \circ f_0$ is said to be a deformation. Evidently, any deformation f has a uniquely defined homotopy type. If K contains a subcomplex M and $f = f_{j-1} \circ \cdots \circ f_0: K \to L$ is a deformation such that $f_i|_M = \mathrm{id}$, then we say that f is a deformation with respect to the subcomplex M. It is known that a continuous mapping $f: K \to L$ is a simple homotopy equivalence with respect to a subcomplex M if f is homotopy equivalent to some deformation with respect to M. Since with each cellular mapping $f: K \to L$ one can associate the CW-complex M_f, i.e., the cylinder of the mapping f, let us discuss the connection between formal deformations and M_f. It is clear that if the mapping $f: K \to L$ is a cellular one, then M_f is contractible on L. The following statements are equivalent:

(1) f is a simple homotopy equivalence;
(2) there exists a cellular mapping $g: K \to L$ homotopy equivalent to f and such that $M_g \curvearrowright K \operatorname{rel} K$;
(3) every cellular mapping $g: K \to L$ homotopy equivalent to f has the property that $M_g \curvearrowright K \operatorname{rel} K$.

Fix a CW-complex L. Consider the set of pairs (K, L) of CW-complexes such that L is a strong deformation retract of K. Define an equivalence relation on this set as follows: $(K, L) \sim (K', L')$ if and only if $K \curvearrowright K' \operatorname{rel} L$. Denote the equivalence class containing the pair (K, L) by $[K, L]$. An operation of addition is defined on the set of equivalence classes

$$[K, L] + [K', L] = [K \cup_L K', L].$$

It can be shown that this operation turns the set of equivalence classes into an abelian group which is called the Whitehead group of the group L and denoted by $\mathrm{Wh}(L)$. This construction evidently yields a covariant functor from the category of CW-complexes and cellular mappings into the category of abelian groups and homomorphisms $L \to \mathrm{Wh}(L)$: $(f: L_1 \to L_2) \to (f_*: \mathrm{Wh}(L_1) \to \mathrm{Wh}(L_2))$. If $f \sim g$, then $f_* = g_*$. Therefore we can define the torsion of the homotopy equivalence $f: L_1 \to L_2$ as follows:

$$\tau(f) = f_*[M_f, L_1] = [M_f \cup_{L_1} M_f, L_2] \in \mathrm{Wh}(L_2).$$

We consider the cellular structure of the pair (K, L), where $K \rightsquigarrow L$, in more detail. The following statements are well known [25].

Proposition 6.1. *If $K_0 = L \cup e_0^n$ and $K_1 = L \cup e_1^n$, where L is a CW-complex, and e_i $(i = 0, 1)$ are n-dimensional cells with the characteristic mappings $\varphi_i: D^n \to K_i$ such that $\varphi_0|\partial D^n$ and $\varphi_1|\partial D^n$ are homotopy equivalent in L, then $K_0 \curvearrowright K_1 \operatorname{rel} L$.*

Proposition 6.2. *If (K, L) is a pair of connected CW-complexes, r is an integer such that $\pi_r(K, L) = 0$, and*

$$K = L \cup \left(\bigcup_{i=1}^{k_r} e_i^r\right) \cup \left(\bigcup_{i=1}^{k_{r+1}} e_i^{r+1}\right) \cup \cdots \cup \left(\bigcup_{i=1}^{k_n} e_i^n\right),$$

then $K \curvearrowright M \operatorname{rel} L$, where M is a CW-complex of the form

$$M = L \cup \left(\bigcup_{i=1}^{k_{r+1}} \tilde{e}_i^{r+1}\right) \cup \left(\bigcup_{i=1}^{k_r + k_{r+2}} \tilde{e}_i^{r+2}\right) \cup \left(\bigcup_{i=1}^{k_{r+3}} \tilde{e}_i^{r+3}\right) \cup \cdots \cup \left(\bigcup_{i=1}^{k_n} \tilde{e}_i^{n}\right). \qquad \square$$

Proposition 6.3. *Suppose that (K, L) is a pair of connected CW-complexes such that L is a deformation retract of K. Let $n = \dim(K \setminus L)$ and*

$r \geq n-1$ be a positive integer. Denote by e^0 the 0-dimensional cell in L. Then $K \curvearrowright M \operatorname{rel} L$, where

$$M = L \cup \left(\bigcup_{j=1}^{s} e_j^r\right) \cup \left(\bigcup_{i=1}^{s} e_i^{r+1}\right)$$

and the cells e_i^{r+1} and e_j^r have characteristic mappings $\varphi_j \colon D^{r+1} \to M$ and $\psi_j \colon D^r \to M$ such that $\psi_j(D^r) = e^0 = \varphi_i(J^r)$ $(J^r = \partial D^{r+1} \setminus D^r)$. □

Consider a pair of CW-complexes (K, L), where K is of the form

$$K = L \cup \left(\bigcup_{j=1}^{k_r} e_j^r\right) \cup \left(\bigcup_{i=1}^{k_r} e_i^{r+1}\right).$$

As is known, the fundamental group $\pi_1 = \pi_1(L, e^0)$ acts on the group $\pi_n(K, L, e^0)$ which can be considered as a $\mathbb{Z}[\pi_1(L, e^0)]$-module. Denote by K' the CW-complex

$$K' = L \cup \left(\bigcup_{j=1}^{k_r} e_j^r\right).$$

Let $\varphi_j \colon D^r \to K'$ be characteristic mappings of e_j^r such that if $j = 2$, then $\varphi_j(\partial D^2) = e^0$. Then $\pi_r(K', L, e^0)$ is a free $\mathbb{Z}[\pi_1(L, e^0)]$-module with the basis $[\varphi_1], \dots, [\varphi_{k_r}]$. Let ψ_i be characteristic mappings of the cells e_i^{r+1}. Consider the free $\mathbb{Z}[\pi_1(L, e^0)]$-modules $\pi_r(K', L, e^0)$ and $\pi_{r+1}(K, K', e^0)$. Making use of the boundary homomorphism $\partial \colon \pi_{r+1}(K, K', e^0) \to \pi_r(K, K', e^0)$ in the exact homotopy sequence for the triple (K, K', L), we can, for each pair of complexes (K, L) and the bases given by the characteristic mappings φ_i and ψ_j, define the matrix (a_{ij}) by the relation $\partial[\varphi_j] = \sum_{i,j} a_{i,j}[\psi_i]$. Note that this matrix is invertible if $K \rightsquigarrow L$, since in this case $\pi_r(K, L, e^0) = \pi_{r+1}(K, L, e^0) = 0$ and consequently ∂ is an isomorphism. This matrix defines an element of the Whitehead group $\operatorname{Wh}(\pi_1(L, e^0))$. It can be shown that if two pairs of CW-complexes (K, L) and (M, L) are of the form

$$K = L \cup \left(\bigcup_{j=1}^{k_r} e_j^r\right) \cup \left(\bigcup_{i=1}^{k_r} e_i^{r+1}\right), \qquad M = L \cup \left(\bigcup_{j=1}^{s} e_j^r\right) \cup \left(\bigcup_{i=1}^{s} e_i^{r+1}\right)$$

and the mappings $\partial \colon \pi_{r+1}(K, K', e^0) \to \pi_r(K', L, e^0)$, $\partial \colon \pi_{r+1}(M, M', e^0) \to \pi_r(M', L, e^0)$ define matrices (a_{ij}), $(\overline{a}_{ij})$ with respect to characteristic mappings of the cells determining the same element in the Whitehead group $\operatorname{Wh}(\pi_1(L, e^0))$, then $K \curvearrowright M \operatorname{rel} L$. Now let G be a finitely representable group and A an invertible matrix over the ring $\mathbb{Z}[G]$. Then there exists a connected CW-complex L with $\pi_1(L, e^0) = G$ and a complex K of the form

$$K = L \cup \left(\bigcup_{j=1}^{k} e_j^r\right) \cup \left(\bigcup_{i=1}^{k} e_i^{r+1}\right),$$

such that $\pi_i(K, L) = 0$ and the matrices of the pair (K, L) with respect to the characteristic mappings of the cells coincide with A. Thus there is an isomorphism of groups $\mathrm{Wh}(L) \to \mathrm{Wh}(\pi_1(L, e^0))$. The element of the Whitehead group $\mathrm{Wh}(\pi_1(L, e^0))$ corresponding to the matrix A for the pair of complexes (K, L) is called the Whitehead torsion and denoted $\tau(K, L)$.

There is an equivalent way to define the Whitehead torsion for a pair of CW-complexes (K, L) such that $K \rightsquigarrow L$. It involves constructing an algebraic chain complex corresponding to the geometric situation. Let (K, L) be a pair of CW-complexes such that $\pi_1(L, e^0) \to \pi_1(L, e^0)$ is an isomorphism. For each pair of CW-complexes (K, L) the ith chain group is defined as

$$C_i(K, L, \mathbb{Z}) = H_i(K^i \cup L, K^{i-1}, L, \mathbb{Z}).$$

This chain group is a free abelian group with an i-dimensional cell belonging to $K^i \cup L \setminus K^{i-1} \cup L$ as a generator. Consider the universal covering space $p: \hat{K} \to K$ and put $\pi = \pi_1(K, e^0)$. Let us identify the fundamental group π with the group of automorphisms of the covering. Therefore each element $\alpha \in \pi$ defines the chain mapping $\alpha_*: C_i(\hat{K}, \hat{L}, \mathbb{Z}) \to C_i(\hat{K}, \hat{L}, \mathbb{Z})$, and $C_i(\hat{K}, \hat{L}, \mathbb{Z})$ is a free $\mathbb{Z}[\pi]$-module, with cells from $K \setminus L$ as generators. We obtain a free chain complex

$$C(\hat{K}, \hat{L}): C_0(\hat{K}, \hat{L}, \mathbb{Z}) \xleftarrow{\partial_1} C_1(\hat{K}, \hat{L}, \mathbb{Z}) \leftarrow \cdots \xleftarrow{\partial_n} C_n(\hat{K}, \hat{L}, \mathbb{Z})$$

over $\mathbb{Z}[\pi]$. The homology modules of this chain complex coincide with $H_i(\hat{K}, \hat{L}, \mathbb{Z})$. The geometrical nature of the situation yields a class of distinguished bases in $C_i(\hat{K}, \hat{L}, \mathbb{Z})$. Let $e_1^i, \dots, e_r^i$ be i-dimensional cells in $K \setminus L$. For each e_j^i choose a representative $\hat{e}_j^i$, i.e., a cell of the complex $\hat{K}$ lying over e_j^i, and fix an orientation such that $\hat{e}_j^i$ defines an element of the module $C_i(\hat{K}, \hat{L}, \mathbb{Z})$ which we will also denote by $\hat{e}_j^i$. Then $\hat{e}^i, \dots, \hat{e}_r^i$ is the required basis in $C_i(\hat{K}, \hat{L}, \mathbb{Z})$. However, there is an arbitrariness in the choice of the cell $\hat{e}_j^i$. It is removed when we go to the group $\mathrm{Wh}(\pi)$. The (algebraic) simple homotopy type of the chain complex $C(\hat{K}, \hat{L})$ is defined uniquely and does not depend on the partition of the topological spaces K and L into cells. If L is a deformation retract of K, then $H_i(\hat{K}, \hat{L}, \mathbb{Z}) = 0$, and therefore $C(\hat{K}, \hat{L})$ is an acyclic chain complex. We define the Whitehead torsion $\tau(\hat{K}, L)$ as the torsion of the chain complex $C(\hat{K}, L)$. No distinguished points were taken into account in the computation of $\tau(K, L)$, since any inner automorphism of the group π induces the identity automorphism of $\mathrm{Wh}(\pi)$. If $K \supset L \supset M$, where L and M are deformation retracts of K, then

$$\tau(K, M) = \tau(K, L) + \tau(L, M).$$

Let $f: K \to L$ be a cellular homotopy equivalence between two finite CW-complexes K and L. The torsion of the mapping f is the element $\tau(f) = f_*(\tau(M_f, K))$. We list the basic properties of the torsion $\tau(f)$:

(1) If $f, g: K \to L$ are homotopy equivalences that are cellular mappings, and f is homotopy equivalent to g, then $\tau(f) = \tau(g)$.

(2) A cellular homotopy equivalence $f: K \to L$ is a simple homotopy equivalence if and only if $\tau(f) = 0$.

(3) If $i: L \to K$ is an inclusion, then $\tau(i) = \tau(K, L)$.

(4) $\tau(g \circ f) = \tau(g) + g_*(\tau(f))$.

(5) If M, K, L are finite CW-complexes, where M is connected and L is a deformation retract of K, then $\tau(K \times M, L \times M) = \chi(M) \times i_*(\tau(K, L))$, where $i: L \to L \times M$, $i(x) = (x, y)$ for some fixed point $y \in M$, and $\chi(M)$ denotes the Euler characteristic of M.

(6) If $f \times g: K \times M \to L \times N$, where f and g are homotopy equivalences between connected CW-complexes K and L, and M and N, respectively, and if $i: L \to L \times N$ and $j: N \to L \times N$ for some fixed points $x \in N$, $y \in L$, then

$$\tau(f \times g) = \chi(N) \times i_*(\tau(f)) + \chi(L) \times j_*(\tau(g)).$$

(7) If $\tau_0 \in \mathrm{Wh}(L)$, then there exist a CW-complex K and a cellular homotopy equivalence $f: K \to L$ with $\tau(f) = \tau_0$.

In the remainder of this section we consider the connection between homotopy equivalence and simple homotopy equivalence and the question of topological invariance of simple homotopy type.

Denote by $\varepsilon(K)$ the group of homotopy self-equivalences of a CW-complex K. There is a natural mapping $\tau: \varepsilon(K) \to \mathrm{Wh}(\pi_1(K, e^0))$ associating with each homotopy self-equivalence $f: K \to K$ the torsion $\tau(f)$. In general, τ is not a surjection. For example, as shown by Metzler, for a two-dimensional CW-complex K^2 with the fundamental group $\pi_1(K^2, e^0) \approx \mathbb{Z} \times \mathbb{Z}_p$ constructed from the presentation $\{a, b, b^p, aba^{-1}b^{-1}\}$, not all elements of the group $\mathrm{Wh}(\mathbb{Z} \times \mathbb{Z}_p)$ can be realized as values of the torsion of a homotopy self-equivalence $f: K^2 \to K^2$ [91]. This implies that there is a CW-complex L homotopy equivalent to K^2 but not simple homotopy equivalent to K^2. Recently Lustig and Metzler constructed examples of two-dimensional CW-complexes that are homotopy equivalent but not simple homotopy equivalent (see [79, 92]). In dimension 3 such examples have been known for a long time. There are three-dimensional lense spaces that are homotopy equivalent but not simple homotopy equivalent. For lense spaces a complete classification up to simple homotopy type is known [25].

Denote by $\mathrm{Wh}(G\{a_i, r_i\})$ the subset of the group $\mathrm{Wh}(G)$ defined as follows. Let G be a presentation of the group G by generators and defining relations, K a two-dimensional CW-complex constructed from this presentation, and $\varepsilon(K)$ its group of homotopy self-equivalences. The image of the mapping $\tau: \varepsilon(K) \to \mathrm{Wh}(G)$ is denoted by $\mathrm{Wh}(G\{a_i, r_i\})$.

A problem in the theory of simple homotopy type that has remained unsolved for a long time is that of topological invariance of simple homotopy type: let $f: K \to L$ be a homomorphism between CW-complexes K and L; does this imply that $\tau(f) = 0$?

This problem was recently solved by Chapman [18]. Chapman's approach is based on the analysis of Hilbert manifolds. We now state his main results.

Let $Q = \prod_{j=1}^{\infty} I_j$, $I_j = [0, 1]$, be the Hilbert cube.

THEOREM 6.1. *Let K and L be finite CW-complexes. A mapping $f: K \to L$ is a simple homotopy equivalence if and only if the mapping $f \times \mathrm{id}_Q: K \times Q \to L \times Q$ is homotopy equivalent to a homeomorphism of $K \times Q$ onto $L \times Q$.* □

THEOREM 6.2. *If $f: K \to L$ is a homeomorphism between finite CW-complexes K and L, then f is a simple homotopy equivalence.* □

THEOREM 6.3. *If K and L are finite CW-complexes, then $K \curvearrowright L$ if and only if $K \times Q$ is homeomorphic to $L \times Q$.* □

The proofs of these results can be found in [18].

§3. Two-dimensional complexes

Let a presentation of a finitely defined group $G = \{a_1, \dots, a_k, r_1, \dots, r_l\}$ be given. It is known that there exists a CW-complex K^2 with one zero-dimensional, k one-dimensional, and l two-dimensional cells, and $\pi_1(K^2, e^0) \approx G$. It is constructed as follows. Consider a wedge of k-spheres $\bigvee_{i=1}^k S_i^1$ and take their common point for the zero-dimensional vertex. Evidently, $\pi_1(\bigvee_{i=1}^k S_i^1, e^0) = F$ is a free group for which the spheres S_i^1 are generators. Each relation r_j defines a word in the group F, and therefore an element of the fundamental group $\pi_1(\bigvee_{i=1}^k S_i^1, e^0)$ which we shall denote by γ_j. Let us attach a two-dimensional sphere S_j^2 to $\bigvee_{i=1}^k S_i^1$ via the mapping $f_j: S_j^1 \to \bigvee_{i=1}^k S_i^1$ realizing the element γ_j, $1 \le j \le l$. By the van Kampen-Seifert theorem, the fundamental group of the two-dimensional CW-complex so obtained is isomorphic to G. As follows from the proposition, the simple homotopy type of this two-dimensional CW-complex does not depend on the choice of the mapping in the homotopy class realizing the element γ_j. This complex is called the cellular model of the presentation of the group G. The converse statement is also true. Every two-dimensional CW-complex with the fundamental group $\pi_1(K^2, e^0)$ has the simple homotopy type of the cellular model of some finite presentation of the group $\pi_1(K^2, e^0)$.

Given a presentation $G = \{g_1, \dots, g_k, r_1, \dots, r_l\}$, denote by F the free group generated by the elements $a_1, \dots, a_k$. There is the canonical epimorphism $\varphi: F \to G$, $\varphi(a_i) = g_i$. Clearly, the elements in F corresponding to the relations r_j belong to the kernel of the epimorphism φ, and the least normal subgroup containing them coincides with $\operatorname{Ker}\varphi = N$. Let a two-dimensional CW-complex K^2 be the cellular model of this presentation. Consider the universal covering space $p: \hat{K}^2 \to K^2$ and construct the chain complex of free $\mathbb{Z}[G]$-modules

$$\mathbb{Z} \xleftarrow{\varepsilon} C_0 \xleftarrow{\partial_1} C_1 \xleftarrow{\partial_2} C_2,$$

where $\varepsilon: C_0 = \mathbb{Z}[G] \to \mathbb{Z}$ is the augmentation mapping. As noted in the preceding section, one can take for the basis elements in the modules C_i certain liftings of i-dimensional cells of the complex on the universal covering space. Let $[e^0]$, $[e_1^1], \dots, [e_k^1]$, $[e_1^2], \dots, [e_l^2]$ be fixed bases in the modules C_0, C_1, C_2, respectively. The boundary homomorphism $\partial_1: C_1 \to C_0$ is given by the equality $\partial_1([e_i^1]) = (g_i - 1)[e^0]$. The description of the boundary homomorphism $\partial_2: C_2 \to C_1$ is more complicated, and we have to recall several facts. Let G be a group and M a $\mathbb{Z}[G]$-module. A crossed homomorphism (derivation) from G into M is a mapping $d: G \to M$ such that $d(gh) = d(g)h + gd(h)$, $g, h \in G$. Whitehead defined a crossed homomorphism $\rho: F \to C_1$ by the equalities $\rho(a_i) = [e_i^1]$, $\rho(a_i^{-1}) = -g_i^{-1}\rho(a_i)$, $\rho(a_i^1 a_i^1) = \rho(a_i^1) + g_i(a_i^1)$, where $g_i = \varphi(a_i)$. The group C_1 is regarded as a $\mathbb{Z}[F]$-module via the epimorphism $F \to G$. Clearly, these three conditions make it possible to find the value of the crossed homomorphism on an arbitrary word $r \in F$. The boundary homomorphism $\partial_2: C_2 \to C_1$ is given by the equality $\partial_2(e_j^2) = \rho(r_j)$. There is an alternative description of the boundary homomorphism ∂_2 in terms of the Fox free differential calculus [53]. Let $\partial/\partial a_i: F \to \mathbb{Z}[F]$ be the only crossed homomorphism taking a_i into $\delta_i j$, and let $\varphi_*: \mathbb{Z}[F] \to \mathbb{Z}[G]$ be the ring homomorphism induced by the mapping $\varphi: F \to G$. Then

$$\rho(r) = \sum_{i=1}^{n} \varphi_*\left(\frac{\partial r}{\partial a_i}\right)[e_i^1].$$

The matrix of the homomorphism ∂_2 with respect to the bases chosen is called the Jacobi matrix of the presentation. Since the universal space $\hat{K}^2$ is simply connected, $H_1(\hat{K}^2, \mathbb{Z}) = 0$ and therefore the chain complex is a segment of a free resolution. As is known, from each presentation of the group G we can canonically construct a free resolution of $\mathbb{Z}[G]$-modules. It is the so-called Lyndon-Fox resolution. This resolution has no relation to CW-complexes in the sense that its construction is purely algebraic and makes no use of CW-complexes. The important fact is that there always exists a CW-complex for which the chain complex of its universal covering space coincides with the Lyndon-Fox resolution. In particular, the CW-complex constructed via the presentation of the group G realizes a segment of the Lyndon-Fox resolution. We shall use this fact in what follows.

The following problem (coming back to Wall) is not yet solved completely. Given a group G with the generators $a_1, \dots, a_k$ and relations $r_1, \dots, r_l$, consider a chain complex of $\mathbb{Z}[G]$-modules

$$(C, \partial): 0 \longleftarrow \mathbb{Z} \overset{\varepsilon}{\longleftarrow} C_0 \overset{\partial_1}{\longleftarrow} C_1 \overset{\partial_2}{\longleftarrow} C_2$$

satisfying the conditions

(1) $H_1(C, \partial) = 0$;
(2) $H_0(C, \partial) = \mathbb{Z}$;
(3) $C_0 = \mathbb{Z}[G], \quad \varepsilon\mathbb{Z}[G] \to \mathbb{Z}$;
(4) the boundary homomorphism ∂_1 is given by the equality $\partial_1(c_i^1) = (g_i - 1)c^0$.

The question is: does there exist a two-dimensional CW-complex K^2 such that the chain complex of its universal covering space $p: \hat{K}^2 \to K^2$ coincides with (C, ∂)? The answer is negative [33]. Dunwoody constructed an example of a chain complex satisfying conditions (1)–(4) but not realizable via a two-dimensional CW-complex. The example is as follows. Let $G = \mathbb{Z}_5 * \mathbb{Z}$. Denote the generator of order 5 by u, and the generator of infinite order by v. The modules C_2 and C_1 are of rank 1 and 2, respectively. The boundary homomorphism $\partial_1: C_1 \to C_0$ is given by the equalities $\partial_1(e_1^1) = (u-1)e^0$, $\partial_1(e_2^1) = (v-1)e^0$. The boundary homomorphism $\partial_2: C_2 \to C_1$ is given by $\partial_2(e^2)UvNe_1^1$, where $U = u + u^2 - u^4$ is the nontrivial unit in the ring $\mathbb{Z}[G]$ and $N = 1 + u + u^2 + u^3 + u^4$ is the norm.

The question arises: what conditions should be imposed on a chain complex in order for it to be realizable geometrically via a two-dimensional CW-complex? This question is of great importance since it arises in the problem of classifying two-dimensional CW-complexes up to homotopy type. Following Cockroft and Moss, we say that a chain complex

$$\mathbb{Z} \longleftarrow C_0 \overset{\partial_1}{\longleftarrow} C_1 \overset{\partial_2}{\longleftarrow} C_2 \longleftarrow \cdots$$

satisfying conditions (1)–(4) is stably realizable if the chain complex

$$\mathbb{Z} \overset{\varepsilon}{\longleftarrow} C_0 \overset{\hat{\partial}_1}{\longleftarrow} C_1 \oplus M \overset{\hat{\partial}_2}{\longleftarrow} C_2 \oplus M$$

can be realized via a two-dimensional CW-complex. Here M is a free module of finite rank. Cockroft and Moss proved that the chain complex constructed by Dunwoody is stably realizable. They conjectured that if $SK_1(\mathbb{Z}[G]) = 0$, then a chain complex satisfying conditions (1)–(4) is stably realizable, and verified that for a finite cyclic group this is indeed the case [24].

The following theorem is a cornerstone in the classification problem of two-dimensional CW-complexes.

THEOREM 6.4 [159]. *Let K and L be two-dimensional complexes with isomorphic fundamental groups $\pi_1(K, e^0) \approx \pi_1(L, \bar{e}^0) \approx G$. Denote by $\hat{K}$ and $\hat{L}$ their universal covering spaces, and by $C(\hat{K})$ and $C(\hat{L})$ the associated chain complexes over the ring $\mathbb{Z}[G]$. Then K and L belong to the same homotopy type if and only if $C(\hat{K})$ and $C(\hat{L})$ are homotopy equivalent.* □

This theorem reduces the classification of two-dimensional CW-complexes to the classification of chain complexes. A question now arises about the connection between elementary operations over CW-complexes and geometric operations over CW-complexes.

Let $G = \{a_1, \dots, a_k, r_1, \dots, r_l\}$ be a presentation of the group G. Consider the following transformations of generators and relations:

(1) replacing a_i by a_i^{-1};
(2) replacing a_i by $a_i a_j$, $i \neq j$;
(3) replacing r_i by r_i^{-1};
(4) replacing r_i by $r_i r_j$, $i \neq j$;
(5) replacing r_i by $x r_i x^{-1}$, where $x \in F$;
(6) renumbering the generators a_i and relations r_j.

An application of these transformations to a presentation of the group G yields a new presentation $G\{\bar{a}_1, \dots, \bar{a}_k, \bar{r}_1, \dots, \bar{r}_l\}$. Let K^2 and L^2 be cellular models realizing these presentations. There is a formal deformation $K^2 = K_0 \to \cdots \to K_q = L^2$ such that the dimension of the CW-complex K_j is not greater than three. Consider the chain complex $C(\hat{K}^2)$ of the universal covering space $\hat{K}^2$. The transformations (1)–(6) correspond to the following transformations of bases in the chain modules:

(1) replacing a_i^1 by $g_i^{-1} e_i$, where $g_i = \varphi(a_i)$;
(2) replacing e_i^1 by $e_i^1 + g_i e_j^1$;
(3) replacing e_i^2 by $-e_i^2$;
(4) replacing e_i^2 by $e_i^2 + e_j^2$;
(5) replacing e_i^2 by $g e_i^2$, where $g = \varphi(x)$;
(6) renumbering the elements of the bases in the modules C_1 and C_2.

This is proved easily by a direct calculation.

We now present major results on the homotopy classification of two-dimensional chain complexes.

Whitehead proved that if two CW-complexes K and L have isomorphic fundamental groups, then there exist positive integers n and m such that the complexes $K \bigvee_{i=1}^n S_i^2$ and $L \bigvee_{j=1}^m S_j^2$ are homotopy equivalent [159].

The following theorem is due to Wall [40, 154].

THEOREM 6.5. *Let K^2 be a two-dimensional CW-complex with the free fundamental group $\pi_1(K^2, e^0) \approx F_k$ and Euler characteristic $\chi(K^2) = l$.*

Then K^2 *is homotopy equivalent to a* CW*-complex of the form*

$$\bigvee_{i=1}^{k} S_i^1 \bigvee_{j=1}^{n} S_j^2, \qquad l = 1 - k + n,$$

where $n = \operatorname{rank} H_2(K^2, \mathbb{Z})$. □

Since $\mathrm{Wh}(F_k) = 0$, the homotopy type and simple homotopy type coincide.

Let T^2 be the two-dimensional torus, i.e., the geometrical realization of the presentation $\{a_1, a_2; a_1 a_2 a_1^{-1} a_2^{-1}\}$ of the group $\mathbb{Z} \oplus \mathbb{Z}$. The following theorems were proved by Jajodia [68].

THEOREM 6.6. *Let* K^2 *be a two-dimensional* CW*-complex with the fundamental group isomorphic to* $\mathbb{Z} \oplus \mathbb{Z}$ *and the Euler characteristic* $\chi(K^2) = l$. *Then* K *is homotopy equivalent to a* CW*-complex of the form* $T^2 \bigvee_{i=1}^{l} S_i^2$, *where* $l \geq 0$. □

In this case the homotopy type and the simple homotopy type coincide, because $\mathrm{Wh}(\mathbb{Z} \oplus \mathbb{Z}) = 0$.

THEOREM 6.7. *If each of two finite* CW*-complexes of dimension two has a single two-dimensional cell, then they belong to the same homotopy type if and only if their fundamental groups are isomorphic.* □

Suppose that K^2 realizes a presentation of the finite cyclic group $\mathbb{Z}_n$. The following theorem is due to Dyer and Sieradski [40].

THEOREM 6.8. *Let* K^2 *be a finite two-dimensional complex with the fundamental group isomorphic to* $\mathbb{Z}_n$. *Then* K^2 *has the same homotopy type as the complex* $S^1 \cup_n e^2 \vee S^2 \vee \cdots \vee S^2$, *where the number of two-dimensional spheres coincide with the rank of* $H_2(K^2, \mathbb{Z})$. *For each element* $\tau_0 \in \mathrm{Wh}(\mathbb{Z}_n)$ *there exists a homotopy equivalence* $f\colon K^2 \to K^2$ *with torsion* $\tau(f) = \tau_0$. □

Therefore, in this case the homotopy type and simple homotopy type coincide. For n prime the first part of the theorem was proved by Cockroft and Swan much earlier [23]. Subsequently Cockroft and Moss proved the second part of the theorem [24]. The case of the fundamental group isomorphic to $\mathbb{Z}_n \oplus \mathbb{Z}_m$ turned out to be similar.

THEOREM 6.9. *Two finite two-dimensional* CW*-complexes with the fundamental groups isomorphic to* $\mathbb{Z}_n \oplus \mathbb{Z}_m$ *have the same homotopy type if their Euler characteristics coincide.* □

This theorem was proved by Dyer [37]. Subsequently Latiolais completed the analysis of simple homotopy type in this case and obtained the following theorems [75].

THEOREM 6.10. *Let K^2 be a CW-complex realizing the presentation*

$$G = \{a_1, a_2; a_1^n, a_2^m, [a_1, a_2]\}.$$

Each element $\tau_0 \in \mathrm{Wh}(\mathbb{Z}_n \oplus \mathbb{Z}_m)$ of the Whitehead group can be realized via a homotopic self-equivalence $f: K^2 \to K^2$ with $\tau(f) = \tau_0$. □

THEOREM 6.11. *Let K^2 be a finite CW-complex with $\pi_1(K^2, e^0) \approx \mathbb{Z}_n \oplus \mathbb{Z}_m$ and Euler characteristic equal to k. Then K^2 is simple homotopy equivalent to a complex of the form $K \vee (k-2)S^2$, where K is a complex of the theorem.* □

The general case of a finite fundamental group turned out to be more complicated. Unlike the two preceding examples, the homotopy type is determined not only by the Euler characteristic of the CW-complex, but also by another invariant introduced by Dyer and Sieradski which they called 'bias'. Let us state the final result. Denote by $\{K_i^2, G\}$ the class of two-dimensional finite CW-complexes with the fundamental group isomorphic to G. Denote by $[K^2]$ the set of all representatives from $\{K_i^2, G\}$ having homotopy type of K^2. Dyer and Sieradski constructed an oriented graph $H(G)$ whose vertices are $K^2 \in \{K_i^2, G\}$. An arrow issues from the vertex $[K_{i_0}^2]$ and enters the vertex $[K_{i_1}^2]$ if $K_{i_1}^2$ is homotopy equivalent to $K_{i_0}^2 \vee S^2$. Denote by $\chi(K^2)$ the Euler characteristic of the CW-complex K^2. Let $\chi_{\min} = \min \chi(K_i^2)$. The level of the vertex $[K_{i_0}^2]$ is the number $\chi(K_{i_0}^2) - \chi_{\min}$. A complex $K_{i_0}^2$ is said to be a root if no arrow enters the vertex. A complex $K_{i_0}^2$ is a minimal root if the vertex $[K_{i_0}^2]$ is of level zero.

DEFINITION 6.1 [40]. A group G has essential height $\leq k$ if any two CW-complexes K^2 and L^2 of dimension two with the fundamental group isomorphic to G and Euler characteristics $\chi(K^2) = \chi(L^2) \geq 1 + k + d(G)$ are homotopy equivalent.

The following theorem was proved by Browning [16].

THEOREM 6.12. *For any finite group the essential height is ≤ 1.* □

For finite abelian groups this result was proved by Dyer [36]. It is known that a CW-complex K^2 realizing the presentation of the abelian group

$$G = \{a_1, \dots, a_k; [a_i, a_j]\}, \quad i < j, \qquad n_i | n_{i+1},$$

where $i = 1, \dots, k$, has minimal Euler characteristic. Thus, Theorem 6.12 implies that a CW-complex with the fundamental group isomorphic to G and Euler characteristic $\chi = \chi(K^2) + n$ has the homotopy type of the CW-complex $K \bigvee_{i=1}^n S_i^2$. Metzler, Sieradski, and Browning provided full descriptions of minimal roots for a finite abelian fundamental group. Besides the given presentation of the group $G \approx \mathbb{Z}_{n_1} \oplus \cdots \oplus \mathbb{Z}_{n_k}$ which is realized by a CW-complex with minimal Euler characteristic, there are so-called twisted presentations of this group $G = \{a_i; a_i^{n_i}, [a_1^{r_1}, a_2^{r_2}], [a_{k-1}^{r_{k-1}}, a_k^{r_k}]\}$, $(r_i, n_i) = 1$, $i = 1, \ldots, k$, $k > 2$, that are also realized by CW-complexes with minimal Euler characteristic. The following theorem follows from a comparison of results obtained by Browning, Metzler, and Sieradski [15, 16, 90, 142].

THEOREM 6.13. *Let* $G = \mathbb{Z}_{n_1} \oplus \mathbb{Z}_{n_2} \oplus \cdots \oplus \mathbb{Z}_{n_k}$. *Then any two-dimensional CW-complex with the fundamental group isomorphic to* G *and minimal Euler characteristic has the homotopy type of the CW-complex constructed via the presentation* $\{a_i; a_i^{n_i}, [a_1^{r_1}, a_2^{r_2}], \ldots, [a_{k-1}^{r_{k-1}}, a_k^{r_k}]\}$, $i = 1, \ldots, k$, $(r_i, n_i) = 1$. *Two CW-complexes* K^2 *and* L^2 *constructed via the presentations* $G = \{a_i; a_i^{n_i}, [a_i^{r_i}, a_j^{r_j}]\}$, $G = \{a_i; a_i^{n_i}, [a_i^{\bar r_i}, a_j^{\bar r_j}]\}$ *have the same homotopy type if and only if*

$$\prod_{i,j} r_i r_j \equiv \pm t^{k-1} \pi \bar r_i \bar r_j \mod n_1, \qquad (k, n_1) = 1. \qquad \square$$

The following two theorems were proved by Latiolais and fully explain the structure of finite two-dimensional CW-complexes with finite abelian fundamental group [75].

THEOREM 6.14. *Let* K^2 *be a finite two-dimensional CW-complex with finite abelian fundamental group. Then all elements of* $\mathrm{Wh}(\pi_1(K^2, e^0))$ *can be realized as values of the torsion of a homotopy self-equivalence* $f: K^2 \to K^2$. $\square$

THEOREM 6.15. *For finite CW-complexes of dimension two with finite abelian fundamental group, homotopy equivalence and simple homotopy equivalence coincide.* $\square$

Let $D_m = \{a_1, a_2; a_1^m a_2^{-2}, [a_2, a_1^{m-1/2}]a^{-1}\}$, where m is odd, be a presentation of the dihedral group. Since the order of the group D_m is equal to $2m$, the two-dimensional CW-complex realizing this presentation has minimal Euler characteristic. Jajodia and Magurn obtained the following theorem [69].

THEOREM 6.16. *Let* D_m *be the dihedral group of order* $2m$, *where* m *is odd. Then two finite two-dimensional CW-complexes with the fundamental group isomorphic to* D_m *are simple homotopy equivalent if and only if their Euler characteristics are equal.* $\square$

For infinite groups, there also exist nonminimal roots. Dunwoody constructed examples of finite two-dimensional CW-complexes with fundamental groups isomorphic to the trefoil group $G = \{a_1, a_2; a_1^2 = a_2^3\}$; their level is equal to one, they have the same Euler characteristic but are not homotopy equivalent. The complexes are of the following form. The complex K^2 realizes the presentation of the trefoil group $G = \{a_1, a_2; a_1^2 = a_2^3, 1\}$, and the complex L^2 realizes the presentation of the same group of the form $G = \{a_1, a_2; (a_1^2 a_2^{-3})(a_1^2 a_2^{-3})^{a_1}(a_1^2 a_2^{-3})^{a_1^2}, (a_1^2 a_2^{-3})(a_1^2 a_2^{-3})^{a_2}(a_1^2 a_2^{-3})^{a_2^2}\}$. Note that the complexes $K^2 \vee S^2$ and $L^2 \vee S^2$ are homotopy equivalent $(a^h = h^{-1}ah)$.

Recall that by the defect of the presentation $G = \{a_1, \dots, a_k; r_1, \dots, r_l\}$ we mean the number $l - k = d(a_i, r_j)$. The defect of the group G is, by definition, equal to $d(G) = d(a_i, r_j)$, where (a_i, r_j) runs over all presentations of the group G. The presentation of the form $G = \{a_1, \dots, a_{\mu(G)}; r_1, \dots, r_l\}$, where $\mu(G)$ is the minimal number of generators of the group G, is said to be minimal. Set $d^\mu(G) = \min d(a_i, r_j)$, where (a_i, r_j) runs over all minimal presentations of G. Evidently, $d^\mu(G) \geq d(G)$; however, the author knows of no example of a finitely presented group with $d^\mu(G) > d(G)$.

DEFINITION 6.2. A group G is said to satisfy condition (h, d) if for any presentation of the group G whose defect is equal to d there exists a minimal presentation of G such that the segments of the Lyndon-Fox resolution of length equal to 3, constructed via these presentations, are homotopy equivalent.

DEFINITION 6.3. A group G is said to satisfy condition (h) if G satisfies condition (h, d) for all d, where d is the defect of the presentation.

An equivalent definition can be given in terms of CW-complexes.

DEFINITION 6.4. A group G is said to satisfy condition (h, d) if for any finite two-dimensional CW-complex K^2 constructed via a presentation of G with defect d, there exists a homotopy equivalent finite two-dimensional CW-complex L^2 constructed via a minimal presentation.

A similar definition can be given in terms of CW-complexes for a group satisfying condition (h).

PROPOSITION 6.4. *The following groups satisfy condition* (h):

(1) *free abelian groups*;
(2) $\mathbb{Z} \oplus \mathbb{Z}$;
(3) *finite abelian groups*;
(4) *the dihedral group* D_m (*where* m *is odd*). □

PROOF. Item (1) follows from the Wall theorem (Theorem 6.5), (2) is a consequence of the Jajodia theorem (Theorem 6.6), (3) follows from the results of Dyer, Sieradski, Browning, and Metzler (Theorems 6.8–6.13), and (4) from the Jajodia-Magurn theorem (Theorem 6.16). □

PROPOSITION 6.5. *Finite groups satisfy condition* (h, d) *for* $d \geq d^\mu(G)+1$, *and nilpotent groups for* $d \geq d(G) + 1$.

PROOF. The first assertion follows from the Browning theorem (Theorem 6.12). In the case of nilpotent groups we have to make use of a theorem proved by Rapaport [117].

There is a conjecture that any s-group satisfies condition (h). It is not known whether there exists a positive integer n such that the group G satisfies condition (h, N) for all $N \geq n$. □

§4. Crossed modules

A crossed module is a triple (C, G, d), where C and G are groups, $d: C \to G$ is a homomorphism, and G acts on C on the left (the action will be denoted by gc). Furthermore, the homomorphism d is to satisfy the conditions

$$c_1 + c_2 - c_1 = d(c_1)c_2, \qquad d(gc) = g(d(c))g^{-1}.$$

A morphism of crossed modules (C, G, d) and (C', G', d') is a commutative diagram

$$\begin{array}{ccc} C & \xrightarrow{d} & G \\ {\scriptstyle\varphi}\downarrow & & \downarrow{\scriptstyle\psi} \\ C' & \xrightarrow{d'} & G', \end{array}$$

where $\varphi: C \to C'$, $\psi: G \to G'$ are group homomorphisms satisfying the conditions $\psi \circ d = d' \circ \varphi$, $\varphi(gc)\psi = \psi(g)\varphi(c)$.

The following are immediate consequences of the definition:

(1) $d(C)$ is a normal subgroup of G;

(2) $\operatorname{Ker} d$ belongs to the center of C;

(3) the action of G on C induces the natural structure of a $(G/d(C))$-module on the center of C, and $\operatorname{Ker} d$ is a submodule of this module;

(4) the action of G on C induces the structure of a $(G/d(C))$-module on the group $C^{\mathrm{ab}} = C/[C, C]$.

An important case of a crossed module is the so-called free crossed module defined by Whitehead [159].

DEFINITION 6.5. Let (C, G, d) be a crossed module and $\{c_{i\in I}\}$ a fixed set of elements of C. Then (C, G, d) is called a free crossed module with basis $\{c_{i\in I}\}$ if for every crossed module (C', G', d'), arbitrary set of elements $\{c'_{i\in I}\}$ of C', and homomorphism $\psi: G \to G'$ such that $\psi \circ d(c_i) = d'(c'_i)$, there exists a unique homomorphism $\varphi: C \to C'$ such that $\varphi(c_i) = c'_i$ and (φ, ψ) is a crossed module homomorphism.

A constructive definition of free crossed modules was given by Whitehead. Let us recall the main points of this construction.

Let G be a group, and $\{g_{a\in A}\}$ a set of fixed elements in G. Denote by E the free group generated by the set $G \times A$. Denote by C the quotient of the group E by the normal subgroup W, the closure of the set $\{(x, a)(y, b)(x, a)^{-1}(xg_a x^{-1}y, b)^{-1},\ x, y \in G,\ a, b \in A\}$. The group C is a crossed module with the G-action induced by the action of G on E given by $g(x, a) = (gx, a)$. The boundary homomorphism $\hat{d}: E \to G$

is induced by the boundary homomorphism $\hat{d}: E \to G$ given by the formula $\hat{d}(x, a) = xg_a x^{-1}$. Let $p: E \to C$ be the natural projection and $c_a = p(1, a)$. Whitehead proved that C is a free crossed module with basis $\{c_{a\in A}\}$. Up to an isomorphism, this module is defined just by the choice of basis elements in C. The following fundamental theorem is also due to Whitehead [159].

THEOREM 6.17. *Let X be a path-connected space, and Y a space obtained from X by attaching two-dimensional cells. Then $\pi_2(Y, X, x)$ is a free crossed $\pi_1(X, x)$-module with basis corresponding to the cells so attached.*

Whitehead arrived at the definition of a crossed module while studying the structure of the second relative homotopy group [159]. Peiffer and Reidemeister obtained the same notion independently when they were studying identities between relations in group presentations. We list here several important properties of free crossed modules that we need; their proofs can be found in [63, 65, 77, 112, 118, 119, 162]. If (C, G, d) is a free crossed module with basis $\{c_{i\in I}\}$, then $C^{\mathrm{ab}} = C/[C, C]$ is a free $(G/d(C))$-module with basis elements $c_i[C, C]$.

A set of elements $\{c'_{i\in I}\}$ of C is a basis if and only if

(1) the set of elements $c'_{i\in I}[C, C]$ is a basis for C^{ab};
(2) the normal closure of $d(c'_i)$ in G is $\mathrm{Im}\, d$.

The cardinal number of the basis elements of a crossed module is the same for all bases. The free module C^{ab} will be called the module associated with the crossed module (C, G, d). We recall what is meant by a free crossed submodule of a free crossed module. Let $(C, \mathfrak{F}, d)$ be a free crossed module with basis $(c_1, \dots, c_k)$, and $d(c_i) = f_i$, where $\mathfrak{F}$ is a free group with basis $(f_1, \dots, f_s)$. Suppose $\overline{\mathfrak{F}}$ is the subgroup generated by the part $(f_{i_1}, \dots, f_{i_r})$ of the basis, and $\overline{C}$ is the subgroup of C generated by $(c_{j_1}, \dots, c_{j_t})$ with the action of $\overline{\mathfrak{F}}$. Let, in addition, $d(\overline{C}) \subset \overline{\mathfrak{F}}$. If $\mathfrak{F}/d(C) = \overline{F}/d(\overline{C})$, then, as shown by Whitehead, $(\overline{C}, \overline{\mathfrak{F}}, d)$ is a free crossed module with basis $(c_{j_1}, \dots, c_{j_t})$.

Let $\hat{\mathfrak{F}}$ be a free group with basis $(\hat{f}_1, \dots, \hat{f}_k)$. We can then construct a free crossed module $(\hat{C}, \mathfrak{F} \cdot \hat{\mathfrak{F}}, \hat{d})$ with basis $(c_1, \dots, c_k, \hat{c}_1, \dots, \hat{c}_k)$, where $\hat{d}(c_i) = l_i$, $\hat{d}(\hat{c}_j) = \hat{f}_j$. Obviously there exists a canonical embedding $(C, \mathfrak{F}, d) \to (\hat{C}, \mathfrak{F}\cdot\hat{\mathfrak{F}}, \hat{d})$, i.e., the crossed module $(C, \mathfrak{F}, d)$ is a submodule of the crossed module $(\hat{C}, \mathfrak{F} \cdot \mathfrak{F}, \hat{d})$.

DEFINITION 6.6. The crossed module $(\hat{C}, \mathfrak{F}\cdot\hat{\mathfrak{F}}, \hat{d})$ is called a stabilization of the module of the crossed module $(C, \mathfrak{F}, d)$, and $(C, \mathfrak{F}, d)$ a contraction of $(\hat{C}, \mathfrak{F} \cdot \hat{\mathfrak{F}}, \hat{d})$.

Let M be a free module with basis $(m_1, \dots, m_n)$ over the ring $\mathbb{Z}[\mathfrak{F}/d(C)]$. Consider the group $C \times M$, and put $\hat{d}|_{C\times 0} = d$, $\hat{d}|_{0\times M} = 0$.

Obviously, $(C \times M, \hat{F}, \hat{d})$ is a free crossed module with basis $(c_1, \dots, c_k, m_1, \dots, m_n)$.

DEFINITION 6.7. The crossed module $(C \times M, \mathfrak{F}, \hat{d})$ is called a thickening of the crossed module $(C, \mathfrak{F}, d)$.

Let $(C, \mathfrak{F}, d)$ be a free crossed module with basis $(c_1, \dots, c_k)$. We say that a basis $(c_1', \dots, c_k')$ is obtained from the basis $(c_1, \dots, c_k)$ by elementary transformations if

(1) $c_{i_0}' = fc_{i_0}$, $c_i = c_i'$, $i \neq i_0$; or

(2) $c_{i_0}' = -c_{i_0}$, $c_i = c_i'$, $i \neq i_0$; or

(3) $c_{i_0}' = c_{i_0} + c_j$ $(c_{i_0}' = c_j + c_{i_0})$, $c_i = c_i'$, $i \neq i_0$, $j \neq i_0$.

DEFINITION 6.8. Let $(C, \mathfrak{F}, d)$ be a free crossed module with bases $(c_1, \dots, c_k)$ and $(c_1', \dots, c_k')$. The bases are said to be equivalent if one can be obtained from the other by a sequence of elementary transformations.

Let (C, G, d) be a crossed module and $\psi: G' \to G$ a homomorphism. Consider the diagram

$$\begin{array}{ccc} C' & \xrightarrow{d'} & G' \\ \varphi\downarrow & & \downarrow\psi \\ C & \xrightarrow{d} & G. \end{array}$$

Here $C' = \{c, g'\} \in C \times G'/\{d(C) = \psi(g')\}$. Evidently, (C', G', d') is a crossed G'-module. The group G' acts on C' by the formula $g'(c, g') = (\psi(\overline{g}')c, \overline{g}'g'(\overline{g}'))^{-1}$, and the boundary homomorphism is given by the formula $d(c, g') = g'$. The crossed module (C', G', d') is called the pullback induced from the homomorphism ψ.

Let (C, G, d) be a crossed module defining the exact sequence

$$e \longleftarrow G/d(C) \longleftarrow G \overset{d}{\longleftarrow} C \longleftarrow \operatorname{Ker} d \longleftarrow 0,$$

where $\operatorname{Ker} d$ is a $\mathbb{Z}[G/d(C)]$-module. Suppose that there is a homomorphism of modules $f: \operatorname{Ker} d \to M$, and a commutative diagram

$$\begin{array}{ccccccccc} e & \longleftarrow & G/d(C) & \longleftarrow & G & \overset{d}{\longleftarrow} & C & \overset{i}{\longleftarrow} & \operatorname{Ker} d & \longleftarrow & 0 \\ & & \mathrm{id}\downarrow & & \mathrm{id}\downarrow & & \downarrow g & & \downarrow f & & \\ e & \longleftarrow & G'/d(C') & \longleftarrow & G & \overset{d'}{\longleftarrow} & C' & \longleftarrow & M & \longleftarrow & 0. \end{array}$$

Here $C' = M \times C/N$, $N = \{-f(x), i(x)\}$, $x \in \operatorname{Ker} d$. The action of G on C' is induced by the diagonal action of G on $M \times C$. The crossed module (G, C', d') defined by the exact sequence

$$e \longleftarrow G/d(C) \longleftarrow G \longleftarrow C \longleftarrow \operatorname{Ker} d \longleftarrow 0$$

is called the pushout induced from the homomorphism f.

Consider another important class of crossed modules, viz. projective crossed modules introduced by Ratcliffe [119]. A crossed module (C, G, d) is said to be projective, if for any epimorphism of crossed modules

$$\begin{array}{ccc} D & \xrightarrow{\partial} & H \\ {\scriptstyle\varphi}\downarrow & & \downarrow{\scriptstyle\psi} \\ D' & \xrightarrow{\partial'} & H' \end{array}$$

and any morphism of crossed modules

$$\begin{array}{ccc} C & \xrightarrow{d} & G \\ {\scriptstyle\eta}\downarrow & & \downarrow{\scriptstyle\xi} \\ D' & \xrightarrow{\partial'} & H' \end{array}$$

there exists a morphism

$$\begin{array}{ccc} C & \xrightarrow{d} & G \\ {\scriptstyle\beta}\downarrow & & \downarrow{\scriptstyle\alpha} \\ D & \xrightarrow{\partial} & H \end{array}$$

such that $\eta = \varphi \circ \beta$, $\xi = \psi \circ \alpha$.

Let (C, G, d) be a crossed module, $N = \operatorname{Im} d$, $C^{\mathrm{ab}} = C/[C, C]$. Ratcliffe proved that (C, G, d) is a projective crossed module if and only if C^{ab} is a projective $\mathbb{Z}[G/d[G]]$-module and the mapping of the two-dimensional homology groups

$$d_*: H_2(C) \to H_2(N)$$

induced by the homomorphism $d: C \to G$ is trivial [119]. It is clear that all free crossed modules are projective. The following important theorem is due to Dyer [39].

THEOREM 6.18. *Let L^2 be a connected subcomplex of a two-dimensional CW-complex K^2. Denote by G the group $\pi_1(L^2, e_0)$, $C = \pi_2(K^2, L^2, e_0)$. The triple $(C, G, d: \pi_2(K^2, L^2, e^0) \to \pi_1(L^2, e^0))$ is a crossed projective module.* □

The homomorphism $d: \pi_2(K^2, L^2, e^0) \to \pi_1(L^2, e^0)$ is taken from the exact homotopy sequence for the pair (K^2, L^2).

Every crossed module (C, G, d) determines an exact sequence

$$e \longleftarrow G/d(C) \longleftarrow G \longleftarrow C \longleftarrow \operatorname{Ker} d \longleftarrow 0,$$

called by Mac Lane and Whitehead in [82] a crossed sequence. Put

$$G/d(C) = B, \qquad \operatorname{Ker} d = A.$$

We denote by $\Omega(B, A)$ the set of crossed sequences starting with the $\mathbb{Z}[B]$-module A and ending with the group B. On $\Omega(A, B)$ we define an equivalence relation. We say that a crossed sequence

$$e \longleftarrow B \longleftarrow G \longleftarrow C \longleftarrow A \longleftarrow 0$$

is congruent to a crossed sequence

$$e \longleftarrow B \longleftarrow G' \longleftarrow C' \longleftarrow A \longleftarrow 0$$

if there exists a homomorphism of the corresponding crossed modules

$$\begin{array}{ccccccccccc} e & \longleftarrow & B & \longleftarrow & G & \longleftarrow & C & \longleftarrow & A & \longleftarrow & 0 \\ & & {\scriptstyle \mathrm{id}}\downarrow & & {\scriptstyle \psi}\downarrow & & \downarrow{\scriptstyle \varphi} & & {\scriptstyle \mathrm{id}}\downarrow & & \\ e & \longleftarrow & B & \longleftarrow & G' & \longleftarrow & C' & \longleftarrow & A & \longleftarrow & 0. \end{array}$$

This congruence relation on $\Omega(B, A)$ generates an equivalence relation on $\Omega(B, A)$. We denote by $X\,\mathrm{seq}(B, A)$ the corresponding set of equivalence classes of crossed sequences. Using composition, $X\,\mathrm{seq}(B, A)$ can be turned into a group which is isomorphic to the cohomology group $H^3(B, A)$ (see [118] for details). In what follows we shall need a specific description of this isomorphism $X\,\mathrm{seq}(B, A) \to H^3(B, A)$. It is known that in each equivalence class of crossed sequences

$$e \longleftarrow B \xleftarrow{\;j\;} G \xleftarrow{\;d\;} C \xleftarrow{\;i\;} A \longleftarrow 0$$

there exists a crossed sequence in which G is a free group. It is obtained by choosing an arbitrary free presentation of B,

$$e \longleftarrow B \xleftarrow{\;p\;} \mathfrak{F} \longleftarrow R \longleftarrow e,$$

and constructing the pullback induced from the homomorphism $\psi\colon \mathfrak{F} \to G$, such that $j \cdot \psi = \psi$:

$$\begin{array}{ccccccccccc} e & \longleftarrow & B & \longleftarrow & \mathfrak{F} & \longleftarrow & D & \longleftarrow & A & \longleftarrow & 0 \\ & & \downarrow & & \downarrow & & \downarrow & & \downarrow & & \\ e & \longleftarrow & B & \longleftarrow & G & \longleftarrow & C & \longleftarrow & A & \longleftarrow & 0. \end{array}$$

Now let

$$e \longleftarrow B \xleftarrow{\;p\;} \mathfrak{F} \xleftarrow{\;d\;} C \xleftarrow{\;i\;} A \longleftarrow 0$$

be a crossed sequence for which $\mathfrak{F}$ is a free group. Consider the subsequence

$$e \longleftarrow R \longleftarrow C \longleftarrow A \longleftarrow 0,$$

where $R = \operatorname{Ker} p$, and make it commutative:

$$0 \longleftarrow R^{\mathrm{ab}} \longleftarrow C^{\mathrm{ab}} \longleftarrow A \longleftarrow 0.$$

Since $\mathfrak{F}$ acts on R by conjugation, the group R^{ab} can be turned into a $\mathbb{Z}[B]$-module. It is easy to see that C^{ab} is also a $\mathbb{Z}[B]$-module. Since A is in the center of C, C can be represented as a product $R \times A$. Let $s: R \to C$ be a homomorphism such that $d \cdot s = \mathrm{id}$. Then every element $c \in C$ has a unique representation $c = a_c + s \cdot d(c)$. If $p(f) = b \in B$, we define the action of b on C^{ab} by the formula $bc = ba_c + s \cdot d(f \cdot c) \bmod [C, C]$. From the exact sequence of groups

$$e \longleftarrow B \longleftarrow \mathfrak{F} \longleftarrow R \longleftarrow e$$

we can construct an exact sequence of $\mathbb{Z}[B]$-modules

$$0 \longleftarrow I[B] \longleftarrow I[\mathfrak{F}] \otimes \mathbb{Z}[B] \longleftarrow R^{\mathrm{ab}} \longleftarrow 0,$$

where $I[B]$ and $I[\mathfrak{F}]$ are the augmentation ideals; see [63]. Combining this with the sequence

$$0 \longleftarrow R^{\mathrm{ab}} \longleftarrow C^{\mathrm{ab}} \longleftarrow A \longleftarrow 0,$$

we obtain an exact sequence of $\mathbb{Z}[B]$-modules

$$0 \longleftarrow I[B] \longleftarrow I[\mathfrak{F}] \otimes \mathbb{Z}[B] \longleftarrow C^{\mathrm{ab}} \longleftarrow A \longleftarrow 0.$$

We thus obtain a correspondence between the crossed sequence

$$e \longleftarrow B \longleftarrow G \longleftarrow C \longleftarrow A \longleftarrow 0$$

and the exact crossed sequence of modules

$$0 \longleftarrow I[B] \longleftarrow I[\mathfrak{F}] \otimes \mathbb{Z}[B] \longleftarrow C^{\mathrm{ab}} \longleftarrow A \longleftarrow 0.$$

This correspondence determines the isomorphism $X\,\mathrm{seq}(B, A) \to H^3(B, A)$. Observe that the group C in the crossed sequence is representable as a direct product of groups $A \times R$. The crossed sequence determines the zero element in the group $H^3(B, A)$ if C is representable as a direct product, not just of groups, but of operator $\mathfrak{F}$-groups.

We define now a numerical invariant of crossed sequences, which we shall need in the construction of minimal homotopy systems.

DEFINITION 6.9. Let $e \leftarrow B \leftarrow \mathfrak{F} \leftarrow C \leftarrow A \leftarrow 0$ be a crossed sequence in which $\mathfrak{F}$ is a free group of rank n. By the Euler characteristic of this sequence we mean the number $\chi = \mu(C) - n$, where $\mu(C)$ is the minimal number of generators of C as a crossed $\mathfrak{F}$-module. By the Euler characteristic of an element $a \in H^3(B, A)$ we mean the number $\chi(a) = \min_s \chi(s)$, where s runs over all crossed sequences representing a.

It is evident that the sequence

$$e \longleftarrow B \overset{p}{\longleftarrow} \mathfrak{F} \overset{i}{\longleftarrow} R \longleftarrow e,$$

where $\mathfrak{F}$ is a free group, is an example of a crossed module $(R, \mathfrak{F}, d = i)$. Rapaport constructed two presentations of the same group with different defects. Making use of this example, it is easy to construct two crossed sequences of the form

$$e \longleftarrow B \longleftarrow \mathfrak{F} \longleftarrow R \oplus M \longleftarrow M \longleftarrow 0,$$

where M is a finitely generated $\mathbb{Z}[B]$-module, such that both sequences represent the zero element in the group $H^3(B, M)$ but have different Euler characteristics. We note that, as a crossed $\mathbb{Z}[B]$-module, R has finitely many generators. This follows from the following lemma proved in [117].

LEMMA 6.1. *Let B be a finitely presentable group, $\mathfrak{F}$ a finitely generated free group, and $p: \mathfrak{F} \to B$ an epimorphism. Then $R = \operatorname{Ker} p$ is the normal closure in $\mathfrak{F}$ of a finite set of elements.*

§5. Homotopy systems

A homotopy system or a crossed chain complex is a sequence of groups and homomorphisms

$$e \longleftarrow \pi \xleftarrow{\partial_1} G \xleftarrow{\partial_2} C_2 \xleftarrow{\partial_3} C_3 \longleftarrow \cdots \xleftarrow{\partial_n} C_n$$

such that

(1) (C_2, G, ∂_2) is a free crossed module;

(2) for each $i \geq 3$ the module C_i is a free $\mathbb{Z}[\pi]$-module, ∂_i is a homomorphism of $\mathbb{Z}[\pi]$-modules, ∂_2 commutes with the action of the group G, and $\partial_3(C_3)$ is a $\mathbb{Z}[\pi]$-module;

(3) $\partial_i \circ \partial_{i+1} = 0$.

Obviously, G acts on each C_i, $i \geq 2$. A homotopy system is said to be of dimension n if $C_i = 0$ for $i > n$. A morphism of homotopy systems $f: (C_i, G, \partial_i) \to (\overline{C}_i, \overline{G}, \overline{\partial}_i)$ is a set of homomorphisms $f_1: G \to \overline{G}$, $f_i: C_i \to C_i'$, $i \geq 2$, that preserves the structures on the C_i and under which the corresponding diagrams are commutative. One may define a relation of homotopy equivalence between homotopy systems. Let (C_i, G, ∂_i) and $(\overline{C}_i, \overline{G}, \overline{\partial}_i)$ be homotopy systems, $\pi = \operatorname{coker} \partial_2$, $\overline{\pi} = \operatorname{coker} \overline{\partial}_2$. Suppose that we are given two morphism $f, g: (C_i, G, \partial_i) \to (\overline{C}_i, \overline{G}, \overline{\partial}_i)$. We say that $f = \{f_i\}$ and $g = \{g_i\}$ are homotopy equivalent if there exists a set of mappings $\Sigma = \{\Sigma_k, k \geq 1\}$, $\Sigma_1: G \to \overline{C}_2$, $\Sigma_k: C_k \to \overline{C}_{k+1}$, such that

(1) $\Sigma_1: G \to \overline{C}_2$ is a crossed homomorphism associated with g_1, in other words, $\Sigma_1(xy) = \Sigma_1(x) + (g_1(x)(\Sigma_1(y)))$ and $\overline{\partial}_2(\Sigma_1(x)) = f_1(x) g_1(x)^{-1}$, $x, y \in G$;

(2) $\Sigma_2: C_2 \to \overline{C}_3$ is a G-homomorphism and $\overline{\partial}_3(\Sigma_2(c)) = -g_2(c) - (\Sigma_1 \partial_2(c)) + f_2(c)$, $x, y \in G$;

(3) for $k > 2$ the mapping Σ_k is a $\mathbb{Z}[\pi]$-homomorphism ($\mathbb{Z}[\pi]$ acts on $\overline{C}_k$ via the homomorphism $\pi \to \overline{\pi}$) such that $\overline{\partial}_{k+1} \circ \Sigma_k + \Sigma_{k-1} \circ \partial_k = f_k - g_k$.

Two homotopy systems (C_i, G, ∂_i) and $(\overline{C}_i, \overline{G}, \overline{\partial}_i)$ are said to be homotopy equivalent if there exist morphisms

$$f = \{f_i\}: (C_i, G, \partial_i) \to (\overline{C}_i, \overline{G}, \overline{\partial}_i)$$

and

$$g = \{g_i\}: (\overline{C}_i, \overline{G}, \overline{\partial}_i) \to (C_i, G, \partial_i)$$

such that $g\circ f$ and $f\circ g$ are homotopy equivalent to the corresponding identity mappings of homotopy systems. With any homotopy system (C_i, G, ∂_i) we associate the chain complex of free $\mathbb{Z}[\pi]$-modules

$$(C_i^{\mathrm{ab}}, \partial_i)\colon 0 \longleftarrow \mathbb{Z} \overset{\varepsilon}{\longleftarrow} \mathbb{Z}[\pi] \overset{\partial_1}{\longleftarrow} n\mathbb{Z}[\pi] \overset{\partial_2^{\mathrm{ab}}}{\longleftarrow} C_2^{\mathrm{ab}} \overset{\partial_3}{\longleftarrow} C_3 \longleftarrow \cdots \overset{\partial_n}{\longleftarrow} C_n,$$

where $n = \mu(G)$. Bases in the modules C_i are in one-to-one correspondence with bases of modules of the homotopy system. There is the natural projection of the homotopy system (C_i, G, ∂_i) into the chain complex $(C_i^{\mathrm{ab}}, \partial_i)$ under which bases go into bases $h_i\colon C_i \to C_i^{\mathrm{ab}}$. For $i \geq 3$, the mapping $\mathbb{Z}[\pi]$ is an isomorphism of $\mathbb{Z}[\pi]$-modules, h_2 is the abelianization, and h_1 is a crossed homomorphism. A morphism of homotopy systems induces the morphism of associated chain complexes preserving the relation of homotopy equivalence. The converse statement was proved by Whitehead [159].

THEOREM 6.19. *Let (C_i, G, ∂_i) and $(\overline{C}_i, \overline{G}, \overline{\partial}_i)$ be homotopy systems, and $(C_i^{\mathrm{ab}}, \partial_i)$ and $(\overline{C}_i^{\mathrm{ab}}, \overline{\partial}_i)$ the associated chain complexes. Suppose that $f\{f_i\}\colon (C_i^{\mathrm{ab}}, \partial_i) \to (\overline{C}_i^{\mathrm{ab}}, \overline{\partial}_i)$ is a chain mapping. Then there exists at least one morphism of homotopy systems $\overline{f} = \{\overline{f}_i\}\colon (C_i, G, \partial_i) \to (\overline{C}_i, \overline{G}, \overline{\partial}_i)$ that induces this chain mapping.* □

We observe that not every chain complex of the form

$$(C, \partial)\colon 0 \longleftarrow \mathbb{Z} \overset{\varepsilon}{\longleftarrow} \mathbb{Z}[\pi] \overset{\partial_1}{\longleftarrow} C_1 \overset{\partial_2}{\longleftarrow} C_2 \longleftarrow \cdots \overset{\partial_n}{\longleftarrow} C_n,$$

for which $H_1(C, \partial) = 0$, $\partial(c_j) = g_j - 1$, where g_j are the generators of the group π and $\{c_j\}$ is a basis in the module C_1, admits a homotopy system (C_i, G, ∂_i) with which the chain complex (C, ∂) is associated. A counterexample is provided by the already discussed chain complex constructed by Dunwoody [34].

DEFINITION 6.10. A chain complex (C, ∂) over the ring $\mathbb{Z}[\pi]$ is realizable, if there exists a homotopy system for which the associated chain complex coincides with (C, ∂).

In what follows we shall need the following theorem proved by Maller [84].

THEOREM 6.20. *Let (C, ∂) be a based chain complex such that $H_0(C) = \mathbb{Z}$, $H_1(C) = 0$, $C_0 = \mathbb{Z}[\pi]$. Suppose that $\{c_1^1, \ldots, c_k^1\}$ is a basis in the module C_1 and $\partial_1(c_i^1) = (g_i - 1)c^0$, where $g_i \in \pi$ and c^0 is an element of the basis in C_0. Then there exists an acyclic complex $0 \leftarrow D_2 \overset{d}{\leftarrow} D_3 \leftarrow 0$ such that the chain complex*

$$C_0 \overset{\partial_1}{\longleftarrow} C_1 \longleftarrow C_2 \oplus D_2 \longleftarrow C_3 \oplus D_3 \overset{\partial_4}{\longleftarrow} C_4 \longleftarrow \cdots$$

is realizable. □

As is shown by the above examples, the property of realizability is not an invariant of the simple homotopy type of a complex. The following theorem was proved by Whitehead [159].

THEOREM 6.21. *Two homotopy systems* (C_i, G, ∂_i) *and* $(\overline{C}_i, \overline{G}, \overline{\partial}_i)$ *are homotopy equivalent if and only if the associated chain complexes are homotopy equivalent. A morphism* $f = \{f_i\}: (C_i, G, \partial_i) \to (\overline{C}_i, \overline{G}, \overline{\partial}_i)$ *is a homotopy equivalence if and only if the induced chain mapping* $f^{\mathrm{ab}} = \{f_i^{\mathrm{ab}}\}: (C^{\mathrm{ab}}, \partial) \to (\overline{C}^{\mathrm{ab}}, \overline{\partial})$ *is a homotopy equivalence.* □

For each CW-complex K one can construct a homotopy system $\rho(K) = \{\rho_i, d_i\}$ [159]. We can assume without loss of generality that the 0-skeleton of K consists of a single point e^0. Set $\rho_i = \pi_i(K^i, K^{i-1}, e^0)$ and $\rho_1 = \pi_1(K^1, e^0)$, where K^i is the i-skeleton of K. Using the exact homotopy sequences of pairs, we easily obtain the boundary homomorphisms $d_i: \rho_i \to \rho_{i-1}$ and thereby a homotopy system

$$S \leftarrow \pi_1(K, e^0) = \rho_1/d_2\rho_2 \xleftarrow{p} \rho_1 \xleftarrow{d_2} \rho_2 \longleftarrow \cdots \xleftarrow{d_n} \rho_n$$

which essentially depends on the cellular decomposition. In this case we say that the homotopy system $\rho(K)$ has a geometric realization. Not every homotopy system can be realized. The following is an example of such a system [159]: The group ρ_1 is generated by a single element x, and the module ρ_2 has two basis elements a_2, b_2. The boundary homomorphism $d_2: \rho_2 \to \rho_1$ acts on the basis elements by the formula $d_2(a_2) = x^2$, $d_2(b_2) = 1$. The module ρ_i $(i = 3, 4, 5)$ has a single basis element a_i and $d_3(a_3) = (g-1)b_2$, $d_4(a_4) = (g+1)a_3$, $d_5(a_5) = (g-1)a_4$, where $g = p(x)$, $p: p_1 \to p_1/d_2(p_2)$. It is evident that $d_i \circ d_{i-1} = 0$. The following two important results are also due to Whitehead.

THEOREM 6.22. *A homotopy system* (C_i, G, ∂_i) *admits a geometric realization if its dimension does not exceed four.* □

THEOREM 6.23. *Suppose that a homotopy system* $\rho(K)$ *admits a geometric realization. If a homotopy system* (C_i, G, ∂_i) *is homotopy equivalent to* $\rho(K)$, *then* (C_i, G, ∂_i) *also admits a geometric realization.* □

Consider a homotopy system

$$(\rho_i, d_i): e \longleftarrow \rho_1/d_2(\rho_2) \longleftarrow \rho_1 \xleftarrow{d_2} \rho_2 \longleftarrow \cdots \xleftarrow{d_n} \rho_n$$

and construct the crossed sequence of the form

$$e \leftarrow \rho_1/d_2(\rho_2) \leftarrow \rho_2/d_3(\rho_3) \leftarrow \operatorname{Ker} d_2/d_3(\rho_3) \leftarrow 0.$$

Put $\rho_1/d_2(\rho_2) = B$, $\operatorname{Ker} d_2/d_3(\rho_3) = A$. From what has already been said, this crossed sequence determines an element of the group $H^3(B, A)$. This was how Mac Lane and Whitehead defined the first k-invariant of a homotopy system [82]. If (ρ_i, d_i) is the homotopy system constructed for a CW-complex K, then this crossed sequence takes the form

$$e \leftarrow \pi_1(K, e^0) \leftarrow \pi_1(K^1, e^0) \leftarrow \pi_2(K, K^1, x) \leftarrow \pi_2(K, x) \leftarrow 0. \qquad (1)$$

They called the triple $(\pi_1(K, e^0), \pi_2(K, e^0), k(K))$, where the element $k(K)$ is determined by the crossed sequence (1) and belongs to the group $H^3(\pi_1(K, e^0), \pi_2(K, e^0))$, the algebraic 3-type of the complex K. Later the term settled down to "2-type", and we shall stick to that in the sequel.

An abstract 2-type is by definition a triple (π, π_2, k), consisting of a group π, a $\mathbb{Z}[\pi]$-module π_2, and an element $k \in H^3(\pi, \pi_2)$. Two 2-types $T = (\pi, \pi_2, k)$ and $T' = (\pi, \pi_2', k')$ with the same group π are equivalent (one writes $T \simeq T'$) if there exist isomorphisms $f: \pi \to \pi$ and $f': \pi_2 \to \pi_2'$, such that $f'(xa) = f(x)f'(a)$ $(x \in \pi, a \in \pi_2)$ and $f'_*(k) = f^*(k')$ in the chain of mappings

$$f'_*: H^3(\pi, \pi_2) \to H^3(\pi, (\pi_2')_f) \leftarrow H^3(\pi, \pi_2') : f^*,$$

where $(\pi_2')_f$ means the $\mathbb{Z}[\pi]$-module with the module structure induced by the isomorphism f. Let us show that equivalent 2-types $T = (\pi, \pi_2, k)$ and $T' = (\pi', \pi_2', k')$ have equal Euler characteristics. Suppose the crossed sequence

$$s: e \longleftarrow \pi \xleftarrow{g_1} \mathfrak{F} \xleftarrow{g_2} \Gamma_2 \xleftarrow{g_3} \pi_2 \longleftarrow 0$$

realizes the element $k \in H^3(\pi_1, \pi_2)$ and has Euler characteristic χ_0. Since (π, π_2, k) and (π', π_2', k') are equivalent, there exist isomorphisms $f: \pi_1 \to \pi_1'$ and $f': \pi_2 \to \pi_2'$ such that $f'_*(k) = f^*(k')$. Using the homomorphism f, construct the pullback induced from the crossed sequence s

$$\begin{array}{ccccccccc} sf{:}e & \longleftarrow & \pi' & \xleftarrow{r_1} & \mathfrak{F}' & \xleftarrow{r_2} & \Gamma_2 & \xleftarrow{r_3} & \pi_2 & \longleftarrow & 0 \\ & & \downarrow{\scriptstyle f} & & \downarrow{\scriptstyle \varphi} & & \downarrow{\scriptstyle \mathrm{id}} & & \downarrow{\scriptstyle \mathrm{id}} & & \\ e & \longleftarrow & \pi & \xleftarrow{g_1} & \mathfrak{F} & \xleftarrow{g_2} & \Gamma_2 & \xleftarrow{g_3} & \pi_2 & \longleftarrow & 0, \end{array}$$

where $\mathfrak{F}' = \{(a, b) \in \mathfrak{F} \times \pi', g_1(a) = f(b)\}$. The group $\mathfrak{F}'$ acts on Γ_2 by the formula $(a, b)c = ac$, $c \in \Gamma_2$. Obviously ϕ is an isomorphism, and Γ_2 whether as a crossed $\mathfrak{F}$- or a crossed $\mathfrak{F}'$-module has the same minimal number of generators. Now using f', construct the pushout induced from sf:

$$\begin{array}{ccccccccc} e & \longleftarrow & \pi' & \xleftarrow{r_1} & \mathfrak{F}' & \xleftarrow{r_2} & \Gamma_2 & \xleftarrow{r_3} & \pi_2 & \longleftarrow & 0 \\ & & \downarrow{\scriptstyle \mathrm{id}} & & \downarrow{\scriptstyle \mathrm{id}} & & \downarrow{\scriptstyle \psi} & & \downarrow{\scriptstyle f'} & & \\ e & \longleftarrow & \pi' & \xleftarrow{g_1} & \mathfrak{F}' & \xleftarrow{g_2} & \Gamma_2' & \xleftarrow{g_3} & \pi_2' & \longleftarrow & 0. \end{array}$$

Clearly, ψ is an isomorphism, and the crossed $\mathfrak{F}'$-module Γ_2 has the same minimal number of generators as the $\mathfrak{F}'$-module Γ'_2. By the construction, the crossed sequence

$$e \leftarrow \pi' \leftarrow \mathfrak{F}' \leftarrow \Gamma'_2 \leftarrow \pi'_2 \leftarrow 0$$

also has the Euler characteristic χ_0 and realizes the element k'.

Let $A(\pi)$ be the set of equivalent 2-types (π, π_2, k) with the same group π, and $[T]$ the equivalence class of the 2-type T. It is known that any abstract 2-type can be realized by a three-dimensional CW-complex K such that the crossed sequence

$$e \leftarrow \pi_1(K, e^0) \leftarrow \pi_1(K^1, e^0) \leftarrow \pi_2(K, K^1, e^0) \leftarrow \pi_2(K, e^0) \leftarrow 0,$$

and the groups $\pi_1(K, e^0)$ and $\pi_2(K, e^0)$ determine a fixed abstract 2-type [82].

Two CW-complexes K and L are said to have the same homotopy 2-type if there exist mappings

$$f: K^3 \to L^3, \qquad g: L^3 \to K^3$$

such that $g \circ f/K^2 \sim i$, $f \circ g/L^2 \sim \bar{i}$, $i: K^2 \to K$, $\bar{i}: L^2 \to L$. A theorem due to Mac Lane and Whitehead states that the complexes K and L have the same homotopy 2-type if and only if their algebraic 2-types are equivalent [82]. A theorem due to Whitehead asserts that if the three-dimensional CW-complexes K^3 and L^3 have equivalent algebraic 2-types, then there exist positive integers n and m such that the CW-complexes

$$K^3 \cup \left(\bigvee_{i=1}^{m} S_i^3\right) \quad \text{and} \quad L^3 \cup \left(\bigvee_{i=1}^{n} S_i^3\right)$$

are simple homotopy equivalent [160].

Let M^n be a smooth manifold. A homotopy system can be constructed by using either a cellular structure on M^n, or the handlebody decomposition of M^n, or a Morse function on M^n. For example, let $f: M^n \to [0, 1]$ be an ordered Morse function on M^n with one critical point x of index 0, and ξ an ordered gradient-like vector field. Let $M_\lambda = f^{-1}[0, c_\lambda]$, where $f(x_\lambda) < c_\lambda < f(y_{\lambda+1})$, x_λ $(y_{\lambda+1})$ is a critical point of index λ $(\lambda + 1)$. Set $\rho_1 = \pi_1(M^n, x)$, $\rho_i = \pi_1(M_i, M_{i-1}, x)$. As in the case of CW-complexes, we obtain a homotopy system

$$e \longleftarrow \pi_1(M_n, x) \longleftarrow \rho_1 \xleftarrow{d_2} \rho_2 \longleftarrow \cdots \xleftarrow{d_n} \rho_n.$$

If on the middle spheres of critical points of index λ belonging to the manifold $\partial M_{\lambda-1}$ we choose distinguished points and join them by paths to the critical point x of index 0, then every middle disk of a critical point of index λ determines a generator of the module ρ_i. Thus the pair (M^n, M_1) defines a crossed sequence

$$e \leftarrow \pi_1(M_1, x) \leftarrow \pi_1(M^n, x) \leftarrow \pi_2(M^n, M_1, x) \leftarrow \pi_2(M^n, x) \leftarrow 0$$

which determines an element $k^3(M^n)$ in $H^3(\pi_1(M^n, x), \pi_2(M^n, x))$. Here M_1 is a level submanifold containing all critical points of index 0 or 1. The element $k^3(M^n)$ is called the first k-invariant of M^n. Using a general position argument, it is easy to see that any two ordered Morse functions f and g on M^n have the same homotopy 2-type. This follows from the existence of a diffeomorphism, isotopic to the identity, that takes M_2^f into M_2^g, and vice versa (M_2^f and M_2^g are level submanifolds of f and g containing the critical points of index 0, 1, 2 only). Consequently, by the Mac Lane-Whitehead theorem, f and g determine equivalent algebraic 2-types. Making use of the preceding argument, we can define the Euler characteristic of the first k-invariant of the manifold M^n. Clearly, all that has been said in the case of ordered Morse functions is also true for ordered handlebody decompositions.

We conclude this section by briefly recalling the theory of simple homotopy type for homotopy systems.

Suppose that bases are chosen in the homotopy systems $(C_i, \mathfrak{F}, d_i)$ and $(\overline{C}_i, \overline{\mathfrak{F}}, \overline{d}_i)$ as well as in the groups $\mathfrak{F}$ and $\overline{\mathfrak{F}}$. A homomorphism $f = \{f_i\}: (C_i, \mathfrak{F}, d_i) \to (\overline{C}_i, \overline{\mathfrak{F}}, \overline{d}_i)$ is said to be a simple isomorphism if the image of a basis element is (up to a sign) again a basis element. Let c_i be a basis element in the crossed module C_2 of the homotopy system $(C_i, \mathfrak{F}, d_i)$, and f_i a basis element in the group $\mathfrak{F}$. Suppose that $d_2(c_i) = x f_j y$, where f_j is contained neither in the word x nor in the word y. Construct a new homotopy system $(\overline{C}_i, \overline{\mathfrak{F}}, \overline{d}_i)$ by removing the elements c_i and f_j and adjusting the boundary homomorphisms in dimensions 2 and 3 in the evident manner. We will call this operation an elementary contraction in dimension 2.

DEFINITION 6.11. Two homotopy systems $(C_i, \mathfrak{F}, d_i)$ and $(\overline{C}_i, \overline{\mathfrak{F}}, \overline{d}_i)$ are simple homotopy equivalent ($(C_i, \mathfrak{F}, d_i) \sim (\overline{C}_i, \overline{\mathfrak{F}}, \overline{d}_i)$) if one can transform one into the other by a sequence of the following transformations:

(1) an elementary contraction or extension in dimension 2;
(2) an elementary transformation of the basis in the crossed $\mathfrak{F}$-module C_2;
(3) elementary transformations of the bases in the modules C_i, $i \geq 3$;
(4) a contraction and stabilization of d_i.

THEOREM 6.24. *Let $f: (C_i, \mathfrak{F}, d_i) \to (\overline{C}_i, \overline{\mathfrak{F}}, \overline{d}_i)$ be a morphism of homotopy systems. The mapping f is a simple homotopy equivalence if and only if the induced chain mapping $\overline{f} = \{f_i^{\mathrm{ab}}\}: (C^{\mathrm{ab}}, d) \to (\overline{C}^{\mathrm{ab}}, \overline{d})$ is a simple homotopy equivalence.* □

This theorem was proved in [160]. For homotopy systems there is an analog of the Cockroft-Swan theorem which we shall prove in the next section.

§6. Homotopy type and stable isomorphism

THEOREM 6.25. *Let there be given a homomorphism of homotopy systems*

$$\begin{array}{ccccccccccc} e & \xleftarrow{d_1} & \mathfrak{F} & \xleftarrow{d_2} & C_2 & \xleftarrow{d_3} & C_3 & \longleftarrow & \cdots & \xleftarrow{d_n} & C_n \\ & & \downarrow f_1 & & \downarrow f_2 & & \downarrow f_3 & & & & \downarrow f_n \\ e & \xleftarrow{\partial_1} & G & \xleftarrow{\partial_2} & D_2 & \xleftarrow{\partial_3} & D_3 & \longleftarrow & \cdots & \xleftarrow{\partial_n} & D_n \end{array}$$

inducing the isomorphism of the homologies, where $\mathfrak{F}$ and G are free groups. By stabilizing d_i and ∂_i, $i \geq 2$, these homotopy systems can be made isomorphic.

PROOF. Consider the crossed modules $(C_2, \mathfrak{F}, d_2)$ and (D_2, G, ∂_2). Let $x_1, \dots, x_n$ be a basis in the group $\mathfrak{F}$, and $y_1, \dots, y_m$ a basis in the group G. Let us stabilize the crossed module $(C_2, \mathfrak{F}, d_2)$ via the group G and the crossed module (D_2, G, ∂_2) via the group $\mathfrak{F}$. Denote the resulting crossed modules by $(\hat{C}_2, \mathfrak{F} * G, \hat{d}_2)$ and $(\hat{D}_2, G * \mathfrak{F}, \hat{\partial}_2)$. We assume that $x_1, \dots, x_n, y_1, \dots, y_m$ is a basis in $\mathfrak{F} * G$, $c_1, \dots, c_k, \overline{c}_1, \dots \overline{c}_m$ is a basis in $\hat{C}_2$, and $\hat{d}_2(c_\alpha) = d_2(c_\alpha)$, $\hat{d}_2(\overline{c}_\beta) = y_\beta$ $(1 \leq \alpha \leq k, 1 \leq \beta \leq m)$. Similarly, $y_1, \dots, y_m, x_1, \dots, x_n$ is a basis in $G * \mathfrak{F}$, $d_1, \dots, d_l, \overline{d}_1, \dots, \overline{d}_n$ is a basis in $\hat{D}_2$, $\hat{\partial}_2(d_\gamma) = \partial_2(d_\gamma)$, $\hat{\partial}_2(\overline{d}_\delta) = x_\delta$ $(1 \leq \gamma \leq l, 1 \leq \delta \leq n)$. We also set $\hat{d}_1(x_\delta) = d_1(x_\delta)$, $\hat{d}_1(y_\beta) = e$, $\hat{\partial}_1(y_\beta) = \partial_1(y_\beta)$, $\hat{\partial}_1(x_\delta) = e$. Construct a mapping $\hat{f}_1 : \mathfrak{F} * G \to G * \mathfrak{F}$. Let $f_1(x_\delta) = \hat{y}_\delta$, and let $\hat{x}_\beta$ be an arbitrary element in $d^{-1} \circ \partial_1(y_\beta)$. The elements $x_1, \dots, \hat{x}_1 y_1, \dots, \hat{x}_m y_m$ form a basis in the group $\mathfrak{F} * G$, and the elements $y_1, \dots, y_m, \hat{y}_1 x_1, \dots, \hat{y}_n x_n$ form a basis in the group $G * \mathfrak{F}$. Since $\hat{\partial}_1(\hat{y}_\delta x_\delta) = \partial_1(\hat{y}_\delta) = \partial_1(f_1(x_\delta)) = d_1(x_\delta)$ and $d_1(\hat{x}_\beta) = d_1(d_1^{-1} \circ \partial_1(y_\beta)) = \hat{\partial}_1(y_\beta)$, there exists an isomorphism $\hat{f}_1 : \mathfrak{F} * G \to G * \mathfrak{F}$ that makes the diagram

$$\begin{array}{ccc} \pi & \xleftarrow{\hat{d}_1} & \mathfrak{F} * G \\ \downarrow \mathrm{id} & & \downarrow \hat{f}_1 \\ \pi & \xleftarrow{\hat{\partial}_1} & G * \mathfrak{F} \end{array}$$

commutative. This can be achieved by setting $\hat{f}_1(x_\delta) = \hat{y}_\delta \cdot x_\delta$, $f_1(\hat{x}_\beta y_\beta) = y_\beta$ and extending $\hat{f}_1$ to the entire group. By construction, $f_1(x_\delta) = p_1 \circ \hat{f}_1(x_\delta)$. Let us construct a mapping $\hat{f}_2 : \hat{C}_2 \to \hat{D}_2$ by defining it on the elements of the basis $c_1, \dots, c_k, \overline{c}_1, \dots, \overline{c}_m$. Consider the following commutative diagrams

$$\begin{array}{ccccccccccc}
\pi & \xleftarrow{\hat{d}_1} & \mathfrak{F} * G & \xleftarrow{\hat{d}_2} & \hat{C}_2 & \longleftarrow & \operatorname{Ker}\hat{d}_2 & \xleftarrow{d_3} & C_3 & \longleftarrow & \\
\uparrow \mathrm{id} & & \uparrow i_1 & & \uparrow i_2 & & \uparrow i_2 & & \uparrow \mathrm{id} & & \\
\pi & \xleftarrow{d_1} & \mathfrak{F} & \xleftarrow{d_2} & C_2 & \longleftarrow & \operatorname{Ker} d_2 & \xleftarrow{d_3} & C_3 & \longleftarrow & \\
\downarrow \mathrm{id} & & \downarrow f_1 & & \downarrow f_2 & & \downarrow f_2 & & \downarrow f_3 & & \\
\pi & \xleftarrow{\partial_1} & G & \xleftarrow{\partial_2} & D_2 & \longleftarrow & \operatorname{Ker}\partial_2 & \xleftarrow{\partial_3} & D_3 & \longleftarrow & ,
\end{array}$$

$$\begin{array}{ccccccccccc}
\pi & \xleftarrow{\hat{d}_1} & \mathfrak{F} * G & \xleftarrow{\hat{d}_2} & \hat{C}_2 & \longleftarrow & \operatorname{Ker}\hat{d}_2 & \xleftarrow{d_3} & C_3 & \longleftarrow & \\
\uparrow \mathrm{id} & & \uparrow f_1 & & \uparrow f_2 & & \uparrow f_2 & & \uparrow f_3 & & \\
\pi & \xleftarrow{\hat{\partial}_1} & G * \mathfrak{F} & \xleftarrow{\hat{\partial}_2} & D_2 & \longleftarrow & \operatorname{Ker}\hat{\partial}_2 & \longleftarrow & D_3 & \longleftarrow & \\
\downarrow \mathrm{id} & & \downarrow p_1 & & \downarrow p_2 & & \downarrow p_2 & & \downarrow \mathrm{id} & & \\
\pi & \xleftarrow{\partial_1} & G & \xleftarrow{\partial_2} & D_2 & \longleftarrow & \operatorname{Ker}\partial_2 & \xleftarrow{\partial_3} & D_3 & \longleftarrow & .
\end{array}$$

Here (i_1, i_2) are embedding homomorphisms of the crossed module $(C_2, \mathfrak{F}, d_2)$ in the crossed module $(\hat{C}_2, \mathfrak{F} * G, \hat{d}_2)$, and (p_1, p_2) are projection homomorphisms from $(\hat{D}_2, G * \mathfrak{F}, \hat{\partial}_2)$ to (D_2, G, ∂_2). Clearly, the restrictions of i_2 to $\operatorname{Ker} d_2$ and p_2 to $\operatorname{Ker}\hat{\partial}_2$ are isomorphisms for which $i_2 \circ d_3 = i_2$, $p_2 \circ \partial_3 = \partial_3$. Let $\hat{d}_\alpha = f_2(c_\alpha)$. By the hypothesis, $\partial_2(\hat{d}_\alpha) = f_1 \circ (d_2(c_\alpha))$, and by construction $f_1 \circ d_2(c_\alpha) = p_1 \circ \hat{f}_1 \circ d_2(c_\alpha)$. Consider all elements in D_2 that go into $f_1 \circ d_2(c_\alpha)$. They are of the form $\hat{d}_\alpha + \operatorname{Ker}\partial_2 = \Gamma_\alpha$. Evidently, $d_2^{-1}(\Gamma_\alpha)$ can be presented in the form $p_2^{-1}(\hat{d}_\alpha) + \operatorname{Ker}\hat{\partial}_2 = B_\alpha$, because the restriction of p_2 to $\operatorname{Ker}\hat{\partial}_2$ is an isomorphism. By construction, B_α contains all elements that the homomorphism $\partial_2 \circ p_2 = p_1 \circ \hat{\partial}_2$ takes into $f_1 \circ d_2(c_\alpha)$. Consequently, there is an element d'_α in B_α such that $\hat{\partial}_2(d'_\alpha) = \hat{f}_1 \circ d_2(c_\alpha)$ since $\hat{f}_1 \circ d_2(c_\alpha) \subset \hat{\partial}_2(\hat{D}_2)$. Put $\hat{f}_2(c_\alpha) = d'_\alpha$. By construction, $\hat{d}_2(\overline{c}_\beta) = y_\beta \in \operatorname{Ker}\hat{d}_1$ and consequently $\hat{f}_1(\hat{d}_2(\overline{c}_\beta)) \subset \operatorname{Ker}\overline{\partial}_1$. But since $\operatorname{Ker}\overline{\partial}_1 = \hat{\partial}_2(\hat{D}_2)$, there is an element $\overline{d}'_\beta$ in the module $\hat{D}_2$ such that $\hat{\partial}_2(\overline{d}'_\beta) = \hat{f}_1(\hat{d}_2(\overline{c}_\beta))$. Put $\hat{f}_2(\overline{c}_\beta) = \overline{d}'_\beta$. Since $\hat{c}_2$ and $\hat{D}_2$ are free crossed modules, the mapping $\hat{f}_2$ can be extended to $\hat{C}_2$. Note that $\hat{f}_2 | \operatorname{Ker}\hat{d}_2 = f_2 | \operatorname{Ker} d_2$ since $f_2 = p_2 \cdot \hat{f}_2$. Thus, in the diagram, we have constructed a homomorphism $\hat{f}_2 \colon \hat{C}_2 \to \hat{D}_2$ such that $p_2 \circ \hat{f}_2 \circ i_2 = f_2$.

Denote $\overline{C}_2 = \hat{C}_2^{\mathrm{ab}}$, $\overline{D}_2 = \hat{D}_2^{\mathrm{ab}}$. As noted above, $\overline{C}_2$ and $\overline{D}_2$ are free $\mathbb{Z}[\pi]$-modules of rank $k + m$ and $l + n$, respectively. Stabilize the homomorphism d_3 via the free module $\overline{D}_2$ and the homomorphism ∂_3 via the free module $\overline{C}_2$. Clearly, $\hat{C}_2 \oplus \overline{D}_2$ is a thickening of the crossed module $\hat{C}_2$ via $\overline{D}_2$, and $\hat{D}_2 \oplus \overline{C}_2$ is a thickening of the crossed module $\hat{D}_2$ via $\overline{C}_2$. Let $\mathring{c}_1, \dots, \mathring{c}_{k+m}$ be a basis in $\overline{C}_2$, and $\mathring{d}_1, \dots, \mathring{d}_{l+n}$ a basis in $\overline{D}_2$. Denote by $c_1, \dots, c_k, c_{k+1}, \dots, c_{k+m}, \mathring{d}_1, \dots, \mathring{d}_{l+n}$ the basis of the crossed module $(\hat{C}_2 \oplus \overline{D}_2, \mathfrak{F} * G, \hat{d}_2)$ (redenote the elements $\overline{c}_\beta$ by $c_{k+\beta}$ and set $\hat{d}_2(c_\alpha) = \hat{d}_2(c_\alpha)$, $\hat{d}_2(c_{k+\beta}) = \overline{d}_2(\overline{c}_\beta)$, $\hat{d}_2(\mathring{d}_\eta) = e$ $(1 \le \eta \le l+n)$). Similarly, let $d_1, \dots, d_l, d_{l+1}, \dots, d_{l+n}, \mathring{c}_1, \dots, \mathring{c}_{k+m}$ be a basis in the module $(\hat{D}_2 \oplus \overline{C}_2, G * \mathfrak{F}, \hat{\partial}_2)$ (we set $d_{l+\delta} = \overline{d}_\delta$, $\hat{\partial}_2(d_\gamma) = \hat{\partial}_2(d_\gamma)$, $\hat{\partial}_2(d_{l+\delta}) = \partial_2(\overline{d}_\delta)$, $\hat{\partial}_2(\mathring{c}_\xi) = e$ $(1 \le \xi \le k+m)$). Denote $\overline{d}'_\xi = \hat{f}_2(c_\xi) \in \hat{D}_2$, and by c'_η an arbitrary element in $\hat{d}_2^{-1} \circ \hat{f}_1^{-1} \circ \hat{\partial}_2(d_\eta)$. Here we make use of the fact that f_1 is an isomorphism. Choose a new basis $c_1, \dots, c_{k+m}, \mathring{d}_1 + c'_1, \dots, \mathring{d}_{l+n} + c'_{l+n}$ in the module $\hat{C}_2 \oplus \overline{D}_2$ and a new basis $d_1, \dots, d_{l+n}, \mathring{c}_1 + \overline{d}'_1, \dots, \mathring{c}_{k+m} + \overline{d}'_{k+m}$ in the module $\hat{D}_2 \oplus \overline{C}_2$. Now $\hat{\partial}_2(\mathring{c}_\xi + \overline{d}'_\xi) = \hat{\partial}_2(\overline{d}'_\xi) = \hat{\partial}_2(\overline{f}_2(c_\xi))$ which is equal to $\hat{f}_1 \circ \hat{d}_2(c_\xi)$ because of the commutativity of the diagram. In addition, we have $\hat{f}_1 \circ \hat{d}_2(\mathring{d}_\eta + c'_\eta) = \hat{f}_1 \circ \hat{d}_2(\mathring{d}_\eta) + \hat{f}_1 \circ \hat{d}_2(c'_\eta) = \hat{\partial}_2(d_\eta)$. Put $\hat{\hat{f}}_2(c_\xi) = \mathring{c}_\xi + d'_\xi$, $\hat{\hat{f}}_2(\mathring{d}_\eta + c'_\eta) = d_\eta$ and extend the mapping $\hat{\hat{f}}$ to the free crossed module $\hat{C}_2 \oplus \overline{D}_2$. It is evident that $\hat{\hat{f}}$ is an isomorphism. Since $\operatorname{Ker} \hat{d}_2 = \operatorname{Ker} \hat{\partial}_2$ belongs to the centers of the modules $D_2^{\text{ab}} \oplus \hat{C}_2$ and $\hat{D}_2 \oplus C^{\text{ab}}$, consider the commuted segments of the homotopy systems

$$\begin{array}{ccccc}
\operatorname{Ker} \hat{d}_1^{\text{ab}} & \longleftarrow & \hat{C}_2^{\text{ab}} \oplus D_2^{\text{ab}} & \xleftarrow{d_3 \oplus \text{id}} & C^3 \oplus D_2^{\text{ab}} \\
 & & \downarrow \hat{f}_1^{\text{ab}} & & \downarrow \hat{\hat{f}}_2^{\text{ab}} \\
\operatorname{Ker} \partial_1^{\text{ab}} & \longleftarrow & \hat{D}_2^{\text{ab}} \oplus C_2^{\text{ab}} & \xleftarrow{\partial_3 \oplus \text{id}} & D_3 \oplus C_2^{\text{ab}}.
\end{array}$$

Since both $\hat{C}_2^{\text{ab}}$ and $\hat{D}_2^{\text{ab}}$ are free modules, the mapping $\hat{f}_2^{\text{ab}}$ has a homotopy inverse mapping $g: \hat{D}_2^{\text{ab}} \to \hat{C}_2^{\text{ab}}$ such that $\text{id} - f_2^{\text{ab}} \circ g = d_3 \circ S$, where $S: \hat{D}_2^{\text{ab}} \to D_3$ is a homotopy operator. Define the mapping $\hat{f}_3: C_3 \oplus D_2^{\text{ab}} \to D_3 \oplus C_2^{\text{ab}}$ by the formula $\hat{f}_3(c, d) = (d_3(c) - g(d), f_3(c) + S(d))$. The argument now proceeds on the same lines. One can however make use of the Cockroft-Swan theorem applied to the segments of the homotopy systems

$$\begin{array}{ccccccccc}
\longleftarrow & d_3(C_3 \oplus D_2^{\text{ab}}) & \xleftarrow{d_3} & C_3 \oplus D_2^{\text{ab}} & \longleftarrow & C_4 & \longleftarrow & \\
 & \downarrow \hat{f}_2 & & \downarrow \hat{f}_3 & & \downarrow f_4 & & \\
\longleftarrow & \partial_3(D_3 \oplus C_2^{\text{ab}}) & \xleftarrow{\partial_3} & D_3 \oplus C_2^{\text{ab}} & \longleftarrow & D_4 & \longleftarrow & \cdots,
\end{array}$$

which can be regarded as chain complexes of free modules. □

§7. Minimal homotopy systems

As for chain complexes, one may define a minimal homotopy system in a given homotopy type.

DEFINITION 6.12. A homotopy system

$$(\rho_i, d_i): e \longleftarrow \pi \xleftarrow{d_1} \rho_1 \xleftarrow{d_2} \rho_2 \longleftarrow \cdots \xleftarrow{d_n} \rho_n,$$

where ρ_1 is a free group, is said to be minimal if for any other homotopy system

$$(\overline{\rho}_i, \overline{d}_i): e \longleftarrow \pi \xleftarrow{\overline{d}_1} \overline{\rho}_1 \xleftarrow{\overline{d}_2} \overline{\rho}_2 \longleftarrow \cdots \xleftarrow{\overline{d}_n} \overline{\rho}_n$$

($\overline{\rho}_1$ is a free group) homotopy equivalent to (ρ_i, d_i), we have $\mu(\rho_i) \leq \mu(\overline{\rho}_i)$.

A homotopy system minimal in a given dimension is defined in an obvious manner.

Take a fixed group π and consider the class of all finite CW-complexes whose fundamental group is isomorphic to π. We can suppose without loss of generality that the 0-skeleton of any such complex consists of a single point. Denote by $\rho(K)$ the class of homotopy systems associated with these CW-complexes.The following theorem gives an indication of when, in the homotopy type of a homotopy system $(\rho_i, d_i) \in \rho(K)$, there is a minimal homotopy system. The question is well posed, since by a theorem of Whitehead, if $(\rho_i, d_i) \in \rho(K)$ and a homotopy system $(\overline{\rho}_i, \overline{d}_i)$ is homotopy equivalent to (ρ_i, d_i), then $(\overline{\rho}_i, \overline{d}_i) \in \rho(K)$.

THEOREM 6.26. *In order that there exist for every homotopy system* $(\rho_i, d_i) \in \rho(K)$ *a homotopy equivalent minimal homotopy system* $(\overline{\rho}_i, \overline{d}_i)$, *it is necessary and sufficient that* $\pi = \rho_1/d_2(\rho_2)$ *satisfy condition* (h) *and be an* s*-group.*

PROOF. *Necessity* of the condition that π be an s-group follows from Theorem 4.2. We show now the necessity of the condition (h). Assume the contrary. Then there exists a presentation $\pi = \{a_1, \dots, a_k, r_1, \dots, r_l\}$ such that $k > \mu(\pi)$ but for which there is no minimal presentation of π such that the chain complexes constructed from the two presentations are homotopy equivalent. For this presentation of π, construct an n-dimensional manifold W^n, $n \geq 6$, having one handle of index 0, k handles of index 1, and l handles of index 2. In order to obtain such a manifold, it is sufficient to construct a two-dimensional CW-complex realizing this presentation of the group π, embed it in a Euclidean space of dimension ≥ 6, and take a tubular neighborhood of this complex. As a result, we obtain a manifold with boundary admitting a handlebody decomposition such that each handle of index λ corresponds to a cell of dimension λ of the two-dimensional CW-complex. At the same time, the manifold W^n can also be given a handle decomposition for which there is one handle of index 0, $\mu(\pi)$ handles of 1, and some number of handles of higher index. This handle decomposition is constructed as follows. Let $\xi_1, \dots, \xi_n$ be the minimal number of generators of the fundamental group $\pi_1(W^n, x)$. Realize the elements ξ_i by smooth embeddings $f_i: (S^1, s_0) \to \mathrm{Int}(W^n)$ such that $f_i(S^1, s_0) \cap f_j(S^1, s_0) = x$. For the handle of index 0 we take a smooth disk $D^n \subset \mathrm{Int}\, W^n$ containing the point x. Suppose that the intersection of $f_i(S^1, s_0)$ with ∂D^n is transversal and consists of exactly two points. Clearly, this can always be achieved. Consider disjoint tubular neighborhoods of the arcs $\overline{f_i(S^1, s_0) \setminus f_i(S^1, s_0) \cap D^n}$ in the manifold $\overline{W^n \setminus D^n}$ and denote them by U_i. We assume that $U_i \cap D^n = D^{n-1}$ (where D^{n-1} is an $(n-1)$-disk). Obviously, we can consider U_i to be handles of index 1 attached to the disk D^n. Denote the submanifold $D^n \cup U_1 \cup \dots \cup U_n$ by W_1^n. By construction, $\pi_1(W_1^n, x) = \mathbb{Z} * \dots * \mathbb{Z}$ and the embedding $W_1^n \to W^n$ induces an epimorphism $\varphi: \pi_1(W_1, x) \to \pi_1(W^n, x)$ whereby $\pi_1(W^n, W_1^n) = 0$. A lemma from [156] implies that on the cobordism $(\overline{W^n \setminus W_1^n}, \partial W_1^n, \partial W^n)$ there exists a handle decomposition having no handles of index 0 or 1. By combining this decomposition with the handles of the manifold W_1^n, we obtain a handle decomposition of the manifold W^n with $\mu(\pi_1(W^n, x))$ handles of index 1. Denote this handle decomposition by P, and let W_2^n be the submanifold of W^n formed by the handles of index 0, 1, and 2 in the decomposition P. Obviously, the mapping $\pi_1(W_2^n, x) \to \pi_1(W^n, x)$ induced by the embedding $W_2^n \to W^n$ is an isomorphism. Consider the exact homology sequence for the pair $(\hat{W}^n, \hat{W}_2^n)$ which is of the form

$$0 \to H_3(\hat{W}^n, \hat{W}_2^n, \mathbb{Z}) \to H_2(\hat{W}_2^n, \mathbb{Z}) \to H_2(\hat{W}^n, \mathbb{Z}) \to 0,$$

since $H_2(\hat{W}^n, \hat{W}_2^n) = 0$ by the Hurewicz theorem.

If $H_3(\hat{W}^n, \hat{W}_2^n) = 0$, then $\overline{W^n \setminus W_2^n} = H$ is an h-cobordism; consequently the manifold W_2^n has the same homotopy type as W^n. But this would mean that the condition $(h, \mu(\pi))$ is satisfied for the given presentation of π, which contradicts our assumption. Therefore $H^3(\hat{W}^n, \hat{W}_2^n) \neq 0$, and the decomposition P of the manifold W^n contains handles of index at least 3. This implies that the homotopy systems associated with the decomposition P of W^n cannot be minimal. This proves necessity.

Sufficiency. Suppose that the homotopy system $\{\rho_i, d_i\}$ is associated with an n-dimensional CW-complex K^n. The homotopy system $(\overline{\rho}_i, \overline{d}_i)$ is minimal in dimension 2 and homotopy equivalent to (ρ_i, d_i). The homotopy system $(\hat{\rho}_i, \hat{d}_i)$ is minimal in dimension 3 and also homotopy equivalent to (ρ_i, d_i). Let K_1^m and K_2^l be CW-complexes providing geometric realizations of the homotopy systems $(\overline{\rho}_i, \overline{d}_i)$ and $(\hat{\rho}_i, \hat{d}_i)$. Embed the complexes K_1^m and K_2^l in a Euclidean space of sufficiently high dimension, and consider tubular neighborhoods for them. Denote the resulting manifolds by W_1^N and W_2^N. The cellular structures on K_1^m and K_2^l induce on W_1^N and W_2^N handle decompositions such that corresponding to every cell of dimension i is a handle of index i. It is easily shown that $\pi_1(\partial W_1^N, x) \to \pi_1(W_1^N, x) = \pi$, $\pi_1(\partial W_2^N, y) \to \pi_1(W_2^N, y)$ are isomorphisms. Denote by $\overline{W}_2^N$ the submanifold in W_2^N formed by the handles of index 0, 1, and 2. The handle decomposition of $\overline{W}_2^2$ gives a presentation of the fundamental group $\pi_2(\overline{W}_2^N, y) = \pi$. Since π satisfies condition (h), there exists a minimal presentation of π such that the corresponding segments of the Lyndon-Fox resolutions are homotopy equivalent. Realize this presentation by a two-dimensional CW-complex K^2 and, using (h), embed this complex in $\overline{W}_2^N$. A tubular neighborhood of the image is a submanifold $\Gamma_2 \subset \overline{W}_2^N$, containing one handle of index 0, $\mu(\pi)$ handles of index 1, and a certain number of handles of index 2. Condition (h) guarantees that the embedding $\Gamma_2 \subset \overline{W}_2^N$ is a homotopy equivalence. Therefore $\overline{\overline{W}_2^N \setminus \Gamma_2}$ is an h-cobordism, in which, as is well known, there is a handle decomposition with handles of index greater than 3. Shift the handles from h-cobordism in the direction of the boundary of the manifold W_2^N, i.e., construct a new handle decomposition of the manifold W_2^N with $\mu(\pi)$ handles of index 1 and the number of handles of index 3 remaining unchanged. We show now that the homotopy system constructed for this decomposition of W_2^N, which we denote by $(\overline{p}_i, \overline{d}_i)$, is minimal in dimensions 1, 2, and 3. Let

$$C(W_1^N)\colon \mathbb{Z} \xleftarrow{\ \varepsilon\ } \mathbb{Z}[\pi] \xleftarrow{\ \overline{d}_1\ } \overline{C}_1 \xleftarrow{\ \overline{d}_2\ } \overline{C}_2 \xleftarrow{\ \overline{d}_3\ } \overline{C}_3 \xleftarrow{\ \overline{d}_4\ } \overline{C}_4 \leftarrow$$

be the chain complex associated with the homotopy system $(\overline{p}_i, \overline{d}_i)$. Without loss of generality we can assume that $f\text{-rank}(\tilde{d}_4(\tilde{C}_4), \tilde{C}_3) = 0$ and is additive, since elementary operations over handles in this dimension are realized geometrically. Let

$$C(W_1^N)\colon \mathbb{Z} \xleftarrow{\ \varepsilon\ } \mathbb{Z}[\pi] \xleftarrow{\ \tilde{d}_1\ } \tilde{D}_1 \xleftarrow{\ \tilde{d}_2\ } \tilde{D}_2 \xleftarrow{\ \tilde{d}_3\ } \tilde{D}_3 \xleftarrow{\ \tilde{d}_4\ } \tilde{D}_4 \leftarrow$$

be the chain complex associated with the homotopy system $(\tilde{p}_i, \tilde{d}_i)$. The condition that $f\text{-rank}(\tilde{d}_4(\tilde{C}_4)\tilde{C}_3) = 0$ and be additive is satisfied automatically, since the homotopy system is minimal in dimension 3. By Lemma 4.7, we have the equality

$$1 - \mu(C_1) + \mu(C_2) - \mu(C_3) = 1 - \mu(\tilde{C}_1) + \mu(\tilde{C}_2) - \mu(\tilde{C}_3).$$

But by construction, $\mu(C_1) \geq \mu(\tilde{C}_1)$ and $\mu(C_3) \geq \mu(\tilde{C}_3)$; hence $\mu(C_2) = \mu(\tilde{C}_2)$. Thus the homotopy system $(\tilde{p}_i, \tilde{d}_i)$ is minimal in dimensions 1, 2, and 3. Using stabilization and contraction of the boundary homomorphisms $(\tilde{d}_i)$, we can arrange that $f\text{-rank}(\tilde{d}_i(\tilde{D}_i), \tilde{D}_{i-1}) = 0$ and is additive for $i \geq 5$. We claim that the homotopy system so obtained is minimal. Consider the chain complex associated with it. Performing contractions and stabilizations on the boundary homomorphisms $\tilde{d}_2$ and $\tilde{d}_3$, we can arrange that f-rank$(\tilde{d}_i(\tilde{D}_i), \tilde{D}_{i-1}) = 0$ and is additive for $i = 2, 3$. By Lemma 4.5 the resulting chain complex is minimal. Therefore the original chain complex was minimal in dimensions greater than 3. □

The proof of the theorem immediately implies that the condition of geometrical realizability for homotopy systems was used only up to dimension 3. However the Whitehead theorem guarantees realizability up to and including dimension 4. The following more general theorem holds.

THEOREM 6.27. *In order that there exist for every homotopy type of the homotopy system* (ρ_i, d_i) *a minimal homotopy system, it is necessary and sufficient that* $\rho_1/d_2(\rho_2) = \pi$ *satisfy condition* (h) *and be an s-group.*

PROOF. *Necessity.* The proof of necessity repeats that of the preceding theorem.

Sufficiency. Apply the argument of the preceding theorem to segments of homotopy systems of length 3, and then, after discarding the resulting h-cobordism, adjust the homotopy system that is minimal in dimension 2 in such a way that the homotopy type remains unchanged. The rest of the argument repeats the preceding one word for word. □

COROLLARY 6.1. *Suppose that the homotopy system* (ρ_i, d_i) *is such that* $\pi = \rho_1/d_2(\rho_2)$ *is isomorphic to one of the following groups:*

(1) *a free group;*
(2) $\mathbb{Z} \oplus \mathbb{Z}$;
(3) *a finite abelian group;*
(4) *a dihedral group* D_n $(n = 2k + 1)$.

Then in the homotopy type of (ρ_i, d_i) *there always exists a minimal homotopy system.*

The proof follows directly from Theorem 6.26 and Proposition 6.4. □

§8. Minimal homotopy systems in a fixed homotopy type

THEOREM 6.28. *Let* $\rho(L^n)$ *be a homotopy system associated with a CW-complex* L^n. *If* $\pi = \pi_1(L^n, x)$ *satisfies the condition* $(h, \chi(k^3(L^n)))$ *and is an s-group, then there exists a minimal homotopy system* (ρ_i, d_i) *homotopy equivalent to* $\rho(L^n)$.

PROOF. Let K be a CW-complex such that the associated homotopy system $\rho(K)$ is minimal in dimension 3 and homotopy equivalent to $\rho(L^n)$. In addition, we can suppose without loss of generality that the chain complex

$$\mathbb{Z} \xleftarrow{\varepsilon} \mathbb{Z}[\pi] \xleftarrow{d_1} C_1 \xleftarrow{d_2} C_2 \longleftarrow \cdots \xleftarrow{d_n} C_n$$

associated with $\rho(K)$ has the property that $f\text{-rank}(d_i(C_i), C_{i-1}) = 0$ and is additive for $i \geq 4$. Denote $\pi_2(L^n) \approx \pi_2(K) \approx \pi_2$. We show that the homotopy system $\rho(K)$ determines a crossed sequence

$$e \leftarrow \pi \leftarrow \pi_1(K^1, e^0) \leftarrow \pi_2(K, K^1, e^0) \leftarrow \pi_2 \leftarrow 0$$

with Euler characteristic equal to $\chi(k^3(L^n)) = \chi_0$. Let

$$e \longleftarrow \pi \xleftarrow{\overline{d}_1} \rho_1 \xleftarrow{\overline{d}_2} \overline{D}_2 \longleftarrow \pi_2 \longleftarrow 0$$

be a crossed sequence representing the element $k^3(L^n) \in H^3(\pi, \pi_2)$ and having Euler characteristic χ_0. Pick a minimal set of generators $\overline{a}_1, \dots, \overline{a}_k$ for the crossed module $\overline{D}_2$. Denote by (D_2, ρ_1, d_2') a free crossed module with basis $a_1, \dots, a_k$ such that $d_2(\overline{a}_i) = d_2'(a_i)$. There exists an epimorphism of crossed modules

$$\begin{array}{ccccccccc}
 & & & & 0 & \longleftarrow & R & \longleftarrow & R & \longleftarrow & 0 \\
 & & & & & & \downarrow & & \downarrow & & \\
e & \longleftarrow & \pi & \xleftarrow{d_1'} & \rho_1 & \xleftarrow{d_2'} & D_2 & \longleftarrow & Z_2 & \longleftarrow & 0 \\
 & & & & \downarrow \mathrm{id} & & \downarrow \mathrm{id} & & \downarrow f_1 & & \downarrow f_2 \\
e & \longleftarrow & \pi & \xleftarrow{\overline{d}_1} & \rho_1 & \longleftarrow & \overline{P}_2 & \longleftarrow & \pi_2 & \longleftarrow & 0.
\end{array}$$

Assume that R is a finitely generated $\mathbb{Z}[\pi]$-module. Let $d_3 \colon D_3 \to R$ be an epimorphism of a finitely generated free module onto R and consider the homotopy system

$$e \longleftarrow \pi \xleftarrow{d_1'} \rho_1 \xleftarrow{d_2'} D_2 \xleftarrow{d_3'} D_3.$$

By construction, its crossed sequence has Euler characteristic χ_0. By a theorem of Whitehead (see Theorem 6.22), this homotopy system has a geometric realization. Let N be a CW-complex of dimension 3 whose homotopy system coincides with

$$e \longleftarrow \pi \xleftarrow{d_1'} \rho_1 \xleftarrow{d_2'} D_2 \xleftarrow{d_3'} D_3.$$

Let K^3 be the 3-skeleton of the complex K. By construction, N^3 and K^3 have equivalent 2-types, and therefore the same homotopy 2-type. Consequently, by another theorem of Whitehead [159] there exist positive integers m and n such that the CW-complexes

$$N^3 \cup \left(\bigvee_{i=1}^{m} S_i^3 \right) \quad \text{and} \quad K^3 \cup \left(\bigvee_{j=1}^{n} S_j^3 \right)$$

are simple homotopy equivalent. If we seal up the wedge of spheres $\bigvee_{j=1}^{n} S_j^3$ by four-dimensional disks in the CW-complex $K^3 \cup (\bigvee_{i=1}^{n} S_j^3)$ and attach the remaining cells of K via their previous characteristic mappings, we obtain a complex $\overline{K}$ simple homotopy equivalent to the complex K. Using $\overline{K}$, build up the complex

$$N^3 \cup \left(\bigvee_{i=1}^{m} S_i^3\right)$$

to a CW-complex N simple homotopy equivalent to $\overline{K}$. Let $\rho(N)$ be a homotopy system associated with N. Embed N in a Euclidean space of large dimension and consider its tubular neighborhood. Clearly, the resulting smooth manifold with boundary W admits the handle decomposition corresponding to the cellular structure. Without loss of generality, we can assume that the chain complex of free $\mathbb{Z}[\pi]$-modules

$$\mathbb{Z} \longleftarrow \mathbb{Z}[\pi] \xleftarrow{d_1'} D_1 \xleftarrow{d_2'} D_2 \xleftarrow{d_3'} D_3 \longleftarrow \cdots$$

constructed from this decomposition has the property that

$$f\text{-rank}(\partial_i(D_i), D_{i-1}) = 0$$

and is additive for $i \geq 4$. Compare it with the chain complex of the CW-complex K. By Lemma 4.7,

$$-\mu(C_1) + \mu(C_2) - \mu(C_3) = \overline{\chi}_0 = -\mu(D_1) + \mu(D_2) - \mu(D_3)$$

and therefore

$$-\mu(C_1) + \mu(C_2) = \chi_0 + \mu(C_3), \qquad -\mu(D_1) + \mu(D_2) = \overline{\chi}_0 + \mu(D_3).$$

Since $\mu(C_3) \leq \mu(D_3)$, we have

$$-\mu(C_1) + \mu(C_2) = -\mu(D_1) + \mu(D_2) = \chi_0.$$

Thus, the homotopy system determined by the homotopy system $\rho(K)$ has Euler characteristic equal to χ_0.

We show now how to use the homotopy system $\rho(K)$ to construct a minimal homotopy system $(\hat{\rho}_i, \hat{d}_i)$ homotopy equivalent to $\rho(L^n)$. The argument is very similar to that in the proof of sufficiency in the theorem of the preceding section. Let $\rho(T)$ be a homotopy system, associated with a CW-complex T, that is homotopy equivalent to $\rho(K)$ and minimal in dimension 2. Embed the CW-complexes T and K in Euclidean spaces of high enough dimension, and consider their tubular neighborhoods, obtaining in them smooth manifolds with boundary W_1 and W_2 respectively. The cellular structures on T and K induce handle decompositions on W_1 and W_2, so that to each cell of dimension i there corresponds a handle of index i. In addition, in view of the restrictions on dimension, we can suppose that the mappings

$$i_{1*}\colon \pi_1(\partial W_1, x) \to \pi_1(W_1, x) \approx \pi,$$
$$i_{2*}\colon \pi_1(\partial W_2, y) \to \pi_1(W_2, y) \approx \pi$$

are isomorphisms. Let Γ be the submanifold in W_2 formed by the handles of index 0, 1, and 2. The decomposition of W_2 gives a presentation of the group $\pi_1(\partial W_2, y) = \pi$ with defect equal to χ_0. Since π satisfies the condition (h, χ_0), there exists a minimal presentation of the group π with defect equal to χ_0. Let $\overline{K}^2$ be a two-dimensional CW-complex constructed from this minimal presentation of the group π. There is an embedding $j: \overline{K}^2 \to \Gamma$ that is a homotopy equivalence. Consider a tubular neighborhood $j: (\overline{K}^2)$ in the manifold Γ and denote it by Ω. Obviously, $H = \overline{\Gamma \setminus \Omega}$ is an h-cobordism. We can suppose that on H there are handles of index greater than 3. Shift them towards the boundary of the manifold W_2. Denote the resulting handle decomposition of the manifold W_2 by P. Consider the chain complexes associated with the handle decomposition of the manifold W_1 and the new decomposition of W_1:

$$\mathbb{Z} \xleftarrow{\varepsilon} \mathbb{Z}[\pi] \xleftarrow{\hat{d}_1} \overline{C}_1 \xleftarrow{\hat{d}_2} \overline{C}_2 \xleftarrow{\hat{d}_3} \overline{C}_3 \longleftarrow \cdots,$$

$$\mathbb{Z} \xleftarrow{\varepsilon} \mathbb{Z}[\pi] \xleftarrow{\partial_1} E_1 \xleftarrow{\partial_2} E_2 \xleftarrow{\partial_3} E_3 \longleftarrow \cdots.$$

We suppose that $f\text{-rank}(\hat{d}_i(\overline{C}_i), \overline{C}_{i-1}) = 0 = f\text{-rank}(\partial_i(E_i), E_{i-1})$ and is additive for $i \geq 4$, since the elementary operations in these dimensions can be realized by operations over handles. By Lemma 4.7, we can write the equality

$$1 - \mu(\overline{C}_1) + \mu(\overline{C}_2) - \mu(\overline{C}_3) = \chi_0 = 1 - \mu(E_1) + \mu(E_2) - \mu(E_3).$$

By construction, $\mu(\overline{C}_1) \leq \mu(E_1)$, $\mu(\overline{C}_3) \leq \mu(E_3)$. Therefore $\mu(\overline{C}_2) = \mu(E_2)$. Thus the homotopy system associated with the handle decomposition P of W_2 is minimal in dimensions 1, 2, and 3. We claim it is minimal also in the remaining dimensions. Indeed, by stabilization and contraction of the boundary homomorphisms $\hat{d}_2$ and $\hat{d}_3$, we can arrange that $f\text{-rank}(d_i(\overline{C}_i), \overline{C}_{i-1}) = 0$ and is additive for $i = 2, 3$. For the remaining dimensions this condition holds automatically. By Lemma 4.5, the chain complex

$$\mathbb{Z} \xleftarrow{\varepsilon} \mathbb{Z}[\pi] \xleftarrow{\hat{d}_1} \overline{C}_1 \xleftarrow{\hat{d}_2} \overline{C}_2 \xleftarrow{\hat{d}_3} \overline{C}_3 \longleftarrow \cdots$$

is minimal and therefore so is the original homotopy system $\rho(W_2)$.

Finally, we prove that the module R is finitely generated. If A and B are two finitely generated modules over the ring $\mathbb{Z}[\pi]$ such that $n\mathbb{Z}[\pi] \oplus A \approx m\mathbb{Z}[\pi] \oplus B$, and $f: F \to A$ is an epimorphism of a finitely generated free module F onto A with $\operatorname{Ker} f$ finitely generated, then for any epimorphism $g: G \to B$, where G is a finitely generated free module, $\operatorname{Ker} g$ is also finitely generated. This follows from Schanuel's lemma, applied to the diagrams

$$0 \leftarrow A \oplus n\mathbb{Z}[\pi] \xleftarrow{f \oplus \mathrm{id}} F \oplus n\mathbb{Z}[\pi] \leftarrow \operatorname{Ker} f \leftarrow 0,$$

$$0 \leftarrow B \oplus m\mathbb{Z}[\pi] \xleftarrow{g \oplus \mathrm{id}} G \oplus m\mathbb{Z}[\pi] \leftarrow \operatorname{Ker} g \leftarrow 0$$

and the composite mapping

$$\operatorname{Ker} f \oplus G \oplus m\mathbb{Z}[\pi] \approx \operatorname{Ker} g \oplus F \oplus n\mathbb{Z}[\pi] \to \operatorname{Ker} g \to 0.$$

Consider the crossed sequences

$$e \longleftarrow \pi \xleftarrow{\ \overline{d}_1\ } \rho_1 \xleftarrow{\ \overline{d}_2\ } \overline{D}_2 \longleftarrow \pi_2 \longleftarrow 0,$$

$$e \longleftarrow \pi \xleftarrow{\ \partial_1\ } \pi_1(K^1, e^0) \xleftarrow{\ \partial_2\ } \pi_2(K, K^1, e^0) \longleftarrow \pi_2 \longleftarrow 0$$

and mappings onto them of free crossed modules:

$$\begin{array}{ccccccccccc}
 & & & & 0 & \longleftarrow & R & \longleftarrow & R & \longleftarrow & 0 \\
 & & & & & & \downarrow & & \downarrow & & \\
e & \longleftarrow & \pi & \xleftarrow{d_1} & \rho_1 & \xleftarrow{d_2'} & D_2 & \longleftarrow & Z_2 & \longleftarrow & 0 \\
 & & \downarrow{\scriptstyle \mathrm{id}} & & \downarrow{\scriptstyle \mathrm{id}} & & \downarrow{\scriptstyle f_1} & & \downarrow{\scriptstyle f_2} & & \\
e & \longleftarrow & \pi & \xleftarrow{\overline{d}_1} & \rho_1 & \xleftarrow{\overline{d}_2} & \overline{D}_2 & \longleftarrow & \pi_2 & \longleftarrow & 0,
\end{array}$$

$$\begin{array}{ccccccccccc}
 & & & & 0 & \longleftarrow & S & \longleftarrow & S & \longleftarrow & 0 \\
 & & & & & & \downarrow & & \downarrow & & \\
e & \longleftarrow & \pi & \xleftarrow{\partial_1'} & \pi_1(K^1, e^0) & \xleftarrow{\partial_2'} & \pi_2(K^2, K^1, e^0) & \longleftarrow & \pi_2(K^2) & \longleftarrow & 0 \\
 & & \downarrow{\scriptstyle \mathrm{id}} & & \downarrow{\scriptstyle \mathrm{id}} & & \downarrow{\scriptstyle i_1} & & \downarrow{\scriptstyle i_2} & & \\
e & \longleftarrow & \pi & \xleftarrow{\partial_1} & \pi_1(K_1, e^0) & \xleftarrow{\partial_2} & \pi_2(K, K^1, e^0) & \longleftarrow & \pi_2 & \longleftarrow & 0
\end{array}$$

Obviously the module $S = d_3(C_3)$ is finitely generated. Making the segments commutative, we have:

$$\begin{array}{ccccccccc}
 & & 0 & \longleftarrow & R & \longleftarrow & R & \longleftarrow & 0 \\
 & & & & \downarrow & & \downarrow & & \\
0 & \longleftarrow & \operatorname{Ker} d_1'^{\mathrm{ab}} & \xleftarrow{d_2'^{\mathrm{ab}}} & D_2^{\mathrm{ab}} & \longleftarrow & Z_2 & \longleftarrow & 0 \\
 & & & & \downarrow{\scriptstyle f_1^{\mathrm{ab}}} & & \downarrow{\scriptstyle f_2} & & \\
0 & \longleftarrow & \operatorname{Ker} \overline{d}_1^{\mathrm{ab}} & \xleftarrow{d_2^{\mathrm{ab}}} & \overline{D}_2^{\mathrm{ab}} & \longleftarrow & \pi_2 & \longleftarrow & 0,
\end{array}$$

$$\begin{array}{ccccccccc}
 & & 0 & \longleftarrow & S & \longleftarrow & S & \longleftarrow & 0 \\
 & & & & \downarrow & & \downarrow & & \\
0 & \longleftarrow & \operatorname{Ker}\partial_1'^{\mathrm{ab}} & \longleftarrow & \pi_2(K^2, K^1, e^0) = C_2 & \longleftarrow & \pi_2(K^2, e^0) & \longleftarrow & 0 \\
 & & \downarrow \mathrm{id} & & \downarrow i_1^{\mathrm{ab}} & & \downarrow i_2^{\mathrm{ab}} & & \\
0 & \longleftarrow & \operatorname{Ker}\partial_1 & \xleftarrow{\partial_2^{\mathrm{ab}}} & \pi_2(K, K^1, e^0)^{\mathrm{ab}} & \longleftarrow & \pi_2(K, e^0) & \longleftarrow & 0
\end{array}$$

Here C_2 and $\overline{D}_2^{\mathrm{ab}}$ are finitely generated free modules. As noted in §4 of this chapter, the crossed sequences determine exact sequences of modules

$$0 \leftarrow I[\pi] \leftarrow \mathbb{Z}[\rho_1] \otimes \mathbb{Z}[\pi] \leftarrow \overline{D}_2^{\mathrm{ab}} \leftarrow \pi_2 \leftarrow 0,$$
$$0 \leftarrow I[\pi] \leftarrow \mathbb{Z}[\pi_1(K^1, e^0)] \otimes \mathbb{Z}[\pi] \leftarrow \pi_2(K, K^1, e_0)^{\mathrm{ab}} \leftarrow \pi_2 \leftarrow 0,$$

which are congruent. By virtue of Corollary 3.6 applied to the modules $\overline{D}_2^{\mathrm{ab}}$ and $\pi_2(K, K^1, e^0)$, there exist positive integers n and m such that the modules $\overline{D}_2^{\mathrm{ab}} \oplus n\mathbb{Z}(\pi)$ and $\pi_2(K, K^1, e^0) \oplus m\mathbb{Z}[\pi]$ are isomorphic. Therefore the module R is finitely generated. This completes the proof of the theorem. □

Observe that we have also proved the following fact: if a crossed sequence $e \leftarrow \pi \leftarrow \rho_1 \leftarrow \rho_2 \leftarrow \pi_2 \leftarrow 0$ is realized by a finite CW-complex, then every congruent crossed sequence is also realized by a finite CW-complex. In addition, we have proved the following statement.

PROPOSITION 6.6. *Let $k \in H^3(\pi, \pi_2)$, $\chi(k) = \chi_0$, and let k be realized by a finite CW-complex. Suppose also that π satisfies condition (h, χ_0) and is an s-group. Then there exists a crossed sequence*

$$(s): e \leftarrow \pi \leftarrow \mathfrak{F} \leftarrow \rho_2 \leftarrow \pi_2 \leftarrow 0$$

representing the element k, where $\chi(s) = \chi_0$ and $\mu(\mathfrak{F}) = \mu(\pi)$. This sequence will be called a minimal crossed sequence. □

Obviously, for any minimal crossed sequence the minimal numbers of generators of the crossed module ρ_2 are the same.

Similarly to the approach used in the preceding section, we obtain from the analysis of the proof of Theorem 6.28 taken together with the Whitehead theorem (see Theorem 6.22) a more general theorem which we do not prove here.

THEOREM 6.29. *Let*

$$e \longleftarrow \pi \xleftarrow{d_1} \rho_1 \xleftarrow{d_2} \rho_2 \longleftarrow \cdots \xleftarrow{d_n} \rho_n$$

be a homotopy system. If π satisfies condition $(h, \chi(\rho_i, d_i))$ and is an s-group, then there exists a minimal homotopy system $(\overline{\rho}_i, \overline{d}_i)$ homotopy equivalent to

$$e \longleftarrow \pi \xleftarrow{d_1} \rho_1 \longleftarrow \cdots \xleftarrow{d_n} \rho_n. \quad \square$$

COROLLARY 6.2. *Suppose that*

$$e \longleftarrow \pi \xleftarrow{d_1} \rho_1 \xleftarrow{d_2} \rho_2 \longleftarrow \cdots \xleftarrow{d_n} \rho_n$$

is a homotopy system, the Euler characteristic of the first k-invariant satisfies the relation $\chi(\rho_i, d_i) \geq 1 + d^{\mu}(\pi)$, and π is an s-group. Then in its homotopy type there exists a minimal homotopy system. In particular, this holds if π is a finite nilpotent s-group.

PROOF. This follows immediately from Theorem 6.29 and Proposition 6.5. $\square$

THEOREM 6.30. *Let (ρ_i, d_i) be a homotopy system with the abelian fundamental group and the first k-invariant zero. Then in its homotopy type there is a minimal homotopy system.*

PROOF. The fact that the first k-invariant is equal to zero means that the crossed sequence representing it, i.e.,

$$e \longleftarrow \pi \xleftarrow{d_1} \rho_1 \longleftarrow \rho_2/d_3(\rho_3) \longleftarrow \operatorname{Ker} d_2 \longleftarrow 0$$

has the property that the crossed ρ_1-module $\rho_2/d_3(\rho_3)$ decomposes into the direct product of operator groups $\operatorname{Ker} d_2 \times N$, where N is the kernel of the epimorphism $d_1 \colon \rho_1 \to \pi$. This is reflected in the structure of the 2-skeleton of the homotopy system. In order to prove the theorem, let us construct a three-dimensional CW-complex that has the same algebraic 2-type as the homotopy system and is minimal in dimensions 1 and 2. Consider the standard minimal presentation of the free abelian group $n\mathbb{Z} = \{a_1, \ldots, a_n; [a_1, a_2], \ldots, [a_{n-1}, a_n]\}$, where $[a_i, a_j]$ are the commutators. Realize this presentation by a CW-complex K^2 of dimension 2. Since every free abelian group is an s-group, the minimal number of generators of its second homotopy group considered as a $\mathbb{Z}[n\mathbb{Z}]$-module is the least possible among all two-dimensional CW-complexes with the fundamental groups isomorphic to $n\mathbb{Z}$. Since the module $\pi_2(K^2, e^0)$ is finitely generated, we can close it up by three-dimensional cells, thus obtaining a complex K^3. Consider the wedge of CW-complexes $K^3 \bigvee_{i=1}^{k} S_i^2 = L^3$, where $k = \mu(\pi_2)$. By attaching cells of dimension 3, we can arrange that the resulting CW-complex $\overline{L}^3$ has the second homotopy group isomorphic to $\operatorname{Ker} d_2$.

Let N^3 be a CW-complex realizing the segment of the homotopy system $\pi \leftarrow \rho_1 \leftarrow \rho_2 \leftarrow \rho_3$. By construction, $\overline{L}^3$ and N^3 have the same 2-type.

Therefore, the homotopy system $\rho(\overline{L}^3)$ associated with the CW-complex L^3 can be built up to a homotopy system

$$\pi \longleftarrow \overline{\rho}_1 \xleftarrow{\overline{d}_2} \overline{\rho}_2 \xleftarrow{\overline{d}_3} \overline{\rho}_3 \longleftarrow \cdots$$

homotopy equivalent to (ρ_i, d_i). Then, using the stabilization and contraction of the boundary homomorphisms $\overline{d}_i$, $i > 3$, we can arrange that f-rank$(d_i(\overline{\rho}_i), \overline{\rho}_{i-1}) = 0$ and is additive. Using Lemma 6.7, we can easily see that the homotopy system so obtained is minimal. This completes the proof of the theorem. □

The author does not know whether this theorem holds in the case of no restriction on the first k-invariant. The essential question is whether a free abelian group of rank ≥ 3 satisfies condition (h) or not.

One can show without difficulty that the homotopy type of a homotopy system is determined by the first k-invariant and the invariants introduced for chain complexes in §2 of the present chapter.

In order to estimate the number of generators of chain modules in homotopy systems, the bounds given in Chapter IV, §6, can be used.

CHAPTER VII

Minimal Morse Functions on Non-Simply-Connected Manifolds

The need to make use of homotopy systems in order to study Morse functions on non-simply-connected closed manifolds or on manifolds with one boundary component arises from the failure of the chain complexes constructed from Morse functions and gradient-like vector fields to capture completely the geometric aspect of the problem. This relates to application of the Whitney lemma to the reduction of the number of points of intersection of manifolds of complementary dimensions. For example, let $f: M^n \to [0, 1]$ be an ordered Morse function, ξ an ordered gradient-like vector field, and $C(f, \xi, M^n)$ the chain complex of $\mathbb{Z}[\pi_1(M^n)]$-modules associated with f and ξ. Suppose that the complex is of the form

$$\begin{array}{ccccccc} \mathbb{Z} \leftarrow \mathbb{Z}[\pi_1(M^n, x)] & \overset{\partial_1}{\leftarrow} & C_1 & \overset{\partial_2}{\leftarrow} & C_2 & \overset{\partial_3}{\leftarrow} & C_3 \leftarrow \cdots \\ & & \oplus & & \oplus & & \\ & \leftarrow & n\mathbb{Z}[\pi_1(M^n, x)] & \overset{E}{\leftarrow} & n\mathbb{Z}[\pi_1(M^n, x)] & \leftarrow 0, & \end{array}$$

where E is the identity matrix. This does not at all imply that the critical points of index 2 and 1 that correspond to the direct summands $n\mathbb{Z}[\pi_1(M^n, x)]$ can be eliminated. A similar observation applies to critical points of index 3 and 2. A suitable algebraic model that permits the determination of whether or not such pairs of critical points can be eliminated is given by the homotopy system constructed from f and ξ. In §1 we give conditions, in terms of homotopy systems, for a pair of critical points of index 2 and 3 to be capable of cancellation.

The main theme of this chapter is the construction of minimal Morse functions on closed manifolds, and manifolds with one boundary component and nontrivial fundamental group.

§1. Homotopy systems and Morse functions

Let $\leftarrow \pi_1 \overset{d_1}{\longleftarrow} \rho_1 \overset{d_2}{\longleftarrow} \rho_2 \longleftarrow \cdots$ be the homotopy system constructed from a Morse function f and vector field ξ defined on the manifold M^n, $n \geq 6$. In what follows, all Morse functions will be assumed to be ordered and to have one critical point of index 0. As distinguished basis elements in each module ρ_i we consider middle disks of critical points. Denote by $[h_i^3]$ the element of ρ_3 corresponding to the middle disk of a critical point of index 3. The element $d_3[h_i^3]$ can be written in the form

$$d_3[h_i^3] = \varepsilon_1 x_1 [h_{j1}^2] + \cdots + \varepsilon_s x_s [h_{js}^2], \qquad 1 \le j \le s,$$

where $\varepsilon_k = \pm 1$, $x_k \in \rho_1$ $(1 \le k \le s)$, and $[h_i^2]$ are the generators corresponding to the middle disks of critical points of index 2.

LEMMA 7.1. *Suppose that* $d_3([h_i^3]) = w \pm x_i[h_j^2] \pm v$, *where* $x_i \in \rho_1$ *and* $[h_i^2]$ *does not enter into the words* w *and* v. *Then the critical points of index* 2 *and* 3 *that correspond to* $[h_j^2]$ *and* $[h_i^3]$ *can be cancelled.*

PROOF. The meaning of the condition in the lemma is as follows: the middle sphere of the critical point of index 3 intersects the comiddle sphere of the critical point of index 2 in just one point. The middle sphere ∂H_i^3 of the critical point of index 3 corresponding to $[h_i^3]$ represents the element $a = w \pm x_i[h_j^2] \pm v$, for which $d_2(a) = 0$. We assume that the corresponding handles are distinguished ones. Realizing the element a by means of a connected sum, consisting of a linear combination of middle disks of critical points of index 2, we obtain a disk D_1^2 embedded in M_2, with $\partial D_1^2 \subset M_1$. Here M_i is the submanifold of M^n containing the critical points with indices $\le i$ $(i = 1, 2)$. The condition $d_2(d) = 0$ implies that ∂D_1^2 bounds a 2-dimensional disk D_2^2 in M_1. Consider the sphere S^2 formed by identifying D_1^2 and D_2^2 along their boundaries. By construction, S^2 is homotopic to the middle sphere ∂h_i^3. By Corollary 5.9 of [123], the sphere ∂h_i^3 is ambient isotopic to S^2. By construction, S^2 intersects the comiddle sphere of the critical point of index 2 corresponding to $[h_j^2]$ in just one point. Carrying out the isotopy from ∂h^3 to S^2, adjust the function f and the vector field ξ on M_2. As a result, we obtain a Morse function for which the pair of critical points of index 2 and 3 has been eliminated. □

We note that since in the crossed module ρ_2 we have the relation $a_1 + a_2 - a_1 = d_2(a_1)a_2$, the representation $d_3([h_i^3]) = \varepsilon_1 x_1[h_{j1}^2] + \cdots + \varepsilon_s x_s[h_{js}^2]$ is nonunique; it can be varied by means of this relation.

Denote by S the class of manifolds W^n with boundary $\partial W^n = V^{n-1}$ such that $\pi_1(V^{n-1}, y) \to \pi_1(W^n, y)$ is an isomorphism, $n \ge 6$. In this section we shall consider manifolds with boundary of this type.

With each ordered Morse function f and gradient-like vector field ξ defined on W^n is associated a homotopy system $\rho(W^n, f, \xi)$. Similarly to the case of cobordisms, the following lemma holds.

LEMMA 7.2. *Let $\rho(W^n, f, \xi)$ and $\rho(W^n, g, \eta)$ be homotopy systems associated with Morse functions f and g and vector fields ξ and η, defined on a manifold W^n with boundary $\partial W^n = V^{n-1}$. Suppose that W^n belongs to the class S and has dimension > 5. Then $\rho(W^n, f, \xi)$ and $\rho(W^n, g, \eta)$ have the same simple homotopy type. Furthermore, if a homotopy system*

$$(\rho_i, d_i) : e \longleftarrow \pi_1(W^n, x) \overset{d_1}{\longleftarrow} \rho_1 \overset{d_2}{\longleftarrow} \rho_2 \longleftarrow \cdots \overset{d_{n-2}}{\longleftarrow} \rho_{n-2}$$

of length $n-2$ is simple homotopy equivalent to the homotopy system $\rho(W^n, f, \xi)$, then W^n admits an ordered Morse function $\overline{f}$ and gradient-like vector field $\overline{\xi}$ such that $\rho(W^n, \overline{f}, \overline{\xi})$ and (ρ_i, d_i) are simple isomorphic.

PROOF. Retract the manifold W^n to the middle disks of the critical points of the functions f and g, obtaining CW-complexes K_1 and K_2 with the same simple homotopy type. Then the chain complexes constructed on the universal coverings $\hat{K}_1$ and $\hat{K}_2$ are simple homotopy equivalent. It follows by the Whitehead theorem that the homotopy systems $\rho(W^n, f, \xi)$ and $\rho(W^n, g, \eta)$ also have the same homotopy type.

Let us now prove the second part of the lemma. The condition imposed on the boundary of the manifold W^n ensures the existence of a Morse function $\overline{f}: W^n \to [0, 1]$, $f^{-1}(1) = V^{n-1}$, without critical points of index $n-1$ or n. We can therefore suppose that the homotopy system $\rho(W^n, \overline{f}, \overline{\xi})$ has length $n-2$. A theorem of [154] ensures the existence of a sequence of elementary operations taking $\rho(W^n, \overline{f}, \overline{\xi})$ into (ρ_i, d_i) without increasing the length of any intermediate homotopy system. Since any elementary operation can be realized geometrically, we can suitably alter the Morse function and the gradient-like vector field to obtain the conclusion of the lemma. □

For a given manifold W^n, different homotopy systems can be constructed either by means of a Morse function and gradient-like vector field or by using cellular decomposition of W^n; but they all have the same homotopy type. It makes sense, therefore, to speak of the simple homotopy type of the homotopy system of W^n.

COROLLARY 7.1. *Let W^n, $n \geq 6$, be a manifold of class S with boundary $\partial W^n = V^{n-1}$. In order that W^n have a minimal Morse function, it is necessary and sufficient that the homotopy system of the manifold be simple homotopy equivalent to a minimal homotopy system.*

PROOF. Necessity is obvious, and sufficiency follows from the preceding lemma. □

We conclude this section with a conjecture. If the fundamental group $\pi = \pi_1(W^n, x)$ of a manifold W^n is an s-group and $\text{Wh}(\pi) = 0$, then the pair of critical points of index 2 and 3 cancels if its (algebraic) index of intersection is equal to $\pm g$.

For free groups the assertion does hold.

§2. Minimal Morse functions on manifolds of class S

THEOREM 7.1. *Let W^n be a manifold of class S with boundary $\partial W^n = V$, $n \geq 6$, such that $\chi(k^3(W^n)) = \chi_0$. Suppose that $\pi_1(W^n, x) = \pi$ is an s-group satisfying condition (h, χ_0), and* r-dim $\pi \leq k$. *Suppose also that the ith Morse number $\mathscr{M}_i(W^n) \geq k$ ($\lambda \geq 3$; for $n = 6$, $\mathscr{M}_4(W^n) \geq k$). Then the manifold W^n admits a minimal Morse function.*

PROOF. Theorem 6.26 ensures the existence of a minimal homotopy system (ρ_i, d_i) in the homotopy type of the homotopy system of W^n. Realize the homotopy system (ρ_i, d_i) by a CW-complex K. Let K^3 be the 3-skeleton of K. Since K and W^n have the same homotopy type, there exists a mapping $j: K^3 \to \operatorname{Int} W^n$ inducing an isomorphism of the homotopy groups $j_*: \pi_2(K^3, e^0) \to \pi_2(W^n, x)$. In view of the dimensional restrictions, we can suppose that j is an embedding on the 2-skeleton K^2. Consider a tubular neighborhood of $j(K^2) \subset W^n$ and denote it by W_2^n. Using the composite mapping

$$K^2 \xrightarrow{\ i\ } K^3 \xrightarrow{\ j\ } W^n,$$

one can show that the mapping $(j \circ i)_*: \pi_2(W_2^n, x^0) \to \pi_2(W^n, x)$ is an epimorphism. Consider the manifold $\overline{W^n \setminus W_2^n} = \Gamma$. Clearly, $\pi_1(\Gamma, y) \approx \pi_1(\partial W_2^n, y) \approx \pi_1(W^n, y)$. Furthermore, from the exact homology sequence for the pair (W^n, W_2^n) and the excision theorem for homology of a universal covering, it follows that $\pi_2(\Gamma, \partial W_2^n, y) = \pi_1(\Gamma, \partial W_2^n, y) = 0$. The Wall theorem [156] implies that on Γ there exists an ordered Morse function $f_2: \Gamma \to [0, 1]$, $f_2^{-1}(0) = \partial W_2^n$, $f^{-1}(1) = \partial W^n$ without critical points of index $0, 1, 2, n-1, n$. Using the cellular structure of $j(K^2)$, construct on W_2^n an ordered Morse function $f_1: W_2^n \to [-1, 0]$ having only critical points of index $0, 1, 2$ (corresponding to the cells of the subcomplex K^2 of dimension $0, 1, 2$). Consider the Morse function $f = f_1 \cup f_2$ on the manifold W^n and its gradient-like vector field ξ, and from them construct the chain complex

$$C(W^n, f, \xi): \mathbb{Z} \leftarrow \mathbb{Z}[\pi_1(W^n)] \xleftarrow{\ \partial_1\ } C_1 \xleftarrow{\ \partial_2\ } C_2 \longleftarrow \cdots.$$

Performing stabilization and contraction on the boundary homomorphisms ∂_i ($i \geq 4$), we can arrange that f-rank$(\partial_i(C_i), C_{i-1}) = 0$ and is additive. Since in these dimensions all elementary operations can be realized, we can assume, by suitably altering f and ξ, that the chain complex $C(W^n, f, \xi)$ has this property, i.e., that f-rank$(\partial_i(C_i), C_{i-1}) = 0$ and is additive for $i \geq 4$. We now show that f is a minimal Morse function. First, let us prove that f has the minimal number of critical points of index 3. Indeed, applying Lemma 6.7 to the chain complex associated with the homotopy system (ρ_i, d_i) and the chain complex $C(W^n, f, \xi)$, we obtain

$$\chi_0 = -\mu(C_1) + \mu(C_2) - \mu(C_3) = -\mu(\rho_1) + \mu(\rho_2^{\text{ab}}) - \mu(\rho_3)$$

(here ρ_2^{ab} is the abelianization of the crossed module ρ_2). By construction, $-\mu(C_1) + \mu(C_2) = -\mu(\rho_1) + \mu(\rho_2^{\text{ab}})$ whence it follows that $\mu(C_3) = \mu(\rho_3)$. By construction, f has the minimal number of critical points of index 0, 1, 2, and, as we have shown, of index 3. Observe that the chain complex $C(W^n, f, \xi)$ is not, in general, minimal in dimensions 2 and 3. By stabilization and contraction of the boundary homomorphisms ∂_i $(i = 1, 2, 3)$ in the chain complex $C(W^n, f, \xi)$ it can be arranged that $f\text{-rank}(\partial(C_i), C_{i-1}) = 0$ and is additive. By Lemma 4.5 the chain complex so obtained is minimal. Since the modules C_i for $i \geq 4$ have been left undisturbed, their ranks must already have been minimal, and therefore the number of critical points of index i for f has also been minimal. This completes the proof of the theorem. □

COROLLARY 7.2. *Let $\pi = \pi_1(W^n, x)$ be an s-group satisfying the condition (h, χ_0), where $\chi_0 = \chi(k^3(W^n))$ and $\text{Wh}(\pi) = 0$. Then W^n has a minimal Morse function. In particular, this holds for free groups and $\mathbb{Z} \oplus \mathbb{Z}$.*
□

COROLLARY 7.3. *Let $\pi = \pi_1(W^n, x)$ be a finite s-group that has no epimorphic mapping onto any of the groups $\mathbb{Z}_{p^2} \oplus \mathbb{Z}_{p^2}$, $\mathbb{Z}_p \oplus \mathbb{Z}_p \oplus \mathbb{Z}_p$, $\mathbb{Z}_p \oplus \mathbb{Z}_2 \oplus \mathbb{Z}_2 \oplus \mathbb{Z}_2$, $\mathbb{Z}_4 \oplus \mathbb{Z}_2 \oplus \mathbb{Z}_2$, $\mathbb{Z}_4 \oplus \mathbb{Z}_4$. Suppose also that the Euler characteristic $\chi(k^3(W^n))$ of the first k-invariant is $\geq d^\mu(\pi) + 1$ (for nilpotent groups $\chi(k^3(W^n)) \geq d(\pi) + 1$) and that $H_i(\hat{W}^n, \mathbb{Z}) \neq 0$ for some $i \geq 3$ (for $n = 6$, $H_4(\hat{W}^n, \mathbb{Z}) \neq 0$). Then W^n has a minimal Morse function.*

PROOF. In this case, $\text{r-dim}\,\pi \leq 1$, and the condition $(h, \chi(k^3(W^n)))$ is satisfied. For nilpotent groups, as noted above, we have $d(\pi) = d^\mu(\pi)$ [117].
□

COROLLARY 7.4. *Let $\pi = \pi_1(W^n, x)$ be a finite s-group. Suppose that the Euler characteristic $\chi(k^3(W^n)) = \chi_0$ of the first k-invariant of W^n is $\geq d^\mu(\pi) + 1$ (for nilpotent groups, $\chi(k^3(W^n)) \geq d(\pi) + 1$), and*

$$\mu(H_i(W^n, \mathbb{Z})) + \mu(\text{Tors}(H_{i-1}(W^n, \mathbb{Z}))) \geq 2$$

for some $i \geq 3$. (If $n = 6$, then $\mu(H_4(W^n, \mathbb{Z})) \geq 2$.) Then W^n has a minimal Morse function.

PROOF. The group π satisfies condition (h, χ_0) and $\text{r-dim}\,\pi \leq 2$. □

THEOREM 7.2. *Let W^n be a manifold of class S with boundary ∂W^n, $n \geq 6$. Suppose*

(1) *π is an s-group, $\pi = \pi_1(W^n, x)$,* r-dim $\pi \leq c$;
(2) *the essential height of π is equal to a;*
(3) *π satisfies condition (h, b) for some b, $d(\pi) \leq b \leq \chi_0 - c$, where $\chi_0 = \chi(k^3(W^n)) \geq a - 1$.*

Then W^n has a minimal Morse function.

REMARK. In Theorem 7.1, the relation between r-dim π and the ith Morse number of the manifold W^n allowed the exclusion of critical points of adjacent indices defining a "torsion block". In the present theorem the "torsion block" can be removed by using the structure of the 2-skeleton of W^n, which has the homotopy type of $K^2 \bigvee_i S_i^2$, where K^2 is a 2-dimensional CW-complex and $\bigvee_i S_i^2$ is a wedge of 2-spheres.

PROOF. The hypothesis of the theorem ensures the existence of a minimal homotopy system (ρ_i, d_i) in the homotopy type of the homotopy system of the manifold W^n. As shown in the proof of the theorem, the crossed sequence of the homotopy system (ρ_i, d_i) has Euler characteristic equal to χ_0. Realize the system (ρ_i, d_i) by a CW-complex K, and consider the 2-skeleton K^2 of K. By assumption, the group π satisfies the condition (h, b). Therefore K^2 is homotopy equivalent to a 2-dimensional CW-complex of the form

$$L^2 = \overline{L}^2 \cup \left(\bigvee_{i=1}^{\chi_0 - b} S_i^2 \right),$$

where $\overline{L}_2^2$ is constructed from a presentation $\pi = \{a_1, \dots, a_\mu(\pi); r_1, \dots, r_t\}$. Using the 3-skeleton $K^3 \subset K$ and Lemma 3.7 of [94], build up L^2 to a 3-dimensional complex L^3 homotopy equivalent to K^3. Since L^3 and W^n have equivalent algebraic 2-types, there exists a mapping $j: L^3 \to W^n$ inducing an isomorphism $j_*: \pi_2(L^3, e^0) \to \pi_2(W^n, x)$. Without loss of generality we can suppose that j is an embedding on K^2. Evidently, $j_*: \pi_2(L^2, e^0) \to \pi_2(W^n, x)$ is an epimorphism. Let W_2^n be a tubular neighborhood of $j(L^2)$ in W^n. Then W_2^n admits a decomposition into handles of index 0, 1, and 2, with one handle of index 0, $\mu(\pi)$ handles of index 1, and $t + \chi_0 - b$ handles of index 2, of which $\chi_0 - b$ are attached by a trivial embedding of the middle spheres. Using this handle decomposition, construct an ordered Morse function on W^n: $f_1: W_2^n \to [0, 1]$, $f_1^{-1}(0) = \partial W_2^n$. Consider the manifold $\overline{W^n \setminus W_2^n} = \Gamma$. Clearly, there exists on the manifold Γ an ordered Morse function f_2 and a gradient-like vector field ξ such that the associated chain complex has the form

$$\begin{array}{ccccccc} & & C_3 & \xleftarrow{\partial_4} & C_4 & \longleftarrow & \cdots \\ & & \oplus & & \oplus & & \\ 0 & \longleftarrow & R_3 & \longleftarrow & R_4 & \longleftarrow & 0, \end{array}$$

where $\mu(R_3) \leq c$ and $f\text{-rank}\partial_{i+1}(C_{i+1})C_i = 0$ and is additive for $i \geq 4$. Consider the Morse function $\overline{f} = f_1 \cup f_2$ on W^n and construct the associated chain complex, which is of the form

$$\begin{array}{ccccccccccccc} \mathbb{Z} & \xleftarrow{\varepsilon} & \mathbb{Z}[\pi] & \xleftarrow{\partial_1} & C_1 & \xleftarrow{\partial_2} & C_2 & \xleftarrow{\partial_3} & C_3 & \xleftarrow{\partial_4} & C_4 & & \\ & & & & & & \oplus & & \oplus & & \oplus & & \\ & & & & & & 0 & \longleftarrow & R_3 & \longleftarrow & R_4 & \longleftarrow & 0. \end{array}$$

By shifting the "torsion block" $0 \leftarrow R_3 \leftarrow R_4 \leftarrow 0$ in the direction of lesser dimension, we obtain the chain complex of the form

$$\begin{array}{ccccccccccc} \mathbb{Z} & \longleftarrow & \mathbb{Z}[\pi] & \xleftarrow{\partial_1} & C_1 & \xleftarrow{\partial_2} & C_2 & \xleftarrow{\partial_3} & C_3 & \xleftarrow{\partial_4} & C_4 \\ & & & & & & \oplus & & & & \\ & & & & & & M & & \oplus & & \\ & & & & & & \oplus & & & & \\ & & & & 0 & \longleftarrow & R_2 & \longleftarrow & R_3 & \longleftarrow & 0. \end{array}$$

Using the submodule M corresponding to $\chi_0 - b$ critical points of index 2, whose middle spheres realize the trivial element of the fundamental group corresponding to a level surface of the function f_1, we can apply the lemma to get rid of the critical points of index 2 and 3 that correspond to the elementary complex $0 \leftarrow R_2 \leftarrow R_3 \leftarrow 0$. Denote the Morse function so obtained by f. It is easily shown that f is a minimal Morse function on the manifold W^n; it suffices to use the argument at the end of the proof of Theorem 7.1. □

COROLLARY 7.5. *If $\pi = \pi_1(W^n, x)$ is a finite s-group and the Euler characteristics of the first k-invariant $\chi(k^3(W^n))$ is $\geq d^\mu(\pi) + 2$, then the manifold W^n has a minimal Morse function.*

THEOREM 7.3. *Let W^n be a manifold of class S such that $\pi_1(W^n, x) \approx n\mathbb{Z}$ $(n \geq 3)$ and the first k-invariant is zero. Then there exists on W^n a minimal Morse function.*

PROOF. See Corollary to Theorem 6.30 and to Lemma 7.2. □

PROPOSITION 7.1. *Let W^n be a manifold of class S. Suppose that $\pi_1(W^n, x) = \pi$ is an s-group and satisfies condition (h, χ_0), where $\chi_0 = \chi(k^3(W^n))$ is the Euler characteristic of the first k-invariant. Then there exists a manifold $\overline{W}^n$ that has a minimal Morse function and is homotopy equivalent to W^n.*

PROOF. Just as in the proof of Theorem 7.1, we find a submanifold W_2^n on which there exists a Morse function with one critical point of index 0, $\mu(\pi)$ critical points of index 1, and the minimal possible number of critical points of index 2. In addition, the embedding $i: W_2^n \to W^n$ induces an isomorphism $\pi_1(W_2^n, x) \to \pi_1(W^n, x)$ and an epimorphism $\pi_2(W_2^n, x) \to \pi_2(W^n, x)$. Since the submanifold $\Gamma = \overline{W^n \setminus W_2^n}$ satisfies the hypothesis of Proposition 5.8, there exists in Γ a submanifold W_1^n such that $\overline{W^n \setminus W_2^n} = \Omega$ is an h-cobordism, $\overline{W}^n = W_2^n \cup W_1^n$, and on W_1^n there exists a minimal Morse function. Obviously the manifold W^n is homotopy equivalent to $\overline{W}^n$ (they differ by an h-cobordism). Using the same argument as in the proof of the theorem, we can show that $f: \overline{W}^n \to [0, 1]$ is a minimal Morse function.
□

§3. A sufficient condition for the existence of a minimal Morse function on a manifold of class S with a fixed fundamental group

Fix a finitely presentable group π and consider all manifolds W^n of class S with the fundamental group isomorphic to π $(n \geq 6)$. In this section we discuss the question of what conditions should be imposed on the group π in order that the existence of minimal Morse functions on W^n be uniquely determined by the fundamental group only.

THEOREM 7.4. *In order for each manifold W^n, $n \geq 6$, of class S with a fixed fundamental group isomorphic to π to admit a minimal Morse function it is sufficient that*

(1) *π be an s-group;*
(2) *the group π satisfy condition (h);*
(3) $\mathrm{Wh}(\pi\{a_i; r_j\}) = \mathrm{Wh}(\pi)$ *for each presentation* $\pi = \{a_i, r_j\}$.

PROOF. Let W^n be an arbitrary manifold of class S satisfying the hypothesis of the theorem. By Theorem 6.26, in the homotopy type of the homotopy system of the manifold W^n there is a minimal homotopy system. Thus the only obstacle in the way of constructing a minimal Morse function on W^n can be a "torsion block". We claim that, with the use of the structure of the 2-dimensional cell complex of the manifold W^n, the "torsion block" can be eliminated.

Let (ρ_i, d_i) be a minimal homotopy system homotopy equivalent to the homotopy system of the manifold W^n. Realize it by a cell complex K. There is an embedding of the 2-skeleton K^2 of this complex in W^n inducing an isomorphism of the fundamental groups and an epimorphism of 2-dimensional homotopy groups. Let W_2^n be a tubular neighborhood of the image of K^2 in W^n.

Using the cellular structure, construct on the submanifold W_2^n a minimal Morse function $f_1: W_2^n \to [0, 1]$, $f_1^{-1}(1) = \partial W_2^n$. Let $\overline{W^n \setminus W_2^n} = \Gamma$. Since

$$\pi_1(\Gamma, \partial W_2^n, x) = \pi_2(\Gamma, \partial W_2^n, x) = 0,$$

there exists on Γ an ordered Morse function $f_2: \Gamma \to [1, 2]$, $f_2^{-1}(1) = \partial W_2^n$, $f_2^{-1}(2) = \partial W^n$ and a gradient-like vector field ξ such that the associated chain complex is of the form

$$\begin{array}{ccccccc} & & C_3 & \xleftarrow{\partial_4} & C_4 & \longleftarrow \cdots \xleftarrow{\partial_n} & C_n \\ & & \oplus & & & & \\ 0 & \longleftarrow & R_2 & \longleftarrow & R_3 & \longleftarrow & 0 \end{array}$$

and f-rank$(\partial_i(C_i), C_{i-1}) = 0$ and is additive. Obviously, the critical points of index 2 and 3 of the function $f = f_1 \cup f_2$ can be regrouped in such a way that the submanifold Ω, $\partial\Omega = \partial W_2^n \cup \overline{V}$ contains only the critical points corresponding to the block $0 \leftarrow R^2 \leftarrow R^3 \leftarrow 0$. Clearly, Ω is an h-cobordism with torsion $\tau_0 \in \mathrm{Wh}(\pi)$. If $\tau_0 = 0$, then Ω is a direct product and therefore the critical points of f on Ω are eliminated. This concludes the proof of the theorem following the scheme of the proof of Theorem 7.1.

Suppose that $\tau_0 = \tau(\Omega, \partial W_2^n) \neq 0$. Using the mapping $\xi(K^2, x) \to \mathrm{Wh}(\pi)$, find an embedding $j: K^2 \to \mathrm{Int}\, W_2^n$ with torsion equal to $-\tau_0$. Consider a tubular neighborhood of $j(K^2)$ in W_2^n. Denote it by $\overline{W}_2^n$. By construction, $\overline{W_2^n \setminus \overline{W}_2^n} = P$ is an h-cobordism with torsion equal to $(-\tau_0)$. Clearly, $P \cup \Omega$ is an h-cobordism. By the addition theorem for h-cobordisms, we have $\tau(P \cup \Omega, \partial\overline{W}_2^n) = 0$, i.e., $\overline{W^n \setminus \overline{W}_2^n} \simeq \partial\overline{W}_2^n \times I$. Let us modify the function f_1 on the manifold W_2^n by using the embedding j. The rest of the argument is similar to that in the proof of Theorem 7.1. □

COROLLARY 7.6. *Let W^n be a manifold of class S, $n \geq 6$. Suppose that the group $\pi_1(W^n, x) = \pi$ is isomorphic to one of the following groups:*

(1) *a finite abelian group;*

(2) *the dihedral group D_n (where n is odd).*

Then there exists a minimal Morse function on W^n.

We conclude this section with a few observations.

If the embedding of the boundary $\partial W^n \to W^n$ does not induce an isomorphism of fundamental groups, then in order to construct minimal Morse functions we must make use of one more homotopy system associated with a Morse function $f: W^n \to [0, 1]$, $f^{-1}(0) = \partial W^n$. But if

$\pi_1(W^n, \partial W^n, x) \neq 0$, the resultant homotopy systems do not easily lend themselves to study.

In all the statements given in §§1–3, the condition that $\pi_1(\partial W^n, x) \to \pi_1(W^n, x)$ be an isomorphism may be replaced by the condition that $\pi_1(\partial W^n, x) \to \pi_1(W^n, x)$ be an epimorphism.

§4. Fundamental group and Morse numbers

As noted above, for a manifold of class (S), the chain complexes associated with Morse functions do not capture completely the critical points of index 1, 2, and 3. Therefore certain adjustments must be made in the universal estimates given in Chapter V, §7.

In this section we give sharper bounds with the use of the fundamental group and the first k-invariant of the manifold.

As noted above if the crossed sequence

$$e \leftarrow \pi_1(K, e^0) \leftarrow \pi_1(K^1, e^0) \leftarrow \pi_2(K, K^1, e^0) \leftarrow \pi_2(K) \leftarrow 0$$

is realized by a finite CW-complex K, then a crossed sequence congruent to it is also realized by a finite CW-complex (K^1, the 1-skeleton of K). Therefore, for smooth manifolds, congruent crossed sequences representing the first k-invariants can be realized by handle decompositions if the dimension of the manifold is greater than 5. According to [85, 12], for 3-manifolds and 4-manifolds this is not the case. However, in any case, given an element $a \in H^3(\pi_1(W^n, x), \pi_2(W^n, x))$ that is the first k-invariant, we can consider all crossed sequences

$$e \leftarrow \pi_1(W^n, x) \leftarrow \mathfrak{F} \leftarrow \pi_2(W^n, W_1^n, x) \leftarrow \pi_2(W^n, x) \leftarrow 0$$

representing it, and speak of the minimal number of generators taken over all crossed modules $\pi_2(W^n, W_1^n, x)$. In what follows we denote this minimum by $\mu(\pi_2(W^n, W_1^n, x))$.

THEOREM 7.5. *Let W^n be a manifold with a single boundary component $\partial W^n = V^{n-1}$. Then*

$$\begin{aligned}
&\mathcal{M}_1(W^n) \geq \mu(\pi_1(W^n, x)), \qquad n \geq 6,\\
&\mathcal{M}_2(W^n) = \mu(\pi_2(W^n, W_1^n, x)), \qquad n \geq 6,\\
&\mathcal{M}_i(W^n) = S_i(W^n) + S_{i+1}(W^n) + \mu(H_i(W^n, \mathbb{Z})) + \mu(\operatorname{Tors} H_{i-1}(W^n, \mathbb{Z})).
\end{aligned}$$

PROOF. The first two inequalities for Morse numbers are evident. The third is proved as follows. We have to estimate the dimension of the "torsion block" that can be present in the chain complex obtained from the handle decomposition of the manifold and formed by the handles of index 2 and 3. Clearly, its dimension is less than $S_i(W^n) + S_{i+1}(W^n)$. The desired conclusion is now obtained by using the inequalities of Chapter V, §7. □

COROLLARY 7.6. * *Let* W^n *be a manifold of class* S *whose fundamental group is isomorphic to one of the following groups*:

(1) *a free group*;
(2) *the dihedral group* D_n, *where* n *is odd*;
(3) $\mathbb{Z} \oplus \mathbb{Z}$;
(4) *a finite abelian group*;
(5) *a free abelian group of rank* $n \geq 3$ *with the first* k*-invariant of the manifold equal to* 0;
(6) *a group satisfying either the hypothesis of Theorem* 7.1 *or that of Theorem* 7.2.

Then, for a minimal Morse function, the number of critical points of index i *is equal to*

$$\mathscr{M}_1(W^n) = \mu(\pi_1(W^n, x)),$$
$$\mathscr{M}_2(W^n) = \mu(\pi_2(W^n, W_1^n, x)),$$
$$\mathscr{M}_i(W^n) = S_i(W^n) + S_{i+1}(W^n) + \mu(H_i(W^n, \mathbb{Z})) + \mu(\operatorname{Tors} H_{i-1}(W^n, \mathbb{Z})).$$

§5. Minimal Morse functions on closed manifolds

Let M^n be a compact manifold without boundary. In this section we discuss the question of the existence of a minimal Morse function on M^n. The homotopy systems constructed for ordered Morse functions f and $-f$, and ordered gradient-like vector fields ξ and $-\xi$, are "dual" to each other; we give a precise definition below. For this reason, closed manifolds produce homotopy systems with a richer structure than those considered above. We shall call them homotopy systems with Poincaré duality. These systems provide an adequate algebraic model for describing Morse functions on closed manifolds.

Let $f: M^n \to [0, 1]$ be an ordered Morse function with one critical point of index 0 and one of index n, and ξ a corresponding gradient-like vector field. From f and ξ (respectively, $-f$ and $-\xi$) construct homotopy systems

$$\rho(M^n, f, \xi): e \longleftarrow \pi_1(M^n, x) \xleftarrow{d_1} \rho_1 \xleftarrow{d_2} \rho_2 \longleftarrow \cdots,$$
$$\rho(M^n, -f, -\xi): e \longleftarrow \pi_1(M^n, \overline{x}) \xleftarrow{\overline{d}_1} \overline{\rho}_1 \xleftarrow{\overline{d}_2} \overline{\rho}_2 \longleftarrow \cdots,$$

where x (respectively, $\overline{x}$) is the critical point of index 0 (respectively, n). On M^n choose a base point m and a fixed isomorphism $\varphi: \pi \to \pi_1(M^n, m)$. Let $p: \hat{M}^n \to M^n$ be the universal covering space, regarded as the set of equivalence classes of paths issuing from the point m. Lifting f and ξ (respectively, $-f$ and $-\xi$) to $\hat{M}^n$, construct a chain complex $C(M^n, f, \xi)$

* *Editor's note.* Misnumbering reflects original Russian.

(respectively, $C(M^n, -f, -\xi)$) of left modules over the ring $\mathbb{Z}[\pi]$:

$$\mathbb{Z} \xleftarrow{\varepsilon} \mathbb{Z}[\pi] \xleftarrow{\partial_1} C_1 \xleftarrow{\partial_2} C_2 \longleftarrow \cdots,$$

$$\mathbb{Z} \xleftarrow{\varepsilon} \mathbb{Z}[\pi] \xleftarrow{d_1} D_1 \xleftarrow{d_2} D_2 \longleftarrow \ldots,$$

We can regard $C(M^n, f, \xi)$ (respectively, $C(M^n, -f, -\xi)$) as associated with the homotopy system $\rho(M^n, f, \xi)$ (respectively, $\rho(M^n, -f, -\xi)$).

LEMMA 7.3. *The homotopy systems* $\rho(M^n, f, \xi)$ *and* $\rho(M^n, -f, -\xi)$ *have the same simple homotopy type.*

PROOF. Denote by K_1 and K_2 the CW-complexes obtained by retracting the manifold M^n to the middle disks of the critical points of the functions f and $-f$, respectively. They are obviously of the same homotopy type. Therefore, $\rho(f, \xi)$ and $\rho(-f, -\xi)$ are also simple homotopy equivalent. □

Let $-: \mathbb{Z}[\pi] \to \mathbb{Z}[\pi]$ be the involution defined by $\overline{g} = w(g)g^{-1}$, $g \in \pi$, where $w: \pi \to \mathbb{Z}_2 = \{-1, 1\}$ is the orientation homomorphism. If A is a left $\mathbb{Z}[\pi]$-module, then, using the involution, we can turn a right $\mathbb{Z}[\pi]$-module into a left one by the formula $\lambda b = b\overline{\lambda}$, where $b \in A^* = \mathrm{Hom}(A, \mathbb{Z}[\pi])$. Applying the functor Hom to the chain complex $C(M^n, f, \xi)$, we obtain a cochain complex

$$C^*(M^n, f, \xi): \mathbb{Z} \longleftarrow \mathbb{Z}[\pi] \xleftarrow{\partial^n} C^n \xleftarrow{\partial^{n-1}} C^{n-1} \longleftarrow \cdots.$$

It is known [95] that the matrix of the homomorphism ∂^i coincides up to sign with the matrix of the homomorphism ∂_{n-i}. We assume that for bases in the modules C_i and D_i we take the corresponding middle and comiddle disks. In the modules C^i we consider the conjugate bases. Thus, the chain complexes $C^*(M^n, f, \xi)$ and $C(M^n, -f, -\xi)$ are simple isomorphic.

As noted above, with any homotopy system (ρ_i, d_i) there is always associated a chain complex (C, ∂). But the converse is false.

DEFINITION 7.1. A chain complex (C, ∂) over the ring $\mathbb{Z}[\pi]$ is said to be realizable if there exists a homotopy system (ρ_i, d_i) for which (C, ∂) is the associated chain complex.

We note that, as shown by examples, the property of realizability is not an invariant of simple homotopy type. Following Maller [84], we give now the definition of a homotopy system with Poincaré duality.

DEFINITION 7.2. Let (ρ_i, d_i) be a homotopy system. We say that (ρ_i, d_i) has Poincaré duality if the cochain complex (C^*, ∂^*) is realizable. Here

$$(C, \partial): \mathbb{Z} \longleftarrow \mathbb{Z}[\pi] \xleftarrow{\partial_1} C_1 \xleftarrow{\partial_2} C_2 \longleftarrow \cdots$$

is the chain complex associated with (ρ_i, d_i).

The homotopy system (ρ_i^*, d_i^*) with which the chain complex (C^*, ∂^*) is associated is called dual to the homotopy system (ρ_i, d_i). Evidently, it also has Poincaré duality, and its dual is the homotopy system (ρ_i, d_i). Thus, ordered Morse functions on closed manifolds give homotopy systems with Poincaré duality. If we perform an elementary operation on the homotopy system associated with the Morse function f, there corresponds an elementary operation on the homotopy system associated with the function $-f$. This fact suggests the following definition.

DEFINITION 7.3. Homotopy systems with Poincaré duality (ρ_i, d_i) and (ρ_i^*, d_i^*) are simple homotopy equivalent if there exist two sequences of homotopy systems (ρ_{ij}, d_{ij}) and (ρ_{ij}^*, d_{ij}^*) such that

(1) $(\rho_{i1}, d_{i1}) = (\rho_i, d_i)$ and $(\rho_{i1}^*, d_{i1}^*) = (\rho_i^*, d_i^*)$;
(2) (ρ_{ij}, d_{ij}) is dual to (ρ_{ij}^*, d_{ij}^*);
(3) $(\rho_{i\,j+1}, d_{i\,j+1})$ (respectively, $(\rho_{i\,j+1}^*, d_{i\,j+1}^*)$) is obtained from the homotopy system (ρ_{ij}, d_{ij}) (respectively, (ρ_{ij}^*, d_{ij}^*)) by an elementary operation.

ASSERTION 7.1. *Let M^n be a closed manifold. All homotopy systems with Poincaré duality associated with ordered Morse functions and gradient-like vector fields on M^n are simple homotopy equivalent. Furthermore, if a homotopy system with Poincaré duality is simple homotopy equivalent to a homotopy system associated with an ordered Morse function and gradient-like vector field ξ, then there exist a Morse function $\overline{f}$ and a field $\overline{\xi}$ such that $\rho(M^n, \overline{f}, \overline{\xi})$ coincides with the given homotopy system.*

The proof of this assertion can be found in Maller [84].

DEFINITION 7.4. Denote by (ρ_{ij}, d_{ij}), $j \in J$, homotopy systems with Poincaré duality that have the same simple homotopy type. A homotopy system $(\overline{\rho}_i, \overline{d}_i) \in (\rho_{ij}, d_{ij})$ is said to be minimal in dimension k if $\mu(\overline{\rho}_k) \leq \mu(\rho_k')$ for all (ρ_i', d_i') in (ρ_{ij}, d_{ij}).

A homotopy system is said to be minimal if it is minimal in all dimensions.

Note that homotopy systems with Poincaré duality need not be minimal as homotopy systems.

Below we establish when there exist minimal homotopy systems with Poincaré duality associated with ordered Morse functions on closed manifolds, or, in other words, determine when there exists a minimal Morse function on a closed manifold.

THEOREM 7.6. *Let M^n be a closed manifold, $n \geq 6$, and $\chi(k^3(M^n)) = \chi_0$. Let $\pi_1(M^n, x) = \pi$ be an s-group satisfying condition (h, χ_0), with* r-dim $\pi \leq k$. *Suppose also that the ith Morse number $\mathcal{M}_i(M^n)$ is $\geq k$, $3 \leq i \leq n-3$ (for $n = 6$, $\mathcal{M}_3(M^n) \geq 2$). Then M^n has a minimal Morse function.*

PROOF. Let f be an arbitrary ordered Morse function, ξ a gradient-like vector field on M^n, and $\rho(M^n, f, \xi)$ the homotopy system associated with f and ξ. The theorem ensures the existence of a minimal homotopy system

(ρ_i, d_i) homotopy equivalent to $\rho(M^n, f, \xi)$. Realize the system (ρ_i, d_i) by a CW-complex K. Since K and M^n have equivalent 2-types, there exists a mapping $j: K^3 \to M^n$ of the 3-skeleton $K^3 \subset K$ into M^n, inducing the isomorphisms

$$j_*: \pi_1(K^3, e^0) \to \pi_1(M^n, x), \qquad j'_*: \pi_2(K^3, e^0) \to \pi_2(M^n, x).$$

Without loss of generality we can suppose that j is an embedding on the 2-skeleton K^2. Since the natural embedding $i: K^2 \to K^3$ induces an epimorphism $i_*: \pi_2(K^2, e^0) \to \pi_2(K^3, e^0)$, the mapping $j'_*: \pi_2(K^2, e^0) \to \pi_2(M^n, x)$ is an epimorphism. In view of dimensional restrictions, there exists another embedding $\bar{j}: K^2 \to M^n$, homotopic to j, such that $j(K^2)$ and $\bar{j}(K^2)$ do not intersect. Let M_2^n and $\overline{M_2^n}$ be nonintersecting tubular neighborhoods of $j(K^2)$ and $\bar{j}(K^2)$, respectively. It is easily shown that

$$\pi_1(\partial M_2^n) \to \pi_1(\Gamma) \leftarrow \pi_1(\partial \overline{M_2^n})$$

are isomorphisms and $\pi_2(\Gamma, \partial M_2^n) = \pi_2(\Gamma, \partial \overline{M_2^n}) = 0$. Consider the manifold $\Gamma = \overline{M^n \setminus (M_2^n \cup \overline{M_2^n})}$. On the manifold Γ there exists an ordered Morse function $f_2: \Gamma \to [1/3, 2/3]$, $f_2^{-1}(1/3) = \partial M_2^n$, $f_2^{-1}(2/3) = \partial \overline{M_2^n}$ with no critical points of index $0, 1, 2, n, n-1, n-2$, if $n \geq 7$. When $n = 6$, it can be arranged that either there are no critical points of index $0, 1, 2, n, n-1$ or none of index $0, 1, n, n-1, n-2$. Using the cellular structures of $j(K^2)$ and $\bar{j}(K^2)$, construct ordered Morse functions

$$f_1: M_2^n \to [0, 1/3], \qquad f_1^{-1}(1/3) = \partial M_2^n,$$

and

$$f_3: \overline{M_2^n} \to [2/3, 1], \qquad f_3^{-1}(2/3) = \partial \overline{M_2^n}.$$

Consider the Morse function $g = f_1 \cup f_2 \cup f_3$ and its gradient-like vector field η. From (g, η) construct a chain complex of $\mathbb{Z}[\pi]$-modules

$$\mathbb{Z} \longleftarrow \mathbb{Z}[\pi] \xleftarrow{\partial_1} C_1 \xleftarrow{\partial_2} C_2 \longleftarrow \cdots \xleftarrow{\partial_{n-1}} C_{n-1} \longleftarrow \mathbb{Z}[\pi].$$

Suppose that $n \geq 7$. Stabilizing and contracting the homomorphisms ∂_i, $4 \leq i \leq n-3$, we can arrange that f-rank$(\partial_i(C_i), C_{i-1}) = 0$ and is additive. Since in these dimensions all elementary operations can be realized, by altering the function g and the field $\bar{\xi}$ we can suppose that g had been chosen in such a way that the segment

$$C_3 \xleftarrow{\partial_4} C_4 \longleftarrow \cdots \xleftarrow{\partial_{n-3}} C_{n-3}$$

already has this property. By Lemma 4.7 applied to the chain complex associated with g and η, and the chain complex associated with the homotopy system (ρ_i, d_i), we have

$$-\mu(C_1) + \mu(C_2) - \mu(C_3) = \mu(\rho_1) + \mu(\rho_2^{\mathrm{ab}}) - \mu(\rho_3).$$

By construction, $-\mu(C_1)+\mu(C_2)=-\mu(\rho_1)+\mu(\rho_2)$, whence it follows that $\mu(\rho_3)=\mu(C_3)$. Consider the function $-g$ and the field $-\eta$. Obviously, the chain complex associated with $(-g, -\eta)$ also has the property that f-rank$(\cdot,\cdot)=0$ and is additive, since the functor Hom preserves the decomposition into direct sum. By a similar argument, using the homotopy system (ρ_i, d_i) and the homotopy system associated with $(-g, -\eta)$ which has the same homotopy type, we show that for $-g$ the number of critical points of index 3 is minimal and therefore for g the number of critical points of index $n-3$ is minimal. Stabilizing and contracting the boundary homomorphisms ∂_i, $1\le i\le 3$, $n-2\le i\le n-1$, in the chain complex

$$\mathbb{Z} \longleftarrow \mathbb{Z}[\pi] \overset{\partial_1}{\longleftarrow} C_1 \longleftarrow \cdots \overset{\partial_{n-1}}{\longleftarrow} C_{n-1} \longleftarrow \mathbb{Z}[\pi],$$

we can arrange that f-rank$(\partial_i(C_i), C_{i-1})=0$ and is additive. By Lemma 4.5, the resulting chain complex is minimal. Since the modules C_j, $4\le j\le n-4$, have been left undisturbed, their ranks were left minimal also.

Consider the case $n=6$. Let f be an arbitrary ordered Morse function on M^n with one critical point of index 0 and one of index 6. It is known that

$$1-\nu_1+\nu_2-\nu_3+\nu_4-\nu_5+1=\chi(M^n),$$

where ν_i is the number of critical points of index i for f. Hence

$$\nu_3=(-\nu_1+\nu_2)+(-\nu_5+\nu_4)+2-\chi(M^n).$$

Therefore ν_3 reaches its minimum when $(-\nu_1+\nu_2)$ and $(-\nu_5+\nu_4)$ are the least possible. Consider the cobordism $(\Gamma, \partial M_2^n, \partial\overline{M_2^n})$. Suppose that the Morse function $\overline{f}:\Gamma\to[0,1]$, $\overline{f}^{-1}(0)=\partial M_2^n$ has critical points of index 2 and 3. By construction, the critical points of the function $\overline{g}=f_1\cup\overline{f}\cup f_3$ of index 3 are concentrated on Γ. By hypothesis, $\mathscr{M}_3(M^n)\ge 2k$ and r-dim $\pi\le k$. Therefore, using Lemma 4.1, we can get rid of the critical points of index 2. Thus, we find on Γ a Morse function $\hat{f}$ having critical points of index 3 only. The functions f_1 and f_2 have been chosen so that $-\nu_1+\nu_2$ and $-\nu_5+\nu_4$ are least possible. Therefore $f_1\cup\hat{f}_2\cup f_3$ is a minimal Morse function. □

We gather here all the consequences of this theorem into a single assertion, whose proof consists in comparing r-dim π and the ith Morse number, which is evaluated by using the homology of the manifold and the invariants $S_i(M^n)$.

COROLLARY 7.7. *Let M^n, $n\ge 6$, be a closed manifold and $\pi_1(M^n, x)=\pi$, where*

(1) *π is a finite s-group, with no epimorphic mapping onto any of the groups*

$$\mathbb{Z}_{p^2}\oplus\mathbb{Z}_{p^2},\ \mathbb{Z}_p\oplus\mathbb{Z}_p\oplus\mathbb{Z}_p,\ \mathbb{Z}_p\oplus\mathbb{Z}_2\oplus\mathbb{Z}_2\oplus\mathbb{Z}_2,\ \mathbb{Z}_4\oplus\mathbb{Z}_2\oplus\mathbb{Z}_2,\ \text{or } \mathbb{Z}_4\oplus\mathbb{Z}_4,$$

$\chi(M^n) \geq d^\mu(\pi)+1$ (*for nilpotent groups,* $\chi(M^n) \geq d(\pi)+1$) *and* $H_i(\hat{M}^n, \mathbb{Z}) \neq 0$, $3 \leq i \leq n-3$; (*for* $n=6$, $\mu(H_3(M^n, \mathbb{Z})) + \mu(\operatorname{Tors} H_2(M^n, \mathbb{Z})) \geq 2$); *or*

(2) π *is a finite s-group,* $\chi(k^3(M^n)) \geq d^\mu(\pi)+1$ (*for nilpotent groups* $\chi(k^3(M^n)) \geq d(\pi)+1$);

if $n=6$, *then* $\mu(H_3(M^n, \mathbb{Z})) + \mu(\operatorname{Tors} H_2(M^n, \mathbb{Z})) \geq 4$;

if $n>6$, *then* $\mu(H_3(M^n, \mathbb{Z})) + \mu(\operatorname{Tors} H_2(M^n, \mathbb{Z})) \geq 2$; *or*

(3) π *is an s-group satisfying the conditions* $\mathrm{Wh}(\pi)=0$ *and* (h, χ_0), *where* $\chi(M^n)=\chi_0$. (*These conditions are satisfied by a free group,* $\mathbb{Z}\oplus\mathbb{Z}$, $\mathbb{Z}_2$, $\mathbb{Z}_3$, $\mathbb{Z}_4$, $\mathbb{Z}_6$.)

Then M^n *has a minimal Morse function.*

THEOREM 7.7. *Let* M^n *be a closed manifold,* $n \geq 6$, *and* $\pi = \pi_1(M^n)$ *satisfy the following conditions:*

(1) π *is an s-group;*
(2) *the essential height of* π *is equal to* a;
(3) $\text{r-dim}\,\pi \leq c$;
(4) *the group* π *satisfies the condition* (h, b), *where* $d(\pi) \leq b < \chi_0 - c$, $\chi_0 = \chi(k^3(M^n)) \geq d(\pi) + \max(c, a)$.

Then M^n *has a Morse function.*

PROOF. We use the argument in the proof of Theorem 7.2. Consider a minimal homotopy system (ρ_i, d_i), which always exists in the homotopy type of the homotopy system associated with an arbitrary ordered Morse function on M^n. Realize this minimal system by a CW-complex K. As shown in the proof of Theorem 7.2, the complex K can be chosen so that its 2-skeleton K^2 is of the form $K^2 = L^2 \cup (\bigvee_i S_i^2)$, where L^2 is constructed from a presentation $\pi = \{a_1, \dots, a_{\mu(\pi)}; r_1, \dots, r_s\}$. There exists an embedding $j: K^2 \to M^n$, inducing an isomorphism of fundamental groups and an epimorphism $\pi_2(K^2, e^0) \to \pi_2(M^n, x)$. Let $\bar{j}: K^2 \to M^n$ be an embedding homotopy equivalent to j such that $j(K^2) \cap \bar{j}(K^2) = \varnothing$. Consider disjoint tubular neighborhoods of $j(K^2)$ and $\bar{j}(K^2)$; denote them by M_1^n and M_2^n, respectively. Using the cellular structure, construct on M_1^n and M_2^n handle decompositions such that there are $\chi_0 - c$ handles of index 2, attached by trivial embeddings of the middle spheres. From these handle decompositions construct Morse functions

$$f_1: M_1^n \to [0, 1/3], \qquad f_1^{-1}(1/3) = \partial M_1^n,$$
$$f_2: M_1^n \to [2/3, 1], \qquad f_1^{-1}(2/3) = \partial M_2^n.$$

Consider the manifold $\Gamma = \overline{M^n \setminus (M_1^n \cup M_2^n)}$. It is easily shown that $\pi_1(\Gamma, \partial M_i^n) = \pi_2(\Gamma, \partial M_i^n) = 0$. Therefore there exist on Γ an ordered Morse function f_2 and a gradient-like vector field ξ such that the associated chain complex is of the form

$$\begin{array}{ccccccc} & & C_3 & \xleftarrow{\partial_4} & C_4 & \longleftarrow \cdots \xleftarrow{\partial_{n-3}} & C_{n-3} \\ & & \oplus & & & & \\ 0 \longleftarrow & R_2 & \xleftarrow{\partial_3} & R_3 & \longleftarrow 0 & & \end{array}$$

and f-rank$\partial_i(C_i)C_{i-1} = 0$ and is additive. Consider the function $\overline{f} = f_1 \cup f_2 \cup f_3$. Using the $\chi_0 - c$ critical points of index 2 of the function f_1, whose middle disks are attached by trivial embeddings, we can apply Lemma 4.1 to get rid of the critical points of index 2 and 3 that correspond to the elementary complex $0 \leftarrow R_2 \overset{\partial_3}{\leftarrow} R_3 \leftarrow 0$. Denote the Morse function so obtained by f. It is easy to see that f is minimal; it suffices to use the argument in the proof of the preceding theorem. □

COROLLARY 7.8. *If $\pi_1(M^n, x) = \pi$ is a finite s-group and the Euler characteristic of the first k-invariant $\chi(k^3(M^n))$ is $\geq d(\pi)$, then M^n has a minimal Morse function.* □

THEOREM 7.8. *For a closed manifold M^n, $n \geq 6$, with fundamental group isomorphic to the group π to have a minimal Morse function, it is sufficient that*

(1) *π be an s-group;*
(2) *π satisfy condition (h);*
(3) $\mathrm{Wh}(\pi\{a_i, r_j\}) = \mathrm{Wh}(\pi)$ *for every presentation $\pi = \{a_i, r_j\}$.*

PROOF. Let K be a CW-complex associated with a minimal homotopy system homotopy equivalent to the homotopy system of the manifold M^n. Consider embeddings of the 2-skeleton $j_i: K^2 \subset K \to M^n$, $i = 1, 2$, inducing an isomorphism of the fundamental groups and an epimorphism of 2-dimensional homotopy groups $\pi_1(K^2, e^0) \to \pi_1(M^n, x)$, $\pi_2(K^2, e^0) \to \pi_2(M^n, x)$. Suppose that $j_1(K^2) \cap j_2(K^2) = \varnothing$. Let M_1^n and M_2^n be disjoint tubular neighborhoods of $j_1(K^2)$ and $j_2(K^2)$. Using the cellular structures, construct ordered Morse functions on M_1^n and M_2^n. Consider the manifold $\Gamma = \overline{M^n \setminus (M_1^n \cup M_2^n)}$. The only obstacle that we can possibly encounter in constructing a minimal Morse function on Γ can be created by a "torsion block", and this can be eliminated with the use of the argument in the proof of the theorem. A minimal Morse function on M^n is now obtained by combining the Morse functions on M_1^n, Γ, and M_2^n. □

COROLLARY 7.9. *Let M be a closed manifold, $n \geq 6$, with $\pi_1(M^n) = \pi$ isomorphic to one of the following:*

(1) *a finite abelian group;*
(2) *the dihedral group D_n, where n is odd.*

Then on M^n there exists a minimal Morse function. □

The next theorem is a direct consequence of Theorems 6.13, 6.15, and 6.16.

THEOREM 7.9. *Let M^n be a manifold without boundary, $n \geq 6$, with fundamental group isomorphic to a free abelian group of rank n $(n \geq 3)$. Suppose that the first k-invariant $k^3(M)$ is equal to zero. Then M^n has a minimal Morse function.*

The following estimates for Morse numbers on closed manifolds are similar to those of the preceding section.

THEOREM 7.10. *Let M^n be a closed manifold. Then the ith Morse numbers satisfy the following relations*:

$$\mathcal{M}_1(M^n) \geq \mu(\pi_1(M^n, x)), \qquad \mathcal{M}_2(M^n) \geq \mu(\pi_2(M^n, M_1^n, x)),$$
$$\mathcal{M}_i(M^n) \geq S_i(M^n) + S_{i+1}(M^n) + \mu(H_i(M^n, \mathbb{Z})) + \mu(\operatorname{Tors} H_{i-1}(M^n, \mathbb{Z})).$$

The case of functions with degenerate critical points as well as applications to Hamiltonian systems is well covered in [89].

CHAPTER VIII

Minimal Round Morse Functions

In the present chapter we study smooth functions on manifolds whose singularities are nondegenerate circles, so-called round Morse functions. These functions were introduced by Thurston in his study of foliations of codimension 1. Then Franks used them to investigate nonsingular Morse-Smale flows [54]. An almost adequate notion of the decomposition of a manifold into round handles was introduced by Asimov who stated an existence condition for such a decomposition [6]. The works of Miyoshi and Morgan should also be mentioned [99, 101]. Recently Fomenko revealed an important role played by these functions in the integrability of Hamiltonian systems. Based on the ideas of S. P. Novikov, V. P. Maslov, I. M. Gelfand, L. D. Faddeev, V. I. Arnold, V. V. Kozlov, Smale, Bott, and Moser, he constructed a topological theory of integrable Hamiltonian systems (see [49]). This line of research was further developed in Fomenko's joint works with Zieshang, S. V. Matveev, A. V. Brailov, and V. V. Sharko [87, 50, 52, 51].

In this chapter we follow [87, 51].

§1. Brief review

Let W^{n+1} be a smooth manifold, $f\colon W^{n+1} \to [0, 1]$ a smooth function, and $x \in W^{n+1}$ its critical point. As is known, a symmetric bilinear functional f_{**}, called the Hessian, can be defined on the tangent space T_x [94]. Let v, w be tangent vectors in T_x. They can be included into vector fields $\tilde{v}, \tilde{w}$. Set $f_{**}(v, w) = \tilde{v}(\tilde{w}(f))$, where $w(f)$ denotes the derivative of f along the vector w. The index of the Hessian is the maximal dimension of the subspace T_x on which f_{**} is negative definite. The index of the critical point x is defined as the index of f_{**}, and the corank of f at the point x is defined as the corank of the Hessian at x. The critical manifold $K(f)$ of the function $f\colon W^{n+1} \to [0, 1]$ is said to be nondegenerate if the Hessian f_{**} is nondegenerate on all planes normal to $K(f)$.

DEFINITION 8.1. A function $f\colon W^{n+1} \to [0, 1]$ is said to be a Bott function if the critical points of f on W^{n+1} form nondegenerate smooth manifolds that do not intersect the boundary ∂W^{n+1}.

General properties of such functions were studied in the well-known works of Bott (see [13]). Consider an important particular case of a Bott function that is called a round Morse function.

Let W^{n+1} be a smooth compact manifold with boundary $\partial W^{n+1} = \partial_+ W^{n+1} \cup \partial_- W^{n+1}$ $(n \geq 2)$, and let $f:(W^{n+1}, \partial_+ W^{n+1}, \partial_- W^{n+1}) \to ([0, 1], 0, 1)$ be a smooth function. Let $K(f)$ be the set of critical points of f.

DEFINITION 8.2. A function f is said to be a round Morse function if

(1) $K(f) \cap \partial W^{n+1} = \varnothing$;
(2) $K(f)$ consists of disjoint circles;
(3) $\operatorname{corank}_{x \in K(f)} f = 1$.

Round Morse functions were introduced by Thurston in his study of foliations of codimension 1. He also observed that the existence of round Morse functions on a manifold W^{n+1} is equivalent to the existence of a decomposition of W^{n+1} into round handles [151]. This assertion was rigorously proved by Miyoshi [99]. Let us recall the facts needed for what follows.

DEFINITION 8.3. A manifold W^{n+1} is said to be obtained from a manifold W_1^{n+1} by attaching a round handle of index λ if $W^{n+1} = W_1^{n+1} \cup_\varphi S^1 \times D^\lambda \times D^{n-\lambda}$, where $\varphi: S^1 \times \partial D^\lambda \times D^{n-\lambda} \to \partial_+ W^{n+1}$ is a smooth embedding.

DEFINITION 8.4. A decomposition of a manifold W^{n+1} into round handles is a filtration $(W_0, W_1, \dots, W_n = W^{n+1})$ of W^{n+1} by submanifolds of codimension 0 such that

$$\partial_- W^{n+1} \times [0, 1] = W_0 \subset W_1 \subset \cdots \subset W_n = W^{n+1},$$

and each manifold W_i is obtained from W_{i-1} by attaching round handles of index i. In the case of $\partial_- W^{n+1} = \varnothing$, the filtration begins with a round handle of index 0.

THEOREM 8.1 (Miyoshi [99]). *Let $f: W^{n+1} \to [0, 1]$ be a round Morse function and C a critical circle of f lying in* $\operatorname{Int} W^{n+1}$. *Then there exists a coordinate chart in a neighborhood of C of one of the following two types:*

(1) *a trivial one: $\nu: S^1 \times D^n(\varepsilon) \to \operatorname{Int} W^{n+1}$, where $D^n(\varepsilon)$ is the disk of radius ε, $\nu(S^1 \times 0) = C$, and*

$$f(\nu(\theta, x)) = -x_1^2 - \cdots - x_\lambda^2 + x_{\lambda+1}^2 + \cdots + x_n^2$$

for $(\theta, x) \in S^1 \times D^n(\varepsilon)$;

(2) *a twisted one: $\tau: [0, 1] \times D^n(\varepsilon)/\sim\, = C$ and*

$$f(\tau(t, x)) = -x_1^2 - \cdots - x_\lambda^2 + x_{\lambda+1}^2 + \cdots + x_n^2$$

for $(t, x) \in [0, 1] \times D^n(\varepsilon)/\sim$. Here $[0, 1] \times D^n(\varepsilon)/\sim$ is diffeomorphic to $S^1 \times D^n(\varepsilon)$ by identifying $(0 \times D^n(\varepsilon))$ and $(1 \times D^n(\varepsilon))$ via the mapping

$$(0, x_1, \dots, x_\lambda, x_{\lambda+1}, \dots, x_n) \sim (1, -x_1, \dots, x_\lambda, -x_{\lambda+1}, \dots, x_n),$$

where the number λ is said to be the index of the critical circle C. □

THEOREM 8.2 [99]. *Let* W^{n+1} *be a compact connected manifold with boundary* $\partial W^{n+1} = \partial_+ W^{n+1} \cup \partial_- W^{n+1}$. *The following two conditions are equivalent*:

(1) *there exists a round Morse function on* W^{n+1};

(2) W^{n+1} *has a decomposition into round handles.*

The case $\partial W^{n+1} = \varnothing$ *is not excluded.* □

Note that, as shown by Asimov [6], in the case when $n \geq 3$, $\partial_- W^n \neq \varnothing$, and $\partial_+ W^{n+1} \neq \varnothing$, there exists a smooth nonsingular vector field on the manifold W^{n+1} issuing from $\partial_- W^{n+1}$ and entering $\partial_+ W^{n+1}$. It is not true that every smooth manifold has a round Morse function.

THEOREM 8.3 (see [6]). *If* W^{n+1} *is a closed manifold and the Euler characteristic* $\chi(W^{n+1}) = 0$, *then* W^{n+1} *has a decomposition into round handles.* □

Therefore there exists a round Morse function on W^{n+1}. The case of 3-dimensional manifolds was analyzed by Morgan [101] and proved to be considerably more complicated. The following trick that makes it possible to construct round handles from ordinary ones is due to Asimov.

LEMMA 8.1 (Asimov [6]). *Let* $W^{n+1} = W_1^{n+1} + h_\lambda + h_{\lambda+1}$, *where* h_λ *and* $h_{\lambda+1}$ *are ordinary handles of index* λ *and* $\lambda + 1$ *that are independent* (*i.e., the comiddle sphere of the handle* h_λ *does not intersect the middle sphere of the handle* $h_{\lambda+1}$). *If* $\lambda \geq 1$, *then* $W^{n+1} = W_1^{n+1} + R_\lambda$, *where* R_λ *denotes a round handle of index* λ. □

The converse statement also holds.

LEMMA 8.2. *Let* $W^{n+1} = W_1^{n+1} + R_\lambda$, *where* R_λ *is a round handle of index* $\lambda \geq 1$. *Then the manifold* W^{n+1} *can be presented in the form* $W^{n+1} = W_1^{n+1} + h_\lambda + h_{\lambda+1}$ (h_λ *are ordinary handles of index* λ).

PROOF. Let us show that a round handle of index λ, i.e., $R_\lambda = S^1 \times D^\lambda \times D^{n-\lambda}$, can be represented as a pair of handles of index λ and $\lambda + 1$. Let $\varphi: S^1 \times \partial D^\lambda \times D^{n-\lambda} \to \partial W_1^{n+1}$ be the attachment mapping. Represent $S^1 \times 0 \times 0$ as the sum of two segments I_1 and I_2 such that $S^1 \times 0 \times 0 = I_1 \cup I_2$, $I_1 \cap I_2 = \varnothing$. Consider the submanifold $h_\lambda = I_1 \times D^\lambda \times D^{n-\lambda} \subset R_\lambda$. Clearly, it can be regarded as a handle of index λ attached to ∂W_1^{n+1} along the set $\partial D^\lambda \times D^{n-\lambda} \times I_1$ via the restriction of the mapping φ. Denote by $h_{\lambda+1}$ the manifold $R_\lambda \setminus I_1 \times D^\lambda \times D^{n-\lambda} = I_2 \times D^\lambda \times D^{n-\lambda}$. We see that $h_{\lambda+1}$ is a handle of index $\lambda + 1$ attached to $\partial W_1^{n+1} + h_\lambda$ along the set $[\partial I_2 \times D^\lambda \cup I_2 \times \partial D^\lambda] \times D^{n-\lambda}$. □

Miyoshi proved [99] that a cobordism $(W^{n+1}, \partial_- W^{n+1}, \partial_+ W^{n+1})$ admits a round Morse function $f: W^{n+1} \to [0, 1]$ with one critical circle C of index λ in $\operatorname{Int} W^{n+1}$, and if the coordinate chart in a neighborhood of C is trivial, then W^{n+1} is diffeomorphic to $\partial_- W^{n+1} \times [0, 1] + R_\lambda$. If the coordinate chart in a neighborhood of C is a twisted one, then W^{n+1} is diffeomorphic to $\partial_- W^{n+1} \times [0, 1] + R_\lambda + R_{\lambda+1}$. In this case there exists a round Morse function on W^{n+1} that has one critical circle of index λ and another one of index $\lambda + 1$ with trivial coordinate charts. Let us make this point more precise.

LEMMA 8.3. *Suppose that $f: (W^{n+1}, \partial_- W^{n+1}, \partial_+ W^{n+1}) \to ([0, 1], 0, 1)$ is a round Morse function with one critical circle C of index λ with a twisted coordinate chart. Let W^{n+1}, $\partial_- W^{n+1}$ be simply connected manifolds and $n \geq 5$. Then there exists a round Morse function on W^{n+1} with two critical circles of indices λ, $\lambda + 1$ or $\lambda - 1$, λ with trivial coordinate charts.*

PROOF. Consider a mapping $\tau: I \times D^n(\delta)/\sim \to W^{n+1}$ such that $f(\tau(\theta, x)) = -x_1^2 - \cdots - x_\lambda^2 + x_{\lambda+1}^2 + \cdots + x_n^2$ for $(\theta, x) \in [0, 1] \times D^n(\delta)/\sim$. Suppose that $f(C) = a$ and choose $\varepsilon > 0$ such that $0 < a - \varepsilon < a < a + \varepsilon < 1$. Then $f^{-1}[0, a] = \partial_- W^{n+1} \times [0, 1]$. Let $0 < \varepsilon' < \varepsilon$. Define a manifold

$$A = \{x \in D^n(\delta) | -x_1^2 - \cdots - x_\lambda^a + x_{\lambda+1}^2 + \cdots + x_n^2 \geq -\varepsilon;\, x_{\lambda+1}^2 + \cdots + x_n^2 \leq \varepsilon'\}.$$

Clearly, A is diffeomorphic to $D^\lambda \times D^{n-\lambda}$. Moving around the circle C, it sweeps a manifold TR_λ such that $f^{-1}[0, a + \varepsilon]$ is diffeomorphic to $f^{-1}[0, a-\varepsilon]+[I\times D^\lambda \times D^{n-\lambda}/\sim]$, where $I\times D^\lambda \times D^{n-\lambda}/\sim$ is obtained from $I\times D^\lambda \times D^{n-\lambda}/\sim$ by means of the identification $(0, x_1, \ldots, x_\lambda, y_1, \ldots, y_{n-\lambda}) \sim (1, -x_1, \ldots, x_\lambda, -y_1, \ldots, y_{n-\lambda})$, and $I \times D^\lambda \times D^{n-\lambda}/\sim$ is attached to $f^{-1}[a - \varepsilon]$ along $I \times \partial D^\lambda \times D^{n-\lambda}/\sim$ via a smooth embedding. Evidently, TR_λ is a nontrivial fiber bundle over the circle with the fiber $D^\lambda \times D^{n-\lambda}$. Let us show that

$$H_i(f^{-1}[0, a+\varepsilon], f^{-1}[0, a-\varepsilon], \mathbb{Z}) = \begin{cases} \mathbb{Z}_2 & \text{for } i = \lambda, \\ 0 & \text{for other dimensions.} \end{cases}$$

Indeed, the excision theorem implies that

$$\begin{aligned} &H_i(f^{-1}[0, a+\varepsilon], f^{-1}[0, a-\varepsilon], \mathbb{Z}) \\ &\quad = H_i(f^{-1}[0, a-\varepsilon] + TR_\lambda, f^{-1}[0, a-\varepsilon], \mathbb{Z}) \\ &\quad = H_i(I \times D^\lambda D^\lambda \times D^{n-\lambda}/\sim, I \times \partial D^\lambda \times D^{n-\lambda}, \mathbb{Z}). \end{aligned}$$

Using the deformation retraction of $I \times D^\lambda \times D^{n-\lambda}/\sim$ to $I \times D^\lambda \times 0/\sim$, we obtain

$$\begin{aligned} &H_i(I \times D^\lambda \times E^{n-\lambda}/\sim, I \times \partial D^\lambda \times D^{n-\lambda}/\sim, \mathbb{Z}) \\ &\quad = H_i(I \times D^\lambda/\sim, I \times \partial D^\lambda/\sim, \mathbb{Z}). \end{aligned}$$

The required fact follows from the exact homology sequence for the pair. Since all our manifolds are simply connected, the Smale theorem [144] implies that there exists a Morse function on $f^{-1}[0, 1] \setminus f^{-1}[0, a - \varepsilon]$ with two critical points of index λ and $\lambda + 1$. Clearly, the handle decomposition constructed from this Morse function has the property that its handles are not independent. Introducing a pair of mutually cancelling handles of index λ and $\lambda+1$, or $\lambda - 1$ and λ, and using the Asimov lemma, we can construct round handles of index λ, $\lambda + 1$ or $\lambda - 1$, λ. It is now easy to construct a round Morse function with two critical circles having trivial coordinate charts. See [99] for details.

In what follows we shall need the following lemma.

LEMMA 8.4. *Let W^{n+1} be a smooth manifold with boundary*

$$\partial W^{n+1} = \partial_- W^{n+1} \cup \partial_+ W^{n+1},$$

where $n \geq 5$, and $\pi_1(\partial_- W^{n+1}) = \pi_1(W^{n+1}) = 0$. Suppose that

$$H_i(W^{n+1}, \partial_- W^{n+1}, \mathbb{Z}) = \begin{cases} \mathbb{Z}_2 & \text{for } i = \lambda, \\ 0 & \text{for } i \neq \lambda. \end{cases}$$

Then on W^{n+1} there exists a round Morse function $f: W^{n+1} \to [0, 1]$ with one critical circle of index λ with a twisted coordinate chart.

PROOF. By the Smale theorem, there exists a Morse function on W^{n+1} having exactly two critical points: one of index λ and another of index $\lambda+1$. Furthermore, we shall assume that the comiddle sphere of the critical point of index λ intersects the middle sphere of the critical point of index $\lambda+1$ at exactly two points, and the intersection index for these spheres is equal to 2. Construct the handle decomposition of the manifold W^{n+1} corresponding to this Morse function:

$$W^{n+1} = (\partial_- W^{n+1} \times I) \cup_\varphi (D^\lambda \times D^{n-\lambda+1}) \cup_\psi (D^{\lambda+1} \times D^{n-\lambda}),$$
$$\varphi: \partial D^\lambda \times D^{n-\lambda+1} \to \partial_- W^{n+1} \times 1,$$
$$\psi: \partial D^{\lambda+1} \times D^{n-\lambda} \to (\partial W_-^{n+1} \times 1) \setminus \varphi(\partial D^\lambda \times \operatorname{Int} D^{n-\lambda+1})$$
$$\cup (D^\lambda \times \varphi(\partial D^{n-\lambda+1})).$$

Here φ and ψ are the embeddings via which the handles are attached.

Consider $\psi(\partial D^{\lambda+1} \times 0) = A$. Clearly, A can be represented in the form $A = A_1 \cup A_2 \cup A_3$, where

$$A_1 \approx D_1^\lambda \approx A_3, \qquad A_2 \approx \partial D^\lambda \times [0, 1],$$
$$A_1 \cap A_2 \approx \partial D_1^\lambda \times 0, \qquad A_3 \cap A_2 \approx \partial D_1^\lambda \times 1, \qquad A_1 \cap A_3 = \varnothing.$$

Here A_1 and A_3 belong to the boundary of the handle of index λ.

Let Γ_1 be a cobordism in $\varphi(D^\lambda \times D^{n-\lambda+1})$ diffeomorphic to $S^{\lambda-1} \times [0, 1]$ with boundaries $\varphi(\partial D^\lambda \times 0)$ and $\partial D_1^\lambda \times 0$, and Γ_2 a similar cobordism in $\varphi(\partial D^\lambda \times D^{n-\lambda+1})$ with boundaries $\varphi(D^\lambda \times 0)$ and $\partial D_1^\lambda \times 1$. By construction, $\Gamma_1 \cap \Gamma_2 = \varphi(D^\lambda \times 0)$. Consider the manifold

$$\Gamma_1 \cup \Gamma_2 \cup A_2 = B.$$

By construction, the manifold B is a fiber bundle over S^1 with the fiber S^{k-1} embedded in $\partial W^{n+1} \times 1$.

Consider a tubular neighborhood of B in $\partial\_W^{n+1} \times 1$ making use of the structure of the direct product on the normal bundle of the manifolds out of which B is constructed. Next, using the middle disk $D^{\lambda+1} \times 0$ of the handle of index λ which seals up $\psi(\partial D^{\lambda+1} \times 0)$, and comiddle disks of the handle of index λ which seal up the manifolds Γ_1 and Γ_2, we can obtain a twisted handle of index λ. For this it is necessary to consider a tubular neighborhood of these sealing up disks in the manifold W^{n+1}. Starting with the twisted handle, we can now construct a round Morse function on W^{n+1} with one critical circle of index λ with a twisted coordinate chart. □

§2. Diagrams

In this section, in order to study round Morse functions, we introduce and explore a notion of the diagram of a round Morse function. A diagram is a set of points in the Euclidean plane joined by edges. As we shall see below, diagrams provide a useful tool for representing the architecture of round Morse functions.

DEFINITION 8.5. By a diagram of length n we mean n sets of points

$$\overset{1}{a}_1, \dots, \overset{1}{a}_{k_1}; \ \overset{2}{a}_1, \dots, \overset{2}{a}_{k_2}; \dots; \ \overset{n}{a}_1, \dots, \overset{n}{a}_{k_n}$$

satisfying the following conditions:

(1) for some i the set $(\overset{i}{a}_1, \dots, \overset{i}{a}_{k_i})$ may be empty;

(2) $k_n - k_{n-1} + k_{n-2} - \dots + (-1)^{n-1}_{k_1} = 0$;

(3) each point of the set $(\overset{i}{a}_1, \dots, \overset{i}{a}_{k_j})$ can be joined with at most one point of the set $(\overset{i-1}{a}_1, \dots, \overset{i-1}{a}_{k_{i-1}})$ or at most one point of the set $(\overset{i+1}{a}_1, \dots, \overset{i+1}{a}_{k_{i+1}})$ by any edge of the following three types:

The set of points $\overset{1}{a}_1, \dots, \overset{1}{a}_{k_1}; \dots; \overset{i}{a}_1, \dots, \overset{i}{a}_{k_i}$ is called the i-skeleton of the diagram. The points $\overset{i}{a}_1, \dots, \overset{i}{a}_{k_i}$ are said to be the points in dimension i. A point joined with no other point is said to be free. If a diagram includes

a fragment of the form

$$a_j^i \Longleftrightarrow a_l^{i+1},$$

then the point a_j^i is said to be semifree. If there is a fragment of the form

$$a_j^i \bullet - - - \bullet\ a_l^{i+1},$$

then the point a_j^i is said to be dependent. A fragment of the form

$$a_j^i \bullet\!\!\longrightarrow\!\!\bullet\ a_l^{i+1}$$

is called an insert in dimension i.

DEFINITION 8.6. A pair of points in dimensions i and $i+1$ is said to be independent in dimension i if either its points are not joined to each other, or the edge joining them forms a fragment of the form $\Longleftrightarrow$.

In what follows we shall decompose a diagram into disjoint pairs of independent points. For fragments of the form

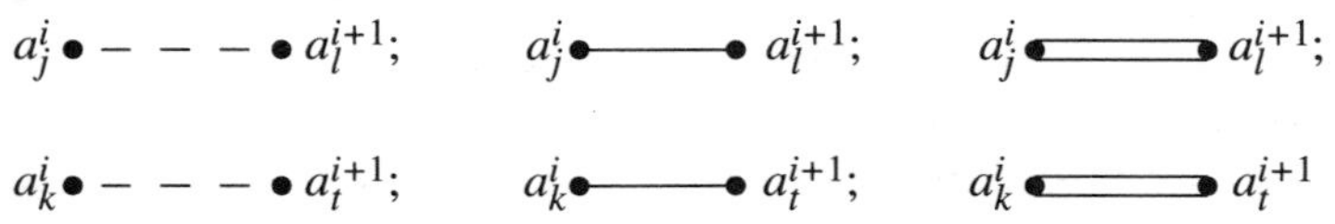

$$a_j^i \bullet - - - \bullet\ a_l^{i+1}; \qquad a_j^i \bullet\!\!\longrightarrow\!\!\bullet\ a_l^{i+1}; \qquad a_j^i \Longleftrightarrow a_l^{i+1};$$

$$a_k^i \bullet - - - \bullet\ a_t^{i+1}; \qquad a_k^i \bullet\!\!\longrightarrow\!\!\bullet\ a_t^{i+1}; \qquad a_k^i \Longleftrightarrow a_t^{i+1}$$

there is the following restriction. It is not allowed to simultaneously decompose any of the fragments into pairs of the form (a_j^i, a_t^{i+1}) and (a_k^i, a_l^{i+1}).

DEFINITION 8.7. If a diagram Ω can be represented as a disjoint union of independent pairs of points, then it is said to admit a decomposition. The pairs of points (a_j^i, a_k^{i+1}) of this decomposition are said to be the vertices of the decomposition in dimension i.

In what follows we denote by (Ω, σ) a fixed decomposition of the diagram Ω. It may occur that Ω does not admit a decomposition because in some dimensions there is not enough points to form independent pairs.

DEFINITION 8.8. A stabilization of a diagram Ω is a diagram of the form $\Omega_s = \Omega \cup_i A_i$, where A_i are new inserts in dimension i.

LEMMA 8.5. *For each diagram Ω there exists a stabilization Ω_s that admits a decomposition.*

PROOF. By introducing inserts in successive dimensions (beginning with dimension 2), we reach dimension $n-1$, where either of the following situations is possible: (1) there is at least one point in dimension $n-1$ (and consequently in dimension n) that forms an independent pair with each point in dimension n (respectively, in dimension $n-1$); (2) there are no such points.

It is evident that in the first case the diagram is decomposed into pairs of independent points without introducing any inserts in dimension $n-1$. In the second case, we introduce an insert (a_j^{n-2}, a_k^{n-1}) and thus reduce the problem to the first case. □

Let k_i be the number of points of the diagram Ω in dimension i.

DEFINITION 8.9. The number $\chi_i(\Omega) = k_i - k_{i-1} + \cdots + (-1)^{k_j - 1} k_1$ is called the ith Euler characteristic of the diagram Ω.

It is clear that an insert in dimension i increments the ith Euler characteristic by 1 and does not change the values of other Euler characteristics.

LEMMA 8.6. *If a diagram Ω admits a decomposition, then the number of vertices in each dimension is the same for all its decompositions.*

PROOF. Fix a decomposition of the diagram Ω into pairs of independent vertices and denote it by σ. Suppose that a decomposition $\overline{\sigma}$ of Ω has the property that in some dimensions the number of the vertices of $\overline{\sigma}$ differs from the number of vertices of σ. Let λ be the minimal dimension in which the numbers of vertices for $\overline{\sigma}$ and σ are different. Obviously, $\lambda > 1$. Denote by d_λ (or $\overline{d}_\lambda$) the number of vertices of σ (respectively, $\overline{\sigma}$) in dimension λ. Suppose that $\overline{d}_\lambda > a_\lambda$. By construction, $\chi_\lambda(\Omega)$ does not depend on how Ω is decomposed into independent pairs of vertices. It is evident that $\chi_\lambda(\Omega) = d_\lambda$, and also that $\chi_\lambda(\Omega) = \overline{d}_\lambda$. The resulting contradiction proves the lemma. □

These lemmas suggest the following definition.

DEFINITION 8.10. Suppose that a diagram Ω admits a decomposition. Then the ith Morse number $\mathscr{M}_i(\Omega, \sigma)$ of the decomposition σ is the number of vertices of the decomposition σ.

By virtue of Lemma 8.6, this number does not depend on the choice of a specific decomposition of σ.

DEFINITION 8.11. The ith Morse number of a diagram Ω is the minimum of the ith Morse numbers taken over all stabilizations of the diagram Ω admitting a decomposition.

In this definition, the diagram Ω is not necessarily a decomposition.

Now consider the question of when inserts must be introduced in order to decompose a diagram Ω into pairs of independent vertices. As it turns out, there are two qualitatively different cases.

DEFINITION 8.12. A dimension λ for a diagram Ω is said to be singular if there are no free points in dimension λ; it consists of semifree points only, and

$$\chi_{\lambda-1}(\Omega) = \chi_{\lambda+1}(\Omega) = 0, \qquad \chi_\lambda(\Omega) = k > 0.$$

Therefore, in dimensions λ and $\lambda + 1$, the diagram Ω has one of the following six forms:

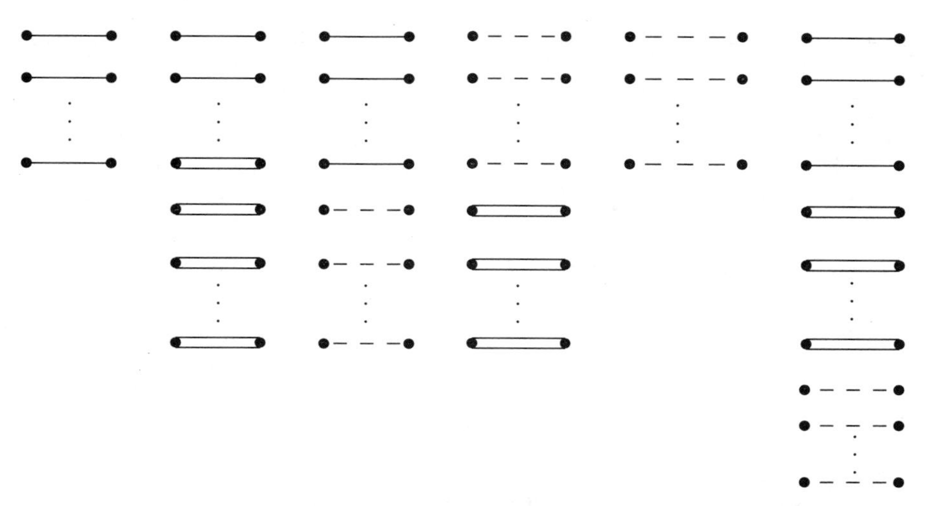

In the process of decomposing Ω into pairs of independent points in this situation, it is necessary to introduce one insert either into dimension $\lambda - 1$ or into dimension $\lambda + 1$ which makes the process nonunique. As a result, we obtain a different number of pairs either in dimension $\lambda - 1$ or in dimension $\lambda + 1$ depending on where the insert was made. The dimension λ does not depend on the position of the insert.

LEMMA 8.7. *The diagram Ω is a decomposition if and only if it has no negative ith Euler characteristics and no singular dimensions. The Morse number of a diagram Ω admitting a decomposition is equal to $\mathscr{M}_i(\Omega, \sigma) = \chi_i(\Omega)$.*

Note that, by virtue of the lemma, σ is an arbitrary decomposition.

PROOF. *Necessity.* The proof of necessity proceeds by induction. For $i = 1$, we evidently have $0 < \mathscr{M}_i(\Omega, \sigma) = \chi_i(\Omega)$. Suppose that the i-skeleton of the diagram can be decomposed into pairs of independent points, making use, if necessary, of points in dimension $i + 1$. Suppose that the number of independent pairs of points in dimension i is equal to $\chi_i(\Omega)$. Let us show that in order for the $(i + 1)$-skeleton of Ω to be decomposed into independent pairs of points, it is necessary that $\chi_{i+1}(\Omega) \geq 0$ and dimension $i + 1$ be nonsingular. If $\chi_i(\Omega) = 0$, then no points in dimension $i + 1$ were used in the decomposition of the i-skeleton. Thus, if there are any points in dimension $i+1$, then in order for them to be joined with points in dimension $i+2$ forming independent pairs it is necessary that dimension $i+1$ be nonsingular. In this case there are exactly $\chi_{i+1}(\Omega)$ pairs of independent points in dimension $i + 1$ because of the equality

$$\chi_{i+1}(\Omega) = k_{i+1} - \chi(\Omega), \tag{1}$$

where k_{i+1} is the number of points in the diagram Ω in dimension $i+1$. If $\chi_i(\Omega) > 0$, then, by the inductive hypothesis, there are $\chi_i(\Omega)$ independent pairs in dimension $i+1$. Using equality (1), we obtain $\chi_{i+1}(\Omega)$ independent pairs in dimension $i+1$.

Sufficiency is obvious. $\square$

It follows directly from the proof of the lemma that if $\chi_i(\Omega) = k < 0$, then it is necessary to introduce k inserts into dimension i in order to make the ith Euler characteristic of the stabilization diagram vanish, thus making it possible to construct a decomposition.

Therefore, inserts must be made in the case of singular dimensions and negative Euler characteristics.

DEFINITION 8.13. A stabilization of a diagram Ω is said to be economical if in the case of $\chi_i(\Omega) = k < 0$ there are k inserts made in dimension i; if i is a singular dimension, then one insert is made either in dimension $i-1$ or in dimension i.

In what follows a decomposition of a diagram into independent pairs will be constructed starting on the left and moving to the right, if necessary.

We introduce the function $\rho(N) = \frac{1}{2}(N + |N|)$. It follows from the proof of the lemma that if $\chi_i(\Omega) < 0$ for some values of i, then the economical stabilization of Ω, viz. Ω_s, admits a decomposition σ and $\mathcal{M}_i(\Omega_s, \sigma) = \rho(\chi_i(\Omega))$. Here we assume that Ω has no singular dimensions.

Suppose that the diagram Ω is a decomposition. Consider the result of introducing an insert in dimension i. Denote the diagram so obtained by Ω_s. There are two possible cases: (1) Ω_s is a decomposition; (2) Ω_s admits no decomposition.

If Ω_s is a decomposition, then the preceding lemmas imply that $\mathcal{M}_j(\Omega, \sigma) = \mathcal{M}_j(\Omega_s, \sigma_s)$ for $j \neq i$, and $\mathcal{M}_i(\Omega, \sigma) + 1 = \mathcal{M}_i(\Omega_s, \sigma_s)$, where σ and σ_s are some decompositions of the diagrams Ω and Ω_s.

If the diagram Ω_s admits no decomposition, this means that dimension i is singular, and in order to decompose Ω_s into independent pairs of points, it is necessary to make one more insert either in dimension $i-1$ or in dimension $i+1$. Denote the resulting stabilization of Ω_s by $\Omega_{\tilde{s}}$. Then

$$\mathcal{M}_{i-1}(\Omega, \sigma) = \mathcal{M}_j(\Omega_{\tilde{s}}, \sigma_{\tilde{s}}) \qquad \text{for } j \neq i-1, i, i+1,$$
$$\mathcal{M}_i(\Omega, \sigma) + 1 = \mathcal{M}_i(\Omega_{\tilde{s}}, \sigma_{\tilde{s}}).$$

Depending on the type of the stabilization of Ω_s, we have either

$$\mathcal{M}_{i-1}(\Omega, \sigma) + 1 = \mathcal{M}_{i-1}(\Omega_{\tilde{s}}, \sigma_{\tilde{s}}), \qquad \mathcal{M}_{i+1}(\Omega, \sigma) = \mathcal{M}_i(\Omega_{\tilde{s}}, \sigma_{\tilde{s}}),$$

or

$$\mathcal{M}_{i-1}(\Omega, \sigma) = \mathcal{M}_{i-1}(\Omega_{\tilde{s}}, \sigma_{\tilde{s}}), \qquad \mathcal{M}_{i+1}(\Omega, \sigma) + 1 = \mathcal{M}_{i+1}(\Omega_{\tilde{s}}, \sigma_{\tilde{s}}).$$

This implies the following proposition.

PROPOSITION 8.1. *If a diagram Ω admits a decomposition, then $\mathcal{M}_i(\Omega) = \mathcal{M}_i(\Omega, \sigma)$, where σ is an arbitrary decomposition of Ω.*

THEOREM 8.4. *Let Ω be an arbitrary decomposition. Then*

$$\mathcal{M}_i(\Omega) = \rho(\chi_i(\Omega)).$$

PROOF. Let Ω be a stabilization of the diagram Ω admitting a decomposition such that $\mathscr{M}_i(\Omega_s, \sigma) = \mathscr{M}_i(\Omega)$. By virtue of Lemma 8.7, we have $\mathscr{M}_i(\Omega_s, \sigma) = \chi_i(\Omega_s, \sigma)$. Since a stabilization can only increase the value of the Euler characteristic, it follows that $\mathscr{M}_i(\Omega) = \rho(\chi_i(\Omega_s, \sigma)) \geq \rho(\chi_i, (\Omega))$. At the same time, as noted above, if Ω has no singular dimensions, then there exists a stabilization of Ω, viz. $\Omega_{\overline{s}}$, admitting a decomposition $\sigma_{\overline{s}}$ such that $\mathscr{M}_i(\Omega_s, \sigma_{\overline{s}}) = \rho(\chi_i(\Omega))$. By virtue of the definition of the ith Morse number of a diagram, we have $\rho(\chi_i(\Omega)) = \mathscr{M}_i(\Omega_s, \sigma_s) \geq \mathscr{M}_i(\Omega)$. Therefore, $\mathscr{M}_i(\Omega) = \rho(\chi_i(\Omega))$. Let $i_1 < i_2 < i_3$ be singular dimensions of the diagram Ω. Denote the i-skeleton of Ω by Ω_i. Since $\chi_{i-1}(\Omega) = 0$, then Ω_{i_1-1} is a diagram without singular dimensions. Therefore, we can apply the above argument, i.e.,

$$\mathscr{M}_j(\Omega_{i-1}) = \mathscr{M}_j(\Omega) = \rho(\chi_j(\Omega)), \qquad 1 \leq j \leq i_1 - 1.$$

Let $\tilde{\Omega}_{i_1-1}$ be a stabilization of Ω_{i_1-1} such that it is a decomposition. Consider Ω_{i_1+1}, the skeleton of Ω. It is evidently a diagram. Replace the skeleton Ω_{i_1-1} in it by the diagram $\tilde{\Omega}_{i_1-1}$. Denote the diagram so obtained by $\overline{\Omega}_{i_1+1}$ and stabilize $\overline{\Omega}_{i_1+1}$ in dimension $i_1 - 1$. Denote the stabilized diagram by $\tilde{\Omega}_{i_1+1}$. Clearly, $\tilde{\Omega}_{i_1+1}$ has no singular dimensions and admits a decomposition. Therefore $\mathscr{M}_{i_1}(\tilde{\Omega}_{i+1}, \sigma_{\tilde{s}}) = \chi_{i_1}(\tilde{\Omega}_{i_1+1})$ because the stabilizations so performed did not change the values of the Euler characteristic in dimensions i_1, i_1+1. Thus $\mathscr{M}_{i_1}(\tilde{\Omega}_{i_1+1}, \tilde{\sigma}) = \chi_{i_1+1}(\tilde{\Omega}_{i_1+1}) = \chi_{i_1}(\Omega) = 0$.

Now consider the (i_2-1)-skeleton of Ω, viz. Ω_{i_2-1}. Replace its (i_1+1)-skeleton by $\tilde{\Omega}_{i_1+1}$ and apply similar reasoning since Ω_{i_2-1} is a diagram without singular dimension. An iterated application of the argument yields the proof of the theorem. □

COROLLARY 8.1. *If Ω_s is a stabilization of a diagram Ω, then $\mathscr{M}_i(\Omega_s) \geq \mathscr{M}_i(\Omega)$.* □

DEFINITION 8.14. The base of a diagram Ω is the diagram obtained from Ω by discarding all inserts.

From Corollary 8.1 we obtain that the Morse number of the base of a diagram does not exceed the Morse number of its stabilization.

DEFINITION 8.15. A diagram Ω is said to be exact if there exists a stabilization of Ω, viz. Ω_s, admitting a decomposition with the ith Morse number equal to $\mathscr{M}_i(\Omega)$ for all i simultaneously.

THEOREM 8.5. *In order for a diagram Ω to be exact, it is necessary and sufficient that it have no singular dimensions.*

PROOF. *Necessity.* Let Ω_s be a stabilization of the diagram Ω admitting a decomposition σ with the ith Morse number $\mathscr{M}_i(\Omega_s, \sigma) = \mathscr{M}_i(\Omega)$. Suppose that i_1 is a singular dimension. Then in this dimension there are $i_1 - 1$

independent pairs of points. Therefore the points in dimension i_1 form independent pairs only with points in dimension $i_1 + 1$, which is possible only if dimension i is not singular. This contradiction proves the necessity.

Sufficiency. As noted above, in the case of no singular dimensions there exists a stabilization of the diagram Ω admitting a decomposition with the ith Morse number equal to $\mathscr{M}_i(\Omega)$. The proof of the theorem is complete. □

§3. Minimal round Morse functions

An ordered Morse function f and a gradient-like vector field ξ on a manifold W^{n+1} define a handle decomposition of W^{n+1}. From the handle decomposition of W^{n+1} we can construct the chain complex

$$C(W^{n+1}, f, \xi): C_0 \xleftarrow{\partial_1} C_1 \longleftarrow \cdots \xleftarrow{\partial_{n+1}} C_{n+1}$$

of free abelian groups with distinguished bases which are in one-to-one correspondence with the handles. The matrices of differentials are defined by the intersection indices of the middle and comiddle spheres of handles of adjacent indices. Using addition of handles and adjusting f and ξ, we can arrange that the matrices of differentials become diagonal:

$$\begin{vmatrix} 1 & & & & & & & \\ & \ddots & & & & & & \\ & & 1 & & & 0 & & \\ & & & n_1 & & & & \\ & & & & \ddots & & & \\ & & & & & n_k & & \\ & & 0 & & & & 0 & \\ & & & & & & & \ddots \\ & & & & & & & & 0 \end{vmatrix},$$

where n_i divides n_{i+1}. With every such chain complex we associate a diagram. Each element of the basis in C_i is assigned a point in dimension i. If the element c_{i+1} of the basis is mapped by the boundary homomorphism into another element c_i of the basis by the formula $\partial_{i+1}(c_{i+1}) = k \cdot c_i$, then the corresponding points on the diagram are joined as follows: •——• if $k = 1$; ⊂═══⊃ if $k = 2$; • – – – • if $k > 2$.

If an insert is made in the diagram in dimension i, this corresponds to the introduction of a pair of mutually cancelling handles of index i and $i + 1$. Note that if the diagram already had an insert, this does not necessarily mean that the corresponding handles cancel each other since the intersection index does not necessarily coincide with the geometric intersection number. This remark is also valid for free and semifree points. However, if W^{n+1}, $n > 5$, is a closed manifold with trivial fundamental group, then the relation between handles correspond to the relation between points on the diagram.

DEFINITION 8.16. By the ith Euler number of a manifold W^{n+1} we mean the number

$$\chi_i(W^{n+1}) = \sum_{j=0}^{i}(-1)^{i+j}\,\mathrm{rk}\,H_j(W^{n+1}, Q).$$

Put

$$\kappa_i(W^{n+1}) = \mu(\mathrm{Tors}\,H_i(W^{n+1}, \mathbb{Z})) + \chi_i(W^{n+1}),$$

where $\mu(H)$ is the minimal number of generators of the group H.

DEFINITION 8.17. The dimension of the manifold W^{n+1} is said to be singular if

$$\chi_{\lambda-1}(W^{n+1}) = \kappa_{\lambda-1}(W^{n+1}) = \kappa_{\lambda+1}(W^{n+1}) = 0,$$
$$\kappa_\lambda(W^{n+1}) = k > 0, \qquad H_\lambda(W^{n+1}, \mathbb{Z}) \neq \underbrace{\mathbb{Z}_2 \oplus \cdots \oplus \mathbb{Z}_2}_{k}.$$

In the rest of this section we consider only manifolds with zero Euler characteristic. By virtue of the Asimov theorem, if the dimension of a manifold is greater than 3, then it admits a round Morse function.

DEFINITION 8.18. The ith Morse S^1-number of a manifold W^{n+1} is the minimum number of critical circles of index i taken over all round Morse functions on W^{n+1}.

Denote the ith Morse S^1-numbers of the manifold W^{n+1} by $\mathscr{M}_i^{S^1}(W^{n+1})$.

DEFINITION 8.19. A round Morse function on a manifold W^{n+1} is said to be minimal if the number of its critical circles of index i is equal to the ith Morse S^1-number of the manifold W^{n+1} for all i.

The next theorem follows from the above proof of Theorem 8.3 and the interrelation between the Euler characteristics of the diagram constructed from the minimal Morse function on a manifold W^{n+1}.

THEOREM 8.6. *In order for a simply connected manifold W^{n+1}, $n \geq 5$, without boundary to admit a minimal round Morse function, it is necessary and sufficient that the manifold W^{n+1} have no singular dimensions. The number of critical circles of a minimal round function is*

$$N_\lambda = \mathscr{M}_\lambda^{S^1}(W^{n+1}) = \rho(\kappa_\lambda(W^{n+1})). \qquad \square$$

THEOREM 8.7. *Let W^{n+1}, $n \geq 2$, be an arbitrary smooth manifold admitting a round Morse function. Then the number of critical circles of index λ for an arbitrary round Morse function on W^{n+1} is not less than*

$$\rho(\kappa_\lambda(W^{n+1}) + S_\lambda(W^{n+1})).$$

PROOF. Let f be an arbitrary round Morse function on W^{n+1}. Construct the associated decomposition of W^{n+1} into ordinary handles, replacing each critical circle of index λ by a pair of handles of index λ and $\lambda+1$. From this handle decomposition, construct a chain complex of free abelian groups. Transform the matrices of the boundary homomorphisms to the diagonal form and construct the diagram which we shall denote by Ω. In the diagram Ω, there always is a block of inserts in dimension i corresponding to the number $S_\lambda(W^{n+1})$. It is not difficult to compute that the number of vertices in the diagram Ω for some and consequently any decomposition is not less than $\rho(\kappa_\lambda(W^{n+1}) + S_\lambda(W^{n+1}))$. □

Bibliography

1. M. K. Agoston, *On handle decompositions and diffeomorphisms*, Trans. Amer. Math. Soc. **137** (1969), 21–26.
2. R. C. Alperin, K. R. Dennis, and M. R. Stein, *The non-triviality of* $SK_1(\mathbf{Z}[\pi])$, Proceedings of the Conference on Orders, Group Rings and Related Topics (Columbus, OH, 1972), Lecture Notes in Math., vol. 353, Springer-Verlag, Berlin and New York, 1973, pp. 1–7.
3. D. V. Anosov, C. Kh. Aranson, I. U. Bronshteĭn, and V. Z. Grines, *Smooth dynamical systems*, Itogi Nauki i Tekhniki: Sovremennye Problemy Mat., vol. 1, VINITI, Moscow, 1985, pp. 151–142; English transl. in J. Soviet Math. **3** (1975), no. 2.
4. V. I. Arnol′d, *Small denominators and problems of stability of motion in classical and celestial mechanics*, Uspekhi Mat. Nauk **18** (1963), no. 6, 91–192; English transl. in Russian Math. Surveys **18** (1963).
5. V. I. Arnol′d, A. N. Varchenko, and S. M. Gusein-Zade, *Singularities of differentiable maps*. I, "Nauka", Moscow, 1982; English transl., Birkhäuser, Boston, 1988.
6. D. Asimov, *Round handles and non-singular Morse-Smale flows*, Ann. of Math. (2) **102** (1975), 41–54.
7. D. Barden, *Simply connected five-manifolds*, Ann. of Math. (2) **82** (1965), 365–385.
8. H. Bass, *Algebraic K-theory*, Benjamin, New York, 1968.
9. H. Bass, A. Heller, and R. Swan, *The Whitehead group of a polynomial extension*, Inst. Hautes Études Sci. Publ. Math. No. 22 (1964), 61–79.
10. N. N. Bogolyubov, Yu. A. Mitropol′skiĭ, and A. M. Samoĭlenko, *Methods of accelerated convergence in nonlinear mechanics*, "Naukova Dumka", Kiev, 1969; English transl., Hindustan Publ., Delhi, and Springer-Verlag, Berlin and New York, 1976.
11. O. I. Bogoyavlenskiĭ, *A "nice" function on manifolds*, Mat. Zametki **8** (1970), no. 1, 77–83; English transl. in Math. Notes **8** (1970).
12. M. Boileau and H. Zieschang, *Heegaard genus of closed orientable Seifert* 3*-manifolds*, Invent. Math. **76** (1984), 455–468.
13. R. Bott, *Lectures on Morse theory, old and new*, Bull. Amer. Math. Soc. (N.S.) **7** (1982), 331–358.
14. K. S. Brown, *Cohomology of groups*, Graduate Texts in Math., vol. 87, Springer-Verlag, Berlin and New York, 1982.
15. W. Browning, *Finite CW-complexes of cohomological dimension* 2 *with finite abelian* π_1, Preprint, Forschungsinstitut für Mathematik ETH, Zürich, 1979.
16. ——, *The homotopy classification of nonminimal* 2*-complexes with finite fundamental group*, Preprint, Forschungsinstitut für Mathematik ETH, Zürich, 1979.
17. J. Cerf, *Stratification naturelle des espaces de functions différentiables réeles et le théorème de la pseudo-isotopie*, Inst. Hautes Études Sci. Publ. Math. No. 39 (1970), 5–173.
18. T. Chapman, *Topological invariance of Whitehead torsion*, Amer. J. Math. **96** (1974), 488–497.
19. A. Chenciner and F. Laudenbach, *Contribution a une théorie de Smale a un parametre dans le cas non simplement connexe*, Ann. Sci. École Norm. Sup. (4) **3** (1970), 409–478.

20. G. S. Chogoshvili, *On level surfaces and domains of smaller values of a function defined on a bounded manifold*, Dokl. Akad. Nauk SSSR **24** (1939), no. 3, 635–639. (Russian)
21. ———, *On level surfaces and regions of smaller values*, Trudy Tbiliss. Mat. Inst. **17** (1949), 203–243. (Russian)
22. W. H. Cockcroft, *Simple homotopy type torsion and the Reidemeister-Franz torsion*, Topology **1** (1962), 143–150.
23. W. H. Cockcroft and R. G. Swan, *On the homotopy type of certain two-dimensional complexes*, Proc. London Math. Soc. (3) **11** (1961), 194–202.
24. W. H. Cockroft and R. Moss, *On two-dimensional realisability of chain complexes*, J. London Math. Soc. (2) **11** (1975), 257–262.
25. M. M. Cohen, *A course in simple-homotopy theory*, Graduate Texts in Math., vol. 10, Springer-Verlag, Berlin and New York, 1973.
26. ———, *Whitehead torsion, group extensions, and Zeeman's conjecture in high dimensions*, Topology **16** (1977), 79–88.
27. P. E. Conner and E. E. Floyd, *Differentiable periodic maps*, Academic Press, New York, 1964.
28. J. Derwent, *Handle decompositions of manifolds*, J. Math. Mech. **15** (1966), 329–345.
29. A. Dold, *Zur Homotopietheorie der kettenkomplexe*, Math. Ann. **140** (1960), 278–298.
30. ———, *Lectures on algebraic topology*, Grundlehren Math. Wiss., vol. 200, Springer-Verlag, Berlin and Heidelberg, 1980.
31. B. A. Dubrovin, S. P. Novikov, and A. T. Fomenko, *Modern geometry. Methods and applications*, "Nauka", Moscow, 1979; English transls., Part I, Graduate Texts in Math., vol. 93, Springer-Verlag, Berlin and New York, 1984; Part II, Graduate Texts in Math., vol. 104, Springer-Verlag, Berlin and New York, 1985.
32. ———, *Modern geometry. Methods and applications.* Part III, "Nauka", Moscow, 1984; English transl., Graduate Texts in Math., vol. 124, Springer-Verlag, Berlin and New York, 1990.
33. M. J. Dunwoody, *Relation modules*, Bull. London Math. Soc. **4** (1972), 151–155.
34. ———, *The homotopy type of a two-dimensional complex*, Bull. London Math. Soc. **8** (1976), 282–285.
35. M. N. Dyer, *Homotopy classification of (π, m)-complexes*, J. Pure Appl. Algebra **7** (1976), 249–282.
36. ———, *On the essential height of homotopy trees with finite fundamental group*, Compositio Math. **36** (1978), 209–224.
37. ———, *An application of homological algebra to the homotopy classification of two-dimensional CW-complexes*, Trans. Amer. Math. Soc. **259** (1980), 505–514.
38. ———, *Simple homotopy types for (G, m)-complexes*, Proc. Amer. Math. Soc. **81** (1981), 111–115.
39. ———, *Subcomplexes of two-complexes and projective crossed modules*, Combinatorial Group Theory and Topology (Alta, UT, 1984), Ann. of Math. Stud., vol. 111, Princeton Univ. Press, Princeton, NJ, 1987, pp. 255–264.
40. M. N. Dyer and A. J. Sieradski, *Trees of homotopy types of two-dimensional CW-complexes.* I, Comment. Math. Helv. **48** (1973), 31–44.
41. ———, *Distinguishing arithmetic for certain stably isomorphic modules*, J. Pure Appl. Algebra **15** (1979), 199–217.
42. A. Z. Dymov, *Homology spheres and contractible compact manifolds*, Izv. Akad. Nauk SSSR Ser. Mat. **35** (1971), no. 1, 72–77; English transl. in Math. USSR-Izv. **5** (1971).
43. S. Eilenberg, *Homological dimension and syzygies*, Ann. of Math. (2) **64** (1956), 328–336.
44. L. Elsholz, *Die Änderung der Bettischen Zahlen der Niveauflächen einer stetigen Funktion, die auf einer Manigfaltigkeit definiert ist*, Mat. Sb. **47** (1939), no. 4, 559–564. (Russian)
45. C. Faith, *Algebra. I. Rings, modules, and categories*, Grundlehren Math. Wiss., vol. 190, Springer-Verlag, New York, 1973.
46. M. Sh. Farber, *Sharpness of the Novikov inequalities*, Funktsional. Anal. i Prilozhen. **19** (1985), no. 1, 49–59; English transl. in Functional Anal. Appl. **19** (1985).
47. A. T. Fomenko, *Variational problems in topology*, Moskov. Gos. Univ., Moscow, 1984; English transl., Gordon and Breach, New York, 1990.

48. ———, *Morse theory of integrable Hamiltonian systems*, Dokl. Akad. Nauk SSSR **287** (1986), no. 6, 1071–1075; English transl. in Soviet Math. Dokl. **33** (1986).
49. ———, *Symplectic geometry. Methods and applications*, Moskov. Gos. Univ., Moscow, 1988; English transl., Gordon and Breach, New York, 1988, and Kluwer, Dordrecht, 1988.
50. A. T. Fomenko and A. V. Brailov, *Topology of integral submanifolds of completely integrable Hamiltonian systems*, Mat. Sb. **134** (1987), no. 3, 375–385; English transl. in Math. USSR-Sb. **62** (1989).
51. A. T. Fomenko and V. V. Sharko, *Morse exact round functions, Morse-type inequalities and integrals of Hamiltonian systems*, Ukrain. Mat. Zh. **41** (1989), no. 6, 723–732; English transl. in Ukrainian Math. J. **41** (1989).
52. A. T. Fomenko and H. Zieschang, *On the topology of the three-dimensional manifolds arising in Hamiltonian mechanics*, Dokl. Akad. Nauk SSSR **294** (1987), no. 2, 283–287; English transl. in Soviet Math. Dokl. **35** (1987).
53. R. H. Fox, *Free differential calculus.* I. *Derivation in the free group ring*, Ann. of Math. (2) **57** (1953), 547–560.
54. J. Franks, *The periodic structure of nonsingular Morse-Smale flows*, Comment. Math. Helv. **53** (1978), 279–294.
55. M. Golubitsky and V. Guillemin, *Stable mappings and their singularities*, Graduate Texts in Math., vol. 14, Springer-Verlag, Heidelberg, 1973.
56. M. L. Gromov, *Smoothing and inversion of differential operators*, Mat. Sb. **88** (1972), 382–441; English transl. in Math. USSR-Sb. **17** (1972).
57. B. Hajduk, *Comparing handle decompositions of homotopy equivalent manifolds*, Fund. Math. **95** (1977), 35–47.
58. R. S. Hamilton, *The inverse function theorem of Nash and Moser*, Bull. Amer. Math. Soc. (N.S.) **7** (1982), 65–222.
59. H. Hatcher and J. Wagoner, *Pseudo-isotopies of compact manifolds*, Astérisque, vol. 6, Soc. Math. France, Paris, 1973.
60. A. Heller, *Homological resolutions of complexes with operators*, Ann. of Math. (2) **60** (1954), 283–303.
61. H. Hendriks, *La stratification naturelle de l'espace des fonctions différentiables réeles n'est pas la bonne*, C. R. Acad. Sci. Paris Sér. A **274** (1972), 618–620.
62. P. Hilton and U. Stammbach, *A course in homological algebra*, Graduate Texts in Math., vol. 4, Springer-Verlag, New York, 1971.
63. D. F. Holt, *An interpretation of the cohomology groups* $H^n(G, M)$, J. Algebra **60** (1979), 307–320.
64. L. Hörmander, *The boundary problems of physical geodesy*, Arch. Rational Mech. Anal. **62** (1976), 1–52.
65. J. Huebschmann, *Crossed n-fold extensions of groups and cohomology*, Comment. Math. Helv. **55** (1980), 302–313.
66. H. Jacobinski, *Genera and decompositions of lattices over orders*, Acta Math. **121** (1968), 1–29.
67. H. Jacobowitz, *Implicit functions theorems and isometric embeddings*, Ann. of Math. (2) **95** (1972), 191–225.
68. S. Jajodia, *Spaces dominated by two-complexes*, Proc. Amer. Math. Soc. **82** (1981), 288–290.
69. S. Jajodia and M. Magurn, *Surjective stability of units and simple homotopy type*, J. Pure Appl. Algebra **18** (1980), 45–58.
70. M. A. Kervaire, *Le théorème de Barden-Mazur-Stallings*, Comment. Math. Helv. **40** (1965), 31–42.
71. ———, *Smooth homology spheres and their fundamental groups*, Trans. Amer. Math. Soc. **144** (1969), 67–72.
72. R. C. Kirby and L. C. Siebenmann, *Foundational essays on topological manifolds, smoothings and triangulations*, Ann. of Math. Stud., vol. 88, Princeton Univ. Press, Princeton, NJ, 1977.
73. A. N. Kolmogorov, *Théorie générale des systèmes dynamiques et méchanique classique*, Proc. Internat. Congr. Math. (Amsterdam, 1954), vol. I, Noordhoff, Groningen, and North-Holland, Amsterdam, 1957, pp. 315–333.

74. J. Kuranishi, *Deformations of isolated singularities and* C_b, J. Differential Geom. **19** (1983), 78–95.
75. M. P. Latiolais, *Simple homotopy type of finite* 2*-complexes with finite abelian fundamental group*, Trans. Amer. Math. Soc. **293** (1986), 655–662.
76. H. Lindel, *On the Bass-Quillen and Suslin conjecture concerning projective modules*, Invent. Math. **65** (1981), 319–323.
77. J.-L. Loday, *Cohomologie et groupe de Steinberg relatif*, J. Algebra **54** (1978), 178–202.
78. L. A. Lyusternik and L. G. Snirel′man, *Méthodes topologiques dans les problèmes variationnels*, Actualités Sci. Indust., No. 118, Hermann, Paris, 1934.
79. M. Lustig, *Nielsen equivalence and simple homotopy type*, Proc. London Math. Soc. (3) **62** (1991), 537–562.
80. R. C. Lyndon and P. E. Shupp, *Combinatorial group theory*, Springer-Verlag, Berlin and Heidelberg, 1977.
81. S. Mac Lane, *Homology*, Springer-Verlag, Berlin and Heidelberg, 1963.
82. S. Mac Lane and J. H. C. Whitehead, *On the* 3*-type of a complex*, Proc. Nat. Acad. Sci. U.S.A. **36** (1950), 41–48.
83. B. Magurn, R. Oliver, and L. Vaserstein, *Units in Whitehead groups of finite groups*, J. Algebra **84** (1983), 324–360.
84. M. Maller, *Fitted diffeomorphisms of nonsimply connected manifolds*, Topology **19** (1980), 395–410.
85. R. Mandelbaum, *Four-dimensional topology: an introduction*, Bull. Amer. Math. Soc. (N.S.) **2** (1980), 1–159.
86. T. Matsumoto, *On the minimal ordered Morse functions on compact simply connected manifolds*, Publ. Res. Inst. Math. Sci. **14** (1978), 673–684.
87. S. V. Matveev, A. T. Fomenko, and V. V. Sharko, *Round Morse functions and isoenergy surfaces of integrable Hamiltonian systems*, Mat. Sb. **135** (1988), no. 3, 325–345; English transl. in Math. USSR-Sb. **63** (1989).
88. S. Maumary, *Contributions à la théorie du type simple d'homotopie*, Comment. Math. Helv. **44** (1969), 410–437.
89. J. Mawhin and M. Willem, *Critical point theory and Hamiltonian systems*, Appl. Math. Sci., vol. 74, Springer-Verlag, Berlin and New York, 1989.
90. W. Metzler, *Über den Homotopietyp zweidimensionalen* CW*-Komplexe und elementartransformationen bei Darstellungen von Gruppen durch Erzeugende und definierende Relationen*, J. Reine Angew. Math. **285** (1976), 7–23.
91. ———, *Two-dimensional complexes with torsion values not realizable by self-equivalences*, Homological Group Theory, London Math. Soc. Lecture Note Ser., vol. 36, Cambridge Univ. Press, Cambridge, 1979, pp. 327–337.
92. ———, *Die Unterscheidung von Homotopietyp und einfachen Homotopietyp bei zweidimensionalen Komplexen*, J. Reine Angew. Math. **403** (1990), 201–219.
93. K. Meyer, *Energy functions for Morse-Smale systems*, Amer. J. Math. **90** (1968), 1031–1040.
94. J. Milnor, *Morse theory*, Ann. of Math. Stud., vol. 51, Princeton Univ. Press, Princeton, NJ, 1963.
95. ———, *Whitehead torsion*, Bull. Amer. Math. Soc. **72** (1966), 358–426.
96. ———, *Lectures on the h-cobordism theory*, Princeton Univ. Press, Princeton, NJ, 1965.
97. ———, *Introduction into algebraic* K*-theory*, Ann. of Math. Stud., vol. 72, Princeton Univ. Press, Princeton, NJ, 1971.
98. E. A. Mikhaĭlyuk, *Equivalent Morse functions*, Akad. Nauk Ukrain. SSR Inst. Mat. Preprint **1986**, no. 79. (Russian)
99. S. Miyoshi, *Foliated round surgery of codimension-one foliated manifolds*, Topology **21** (1982), 245–261.
100. M. S. Montgomery, *Left and right inverses in group algebras*, Bull. Amer. Math. Soc. **75** (1969), 539–540.
101. J. W. Morgan, *Nonsingular Morse-Smale flows on* 3*-dimensional manifolds*, Topology **18** (1979), 41–53.

102. M. Morse, *The calculus of variations in the large*, Amer. Math. Soc. Colloq. Publ., No. 18, Amer. Math. Soc., Providence, RI, 1934.
103. J. Moser, *A new technique for the construction of solutions of nonlinear differential equations*, Proc. Nat. Acad. Sci. U.S.A. **47** (1961), 1824–1831.
104. J. F. Nash, *The imbedding problem for Riemannian manifolds*, Ann. of Math. (2) **63** (1956), 20–63.
105. S. P. Novikov, *Manifolds with free abelian fundamental groups and their applications*, Izv. Akad. Nauk SSSR Ser. Mat. **30** (1966), no. 1, 207–246; English transl. in Amer. Math. Soc. Transl. Ser. 2 **71** (1968).
106. ———, *Hamiltonian formalism and a multivalued analogue of Morse theory*, Uspekhi Mat. Nauk **37** (1982), no. 5, 3–49; English transl. in Russian Math. Surveys **37** (1982).
107. ———, *Bloch homology. Critical points of functions and closed 1-forms*, Dokl. Akad. Nauk SSSR **287** (1986), no. 6, 1321–1324; English transl. in Soviet Math. Dokl. **33** (1986).
108. ———, *Topology*, Itogi Nauki i Tekhniki: Sovremennye Problemy Mat.: Fundamental′nye Napravleniya, vol. 12, VINITI, Moscow, 1986, pp. 5–252; English transl. in Encyclopedia of Math. Sci., vol. 12 (Topology, I), Springer-Verlag, Berlin and New York, 1986.
109. S. P. Novikov and M. A. Shubin, *Morse inequalities and von Neumann* II$_1$*-factors*, Dokl. Akad. Nauk SSSR **289** (1986), no. 2, 289–292; English transl. in Soviet Math. Dokl. **34** (1987).
110. L. V. Ovsyannikov, *A singular operator in a scale of Banach spaces*, Dokl. Akad. Nauk SSSR **163** (1965), no. 4, 819–822; English transl. in Soviet Math. Dokl. **6** (1965).
111. A. V. Pazhitnov, *An analytic proof of the real part of Novikov's inequalities*, Dokl. Akad. Nauk SSSR **293** (1987), no. 6, 1305–1307; English transl. in Soviet Math. Dokl. **35** (1987).
112. R. Peiffer, *Über identitaten zwischen Relationen*, Math. Ann. **121** (1949), 67–99.
113. E. Pitcher, *Inequalities of critical point theory*, Bull. Amer. Math. Soc. **64** (1958), 1–30.
114. M. M. Postnikov, *Introduction to the Morse theory*, "Nauka", Moscow, 1971. (Russian)
115. D. Quillen, *Projective modules over polynomial rings*, Invent. Math. **36** (1976), 167–171.
116. P. H. Rabinowitz, *Periodic solutions of nonlinear hyperbolic partial differential equations*, Comm. Pure Appl. Math. **20** (1967), 145–205.
117. E. Rapaport, *Finitely presented groups: the deficiency*, J. Algebra **24** (1973), 531–543.
118. J. Ratcliffe, *Crossed extensions*, Trans. Amer. Math. Soc. **257** (1980), 73–89.
119. ———, *Free and projective crossed modules*, J. London Math. Soc. (2) **22** (1980), 66–74.
120. ———, *On complexes dominated by a two-complex*, Combinatorial Group Theory and Topology (Alta, UT, 1984), Ann. of Math. Stud., vol. 111, Princeton Univ. Press, Princeton, NJ, 1987, pp. 221–254.
121. G. Reeb, *Sur les variétés niveau d'une fonction numérique*, C. R. Acad. Sci. Paris **224** (1947), 1324–1326.
122. K. Reidemeister, *Überdeckungen von Komplexen*, J. Reine Angew. Math. **173** (1935), 163–173.
123. C. P. Rourke and B. J. Sanderson, *Introduction to piecewise-linear topology*, Springer-Verlag, Berlin and Heidelbeg, 1972.
124. F. Sergeraert, *Un théorème de fonctions implicites sur certains espaces de Fréchet et quelques applications*, Ann. Sci. École Norm. Sup. (4) **5** (1972), 599–660.
125. V. V. Sharko, *On Morse functions depending on a parameter.* I, Complex Analysis and Manifolds, Inst. Mat. Akad. Nauk Ukrain. SSR, Kiev, 1978, pp. 158–166. (Russian)
126. ———, *Smooth functions on manifolds*, Akad. Nauk Ukrain. SSR Inst. Mat. Preprint **1979**, no. 22.
127. ———, *Minimum resolvents and Morse functions*, Dokl. Akad. Nauk Ukrain. SSR Ser. A **1980**, no. 11, 31–33. (Russian)
128. ———, *On equivalence of nice Morse functions*, Dokl. Akad. Nauk Ukrain. SSR Ser. A **1980**, no. 12, 18–20; English transl. in Amer. Math. Soc. Transl. Ser. 2 **134** (1987).
129. ———, *Stable algebra and Morse theory*, Ukrain. Mat. Zh. **32** (1980), no. 5, 711–713; English transl. in Ukrainian Math. J. **32** (1980).
130. ———, *Exact Morse functions on simply connected manifolds a boundary that is not simply connected*, Uspekhi Mat. Nauk **36** (1981), no. 5, 205–206; English transl. in Russian Math. Surveys **36** (1981).

131. ——, *Minimal resolutions and Morse functions*, Trudy Mat. Inst. Steklov. **154** (1983), 306–311; English transl. in Proc. Steklov Inst. Math. **1984**, no. 4.
132. ——, *K-theory and the Morse theory*. I, Akad. Nauk Ukrain. SSR Inst. Mat. Preprint **1986**, no. 39; English transl. in Amer. Math. Soc. Transl. Ser. 2 **149** (1991).
133. ——, *Minimal Morse functions*, Geometry and Topology in Global Nonlinear Problems, Voronezh. Gos. Univ., Voronezh, 1984, pp. 123–141; English transl., Global Analysis – Studies and Applications. I, Lecture Notes in Math., vol. 1108, Springer-Verlag, Berlin and New York, 1984, pp. 218–234.
134. ——, *K-theory and the Morse theory*. II, Akad. Nauk Ukrain. SSR Inst. Mat. Preprint **1986**, no. 40; English transl. in Amer. Math. Soc. Transl. Ser. 2 **149** (1991).
135. ——, *Morse numbers and minimal Morse functions on non-simply connected manifolds*, Ukrain. Mat. Zh. **40** (1988), no. 1, 130–131; English transl. in Ukrainian Math. J. **40** (1988).
136. ——, *Deformation of Morse functions*. I, Ukrain. Mat. Zh. **41** (1989), no. 2, 237–243; English transl. in Ukrainian Math. J. **41** (1989).
137. ——, *Stable algebra in Morse theory*, Izv. Akad. Nauk SSSR Ser. Mat. **54** (1990), no. 3, 607–632; English transl. in Math. USSR-Izv. **36** (1991).
138. ——, *Equivariant Morse function*, Akad. Nauk Ukrain. SSR Inst. Mat. Preprint **1990**, no. 65. (Russian)
139. ——, *Minimal Morse functions*, Global Analysis – Studies and Applications. I, Lecture Notes in Math., vol. 1108, Springer-Verlag, Berlin and New York, 1984, pp. 218–234.
140. ——, *On the equivalence of exact Morse functions*. I, Amer. Math. Soc. Transl. Ser. 2 **134** (1987), 137–146.
141. M. Shub and D. Sullivan, *Homology theory and dynamical systems*, Topology **14** (1975), 109–132.
142. A. J. Sieradski, *A semigroup of simple homotopy type*, Math. Z. **153** (1977), 135–148.
143. S. Smale, *Generalized Poincare's conjecture in dimensions greater than four*, Ann. of Math. (2) **74** (1961), 391–406.
144. ——, *On the structure of manifolds*, Amer. J. Math. **84** (1962), 387–399.
145. I. O. Solopko, *Exact Morse functions in the category* TOP (*dimension* > 5), Ukrain. Mat. Zh. **35** (1983), no. 6, 792–796; English transl. in Ukrainian Math. J. **35** (1983).
146. J. Stallings, *Whitehead torsion of free products*, Ann. of Math. (2) **82** (1965), 354–363.
147. A. A. Suslin, *Projective modules over polynomial rings are free*, Dokl. Akad. Nauk SSSR **229** (1976), no. 5, 1063–1066; English transl. in Soviet Math. Dokl. **17** (1976).
148. ——, *Algebraic K-theory and the norm residue homomorphism*, Itogi Nauki i Tekhniki: Sovremennye Problemy Mat.: Noveĭshie Dostizheniya, vol. 25, VINITI, Moscow, 1984, pp. 115–207; English transl. in J. Soviet Math. **30** (1985), no. 6.
149. R. G. Swan, *Projective modules over group ring and maximal orders*, Ann. of Math. (2) **76** (1962), 55-61.
150. ——, *Vector bundles, projective modules and the K-theory of spheres*, Algebraic Topology and Algebraic K-Theory (Princeton, NJ, 1983), Ann. of Math. Stud., vol. 113, Princeton Univ. Press, Princeton, NJ, 1987, pp. 432–522.
151. W. Thurston, *Existence of codimension-one foliations*, Ann. of Math. (2) **104** (1976), 249–268.
152. I. A. Volodin, *Generalized Whitehead groups and pseudoisotopies*, Uspekhi Mat. Nauk **27** (1972), no. 5, 229–230. (Russian)
153. I. A. Volodin, V. E. Kuznetsov, and A. T. Fomenko, *The problem of discriminating algorithmically the standard three dimensional sphere*, Uspekhi Mat. Nauk **29** (1974), no. 5, 71–168; English transl. in Russian Math. Surveys **29** (1974).
154. C. T. C. Wall, *Formal deformations*, Proc. London Math. Soc. (3) **16** (1966), 342–352.
155. ——, *Surgery on compact manifolds*, London Math. Soc. Monographs, vol. 1, Academic Press, London, 1970.
156. ——, *Geometrical connectivity*. I, J. London Math. Soc. (2) **3** (1971), 597–604.
157. R. B. Warfield, *Stable generation of modules*, Module Theory, Lecture Notes in Math., vol. 700, Springer-Verlag, Berlin and New York, 1979, pp. 16–33.

158. P. J. Webb, *The minimal relation modules of a finite abelian group*, J. Pure Appl. Algebra **21** (1981), 205–232.
159. J. H. C. Whitehead, *Combinatorial homotopy*. II, Bull. Amer. Math. Soc. **55** (1949), 453–496.
160. ———, *Simple homotopy types*, Amer. J. Math. **72** (1950), 1–57.
161. E. Witten, *Supersymmetry and Morse theory*, J. Differential Geom. **17** (1982), 661–692.
162. Y.-G. Wu, *$H^3(G, A)$ and obstructions of group extensions*, J. Pure Appl. Algebra **12** (1978), 93–100.
163. A. V. Zorich, *A quasiperiodic structure of level surfaces of a Morse 1-form close to a rational one – the problem of S. P. Novikov*, Izv. Akad. Nauk SSSR Ser. Mat. **51** (1987), no. 6, 1322–1344; English transl. in Math. USSR-Izv. **31** (1988).

Recent Titles in This Series

(*Continued from the front of this publication*)

(See the AMS catalog for earlier titles)